The Phytochemical Landscape

MONOGRAPHS IN POPULATION BIOLOGY

EDITED BY SIMON A. LEVIN AND HENRY S. HORN

A complete series list follows the index.

The Phytochemical Landscape

Linking Trophic Interactions and Nutrient Dynamics

MARK D. HUNTER

PRINCETON UNIVERSITY PRESS

Princeton and Oxford

Library of Congress Cataloging-in-Publication Data

Names: Hunter, Mark D.
Title: The phytochemical landscape : linking trophic interactions and nutrient
dynamics / Mark D. Hunter.
Description: Princeton : Princeton University Press, [2016] | Series:
Monographs in population biology | Includes bibliographical references and index.
Identifiers: LCCN 2015035498 | ISBN 9780691158457 (hardcover : alk. paper)
Subjects: LCSH: Variation (Biology) | Environmental chemistry. | Botanical chemistry.
| Phytochemicals. | Animal-plant relationships. | Autotrophic bacteria. | Heterotrophic
bacteria.
Classification: LCC QH401 .H86 2016 | DDC 572/.2—dc23 LC record available at
http://lccn.loc.gov/2015035498

British Library Cataloging-in-Publication Data is available

This book has been composed in Times Roman

Printed on acid-free paper. ∞

Printed in the United States of America

10 9 8 7 6 5 4 3 2 1

Contents

Acknowledgments

I am very grateful for OPUS Grant DEB 1144922 from the National Science Foundation, which directly supported the writing of this book. In addition, the ideas that I present were developed in large part with NSF support (DEB Grants 8918083, 9527522, 9615661, 9630347, 9632854, 9815133, 9906366, 0104804, 0214946, 0218001, 0342750, 0404876, 0814340, 1010571, 1019527, 1256115) for which I am also very grateful. I owe a huge debt of gratitude to the program officers at NSF, who manage to facilitate so much research with the resources available to them.

Many of ideas in this book arose through conversations with friends and colleagues, including MaryCarol Hunter, Dick Southwood, Willy Wint, Jack Schultz, Heidi Appel, Jeremy McNeil, Dac Crossley, Dave Coleman, Paul Hendrix, Bob Denno, Pej Rohani, and Peter Price. Many thanks for all of your help. Nora Underwood, Lee Dyer and Angela Smilanich provided helpful input on specific sections of text, and Pej Rohani helped with the analyses illustrated in box 4.1. My graduate students have taught me much more than I've ever taught them, and their work appears throughout; many thanks to Rebecca Forkner, Kitti Reynolds, Rebecca Klaper, Katherine Kearns, Alissa Salmore, Mike Madritch, Sandy Helms, Chris Frost, Carmen Hall, Caralyn Zehnder, Becky Ball, Kyle Wickings, Kristine Crous, Susan Kabat, Rachel Hessler, Nate Haan, Rachel Vannette, Liz Wason, Leiling Tao, Huijie Gan, Holly Andrews, Katherine Crocker, Maria Riolo, Amanda Meier, Leslie Decker, Johanna Nifosi, and Kristel Sanchez. I am also very grateful for the enthusiasm and dedication of the many undergraduate students who have worked with us on our projects over the years.

A special thanks to those who read earlier drafts of chapters, and helped to improve them, especially Oswald Schmitz, Beth Pringle, Katherine Crocker, Hillary Streit, Amanda Meier, Leslie Decker, Johanna Nifosi, Holly Andrews and anonymous reviewers at Princeton University Press. Many thanks also to the PUP editorial team for their patience and guidance.

Finally, our research would not be possible without the support of long-term research facilities, which provide the backbone for much of our work. Many thanks to all those who facilitate long-term research at Wytham Woods (UK), the Coweeta LTER Site, the E.S. George Reserve, and the University of Michigan Biological Station.

The Phytochemical Landscape

CHAPTER 1

Introduction

Imagine a world in which all primary producers have exactly the same chemistry. In this world, all autotrophs are equally palatable to herbivores, decompose at equal rates, and do not respond phenotypically to changes in the biotic and abiotic environment. There is no evolutionary change in plant chemical traits because there is no heritable variation upon which natural selection can act. Succession after disturbance is characterized only by the accumulation of equally palatable plant biomass. Algal blooms in lakes and oceans represent changes in the quantity, but not quality, of primary production. Our current vegetation-based classification of Earth's biomes is redundant.

The full ecological and evolutionary consequences of living in such a world are hard to imagine in part because we take for granted the enormous chemical variation that exists among the primary producers on Earth. Perhaps we are so familiar with the different autotrophs that characterize our forests, grasslands, and rocky intertidal zones, with the varying palatability of the species we can and cannot eat, that we are desensitized to the profound importance of that variation. As ecologists and evolutionary biologists, we encounter two major (and related) problems when we ignore the importance of variation in the chemistry of primary producers. First, our understanding of ecological and evolutionary processes is incomplete at best and misleading at worst. Second, we will be unable to predict the consequences of losing variation in plant chemical traits from our ecological communities. As human activities cause us to lose members of autotroph communities to extinction, we lose so much more than just scientific names. We lose the phenotypic variation that characterizes those scientific names, and their associated ecological and evolutionary interactions. We simplify the phytochemical landscape.

In this book, I describe how variation in the chemical traits of primary producers is of fundamental importance in ecology, and how that variation can be used as an organizing force for synthesis. Specifically, by focusing on variation in autotroph chemistry on the phytochemical landscape, we can better link studies of trophic interactions to those of ecosystem processes. Autotroph chemistry influences, and is influenced by, nutrient dynamics at the ecosystem level.

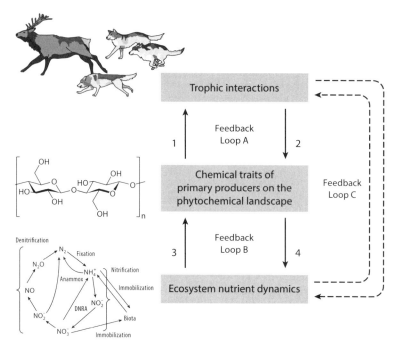

FIGURE 1.1. Links between trophic interactions and nutrient dynamics are mediated by variation in the chemical traits of primary producers on the phytochemical landscape. Trophic interactions such as herbivory and predation vary with the chemical traits of autotrophs (arrow 1). In turn, trophic interactions modify autotroph chemistry (arrow 2) establishing a feedback loop (loop A) between trophic interactions and the chemistry of primary producers. Similarly, nutrient dynamics in the environment influence the chemical traits of autotrophs (arrow 3), which in turn influence the recycling of nutrients (arrow 4), establishing a feedback loop (loop B) between nutrient dynamics and autotroph chemistry. In combination, loops A and B generate a larger feedback loop (loop C) linking trophic interactions with nutrient dynamics through variation on the phytochemical landscape.

Similarly, autotroph chemistry influences, and is influenced by, trophic interactions among organisms. The result is a series of feedback loops that link trophic interactions with the cycling of matter, mediated through the nexus of variation in autotroph chemistry on the phytochemical landscape (figure 1.1). These feedback loops represent powerful trait-mediated indirect effects (Werner and Peacor 2003) in which the chemical traits of primary producers link trophic interactions with nutrient cycling. In chapter 2, I introduce in more detail the concept of the phytochemical landscape, while in chapter 3 I describe the diversity of phytochemistry that exists in our ecological systems. In subsequent chapters (chapters 4–7), I provide evidence to support the importance of each of the feedback loops illustrated in figure 1.1. In chapter 8, I bring the individual feedback loops together into a synthetic whole. Finally, in chapter 9, I discuss priorities for future work and provide some testable predictions that arise from using the concept of the phytochemical landscape to understand the links between trophic interactions and nutrient dynamics.

1.1 A MATTER OF PERSPECTIVE

I doubt that there is a single best way to describe an ecological system. As scientists, we carry our experience, our skill sets, and our biases with us as we try to understand the ecological world. We also target our favorite research questions, which naturally influence the perspective that we take when we try to understand the world around us. Because of these diverse experiences, skill sets, biases, and questions, there are many complementary ways in which we can search for pattern and synthesis in ecology (Sterner and Elser 2002, Brown et al. 2004, Schmitz 2008b, Bardgett and Wardle 2010, Loreau 2010, McCann 2011, Cavender-Bares et al. 2012). I mention this explicitly because the approach that I offer here is one among many ways of seeking synthesis in ecology—it arises from my experience and my biases; it's a matter of perspective. My hope is that there is some strength in this perspective, an opportunity for insight that emerges specifically because of the approach that I describe. Other approaches, other perspectives, will offer their own unique insights into ecological processes. I view these varying perspectives as complementary, and not as alternatives. Moreover, figure 1.1 is meant to illustrate how trophic interactions and nutrient dynamics are linked through the nexus of autotroph chemistry. I do not mean to suggest that these feedback processes are the only factors influencing trophic interactions, the chemistry of primary production, and nutrient cycling. Many edaphic, climatic, and biological forces affect the components of figure 1.1, and can act as external influences on the feedback loops that are the focus of this book. These are discussed in chapter 9.

I also want to stress that this book is not about the primacy of plants. I am not suggesting that variation in the chemistry of primary producers in some way "drives" all of the ecological and evolutionary processes on Earth. Indeed, it is absolutely central to the arguments I present here that autotroph chemistry varies in direct response to nutrient dynamics at the ecosystem scale, and varies in response to trophic interactions. It is now abundantly clear that processes such as trophic cascades (Carpenter et al. 1985) (figure 1.1, arrow 2) and soil nitrogen flux (Aber et al. 1989) (figure 1.1, arrow 3) generate variation in the chemical traits of primary producers at landscape scales. These traits then feed back to influence species interactions and nutrient cycling (figure 1.1, arrows 1 and 4, respectively). Autotrophs are central only in the sense that variation in their chemical traits provides a nexus through which links between trophic interactions and nutrient cycling can take place. Specifically, in no sense am I suggesting that primary producers are more important than herbivores, predators, mutualists, or decomposers in the organization of ecological communities. Rather, I'm suggesting that to understand fully the role of trophic interactions in natural systems, we need to understand how differential consumption generates, and is generated by, variation in the chemistry of primary production.

1.2 THE NATURE OF FEEDBACK

Similarly, this is not a book about the relative importance of top-down and bottom-up forces in ecology. Rather, it is a book about feedback processes. Feedback loops play a central role in ecology at all levels of organization (Odum 1953, Royama 1992) and make irrelevant many arguments about primacy. Figure 1.2A shows an extreme example of a perfectly coupled system in which Consumer B consumes only Producer A. The dynamics of B are driven entirely from the bottom up, and the dynamics of A are driven entirely from the top down. The interaction as a whole cannot be characterized as bottom-up or top-down in any meaningful way. Figure 1.2B extends this to a system in which a variable plant trait (nitrogen content) varies in response to both herbivore damage and environmental nitrogen availability. The presence of feedback loops makes it impossible to characterize any part of the system as driven from the top down or the bottom up.

Why does this approach have merit? As the chemical traits of autotrophs respond to both nutrient cycling and trophic interactions, they establish pathways by which unexpected indirect effects emerge (Hunter et al. 2012). For example, figure 1.2C illustrates a pathway by which spatial patterns of environmental nitrogen availability could influence the dynamics between an herbivore host and its agent of disease. Likewise, disease dynamics between host and parasite may feed back to influence the spatial and temporal availability of nutrients in the environment. As disease ecologists learn more about the contingent nature of parasite-host interactions (Duffy et al. 2012, Orlofske et al. 2012), interactions between nutrient dynamics and agents of disease are starting to appear in the literature (Civitello et al. 2013, Lehahn et al. 2014). These kinds of indirect effects, which are experimentally tractable, can be predicted using the perspective shown in figure 1.1.

Feedback loops have other important properties. In the presence of the time lags inherent in ecological systems, feedback loops can generate complex temporal

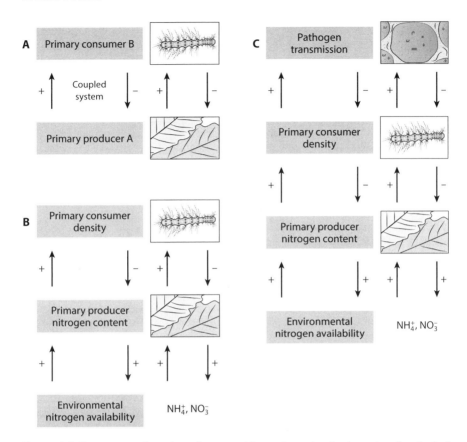

FIGURE 1.2. Bottom-up and top-down forces combine to determine the dynamics of ecological systems. In the simplest 2-level models (A), primary consumer density depends entirely on primary producer density, and vice versa. Extending this to the links between trophic interactions and nutrient dynamics illustrates how primary consumers might reduce the availability of inorganic N (e.g. ammonium NH_4^+ and nitrate NO_3^-) in ecosystems, while N availability simultaneously feeds back to increase the density of primary consumers (B). Note that variation in both consumer density and N availability may be constrained by the overall negative feedback operating between them (multiply the signs on the arrows to determine the overall effect). Adding complexity (C), it becomes apparent how organisms like pathogens both influence, and are influenced by, the availability of nutrients in ecosystems.

dynamics (May 1974). Ecologists still struggle to explain much of the temporal variation in species abundances and ecological processes that we observe in nature. But we should expect temporal variation to emerge in systems such as those illustrated by figure 1.2C. Critically, time lags in the responses of primary producer chemistry to trophic interactions or nutrient dynamics could generate complex temporal dynamics, including cycles and deterministic chaos (Royama

1992). It is a fundamental feature of many phytochemical traits that they are plastic, changing markedly in response to interactions with the biotic and abiotic environment. The timescales of such changes vary from a few seconds to years within individual autotrophs (Liechti and Farmer 2002), and to millennia within whole terrestrial plant communities (Petit et al. 1997). It is a prediction of the perspective described here that temporal variation in trophic interactions and ecosystem processes emerges in part from (a) feedback loops that include variation in autotroph chemistry, and (b) time lags in the responses of autotroph chemistry to abiotic and biotic interactions.

1.3 WHICH AUTOTROPHS AND WHICH TRAITS?

For simplicity, I use some shorthand throughout this book. I use the word "plant" or "autotroph" to include all primary producers, including autotrophic bacteria and archaea, and autotrophic members of the paraphyletic Protista (including diatoms, dinoflagellates, chlorarachniophytes, euglenids, brown algae, and the members of the Plantae that are not characterized as land plants). That being said, the ecological importance of variation in their chemical traits is much better known for some groups (e.g., land plants, cyanobacteria) than for others; the better-known groups will necessarily provide many of the examples in the book. Although the great majority of ecological processes presented here operate in both terrestrial and aquatic environments (chapter 9), there are some specific to each. I discuss relevant differences between terrestrial and aquatic habitats as they arise in each chapter.

I occasionally use the shorthand *autotroph quality* when focusing on the effects of biochemical traits on consumers or microbes. Autotroph quality is a tricky term because quality is a function of the user, not of the plant trait itself. For example, my daily consumption of caffeine and theobromine (in coffee and chocolate, respectively) would be lethal to most organisms, including many mammals (Gans et al. 1980). These particular alkaloid secondary metabolites have no inherent quality beyond their effects (positive or negative) on particular consumers (Scholey et al. 2010, Messerli 2012). However, there are a suite of chemical traits expressed by primary producers that appear to have significant and generalizable effects on nutrient cycling, trophic interactions, or both (Classen et al. 2007a). These traits include mineral nutrient concentrations and ratios (elemental stoichiometry), secondary metabolites (alkaloids, phenolics, terpenoids, etc.), and structural compounds such as lignin, cellulose, and suberin. These are the chemical traits on which I focus here. Some of them are particularly important in mediating interactions between consumers and nutrient cycling because they have broad biological activity. For example, depending on local chemical conditions, hydrolysable tannins can form complexes with proteins, form complexes with iron, and participate in redox reactions (Hartzfeld et al. 2002). Such chemical interactions are important across a broad spectrum of biological processes from nutrient absorption in the gut of mammalian herbivores to the decomposition of terrestrial plant litter.

The chemistry of primary producers varies across a broad range of temporal and spatial scales. In space, autotroph chemistry varies among tissues within single individuals, among individuals (including phytoplankton cells), among species within communities, among landscapes, and among biomes (Denno and McClure 1983, Hunter et al. 1992). This spatial variation generates a complex phytochemical landscape on which trophic interactions and nutrient dynamics take place. The phytochemical landscape might be imagined as a contour map of primary producer chemistry that varies from high to low mineral nutrient content, high to low toxicity to herbivores, and high to low recalcitrance to heterotrophic microbes that mineralize organic chemical compounds into inorganic forms. The phytochemical landscape has both chemical richness and chemical diversity and, as in traditional landscapes, patches of autotroph chemistry are differentially connected to one another through the flow of materials (chapters 2 and 9).

Moreover, the phytochemical landscape is not static over time. Autotroph chemistry can change within individuals over just a few seconds following herbivory (Zagrobelny et al. 2004). Additionally, the phytochemical landscape changes over weeks and years as the result of selective foraging (Pastor and Naiman 1992) and over centuries during the course of ecological succession and recovery from disturbance (Bishop 2002). Finally, the differential fitness of individuals that vary in chemical traits results in evolutionary change in the chemistry of primary producers on the phytochemical landscape (Ehrlich and Raven 1964), with consequences for subsequent trophic interactions and nutrient dynamics (Classen et al. 2007a). Variation in autotroph chemistry at all of these spatial and temporal scales are simultaneously causes and consequences of ecological interactions linked by feedback processes.

1.4 TRAIT VARIATION AND TRAIT DIVERSITY

There are two related yet separate ways by which variation in the chemical traits of primary producers can influence trophic interactions and nutrient cycling. First, key chemical traits in autotrophs can influence (and respond to) trophic interactions or nutrient dynamics. For example, damage by caterpillars to the leaves of many broad-leaved trees results in the induction of polyphenolic

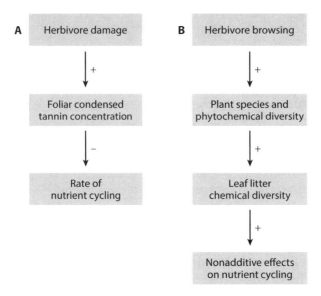

FIGURE 1.3. Both phytochemical trait variation (A) and phytochemical trait diversity (B) mediate the links between trophic interactions and ecosystem processes. For example, herbivore damage can influence variation in the condensed tannin chemistry of tree leaves and litter, thereby modifying the rate at which nutrients are released from that litter (A). Additionally, herbivores may increase the diversity of chemical traits expressed in plant communities, thereby facilitating nonadditive effects during litter decomposition (B). See chapter 6 for further details.

compounds in leaves and decreases in foliar nitrogen (N) levels (Rossiter et al. 1988, Tao and Hunter 2011). In turn, high concentrations of polyphenols and high lignin:N ratios in leaf litter can reduce rates of litter decomposition and nutrient cycling in soils (Madritch and Hunter 2002). In this example, variation in autotroph chemical traits (N, polyphenols) provides a pathway of interaction between insect herbivores and nutrient cycling (Hunter et al. 2012) (figure 1.3A). This is an example of how chemical **trait variation** can link trophic interactions and ecosystem processes.

Additionally, plant **trait diversity** within communities can influence both trophic interactions and nutrient cycling (figure 1.3B). Here, trait diversity refers to the richness of different chemical traits that are expressed simultaneously by communities of autotrophs on the phytochemical landscape. The expanding literature that links species diversity and ecosystem function (Cardinale et al. 2006) is in part a reflection of this. In this body of literature, "species diversity" is an easily measurable variable that serves as a proxy for the complex latent variables that actually determine functional phenotypic differences among taxa (Edwards et al. 2013). The quantification of Latin binomials (species diversity or richness) is convenient shorthand for the variation in phenotype, including variation in phytochemistry, which actually mediates ecological processes (Whitham et al. 2003). Whether described as species diversity, functional diversity, or phenotypic diversity, it is apparent that autotroph trait diversity can influence both trophic interactions (Haddad et al. 2009) and nutrient cycling (Kominoski et al. 2007). Similarly, Latin binomials represent phylogenetic relationships among species that share common ancestors and common ancestral traits. Consequently, a trait-based approach may better serve to explain relationships among the phylogenetic diversity of primary producers, species interactions, and ecosystem processes wherein the assembly of autotroph communities is viewed as the assembly of chemical phenotypes (Asner et al. 2014).

1.5 WHICH TROPHIC INTERACTIONS?

In figure 1.1, arrows 1 and 2 illustrate the feedback loop between trophic interactions and the chemistry of primary producers. What kinds of trophic interactions influence, and are influenced by, variation in the chemistry of autotrophs? The short answer is all of them (see chapters 4 and 5 for detailed coverage). For example, herbivores induce well-known changes in plant chemistry that have subsequent effects on herbivore fitness (Hunter and Schultz 1993, Cronin and Hay 1996b). In turn, the predators, parasites and pathogens of primary consumers can change the densities, distributions, and behaviors of herbivores (Werner et al. 1983, Schmitz 1998), thereby changing patterns of plant consumption and subsequent chemical induction. Omnivores, because they feed at more than one trophic level, may have direct and indirect effects on the chemistry of autotrophs and, in turn, respond to variation in autotroph chemistry (Eubanks and Denno 2000). Mutualisms, because they are based on the exchange of resources, can also be considered as trophic interactions or mutual exploitation (Bronstein 2001). Consequently, many mutualisms can both cause, and respond to, variation in the chemistry of primary producers. For example, mycorrhizal fungi influence the defensive chemistry of terrestrial plants (Vannette and Hunter 2011b), while the effects of recalcitrant phytochemicals on digestion by herbivores may depend critically on nutritional mutualists (Lilburn et al. 2001). Microbial symbionts may even influence the rate and direction of plant succession (Rudgers et al. 2007), with concomitant effects on trophic interactions and ecosystem processes at landscape scales. Finally, there are both direct and indirect mechanisms by which competition for resources can result from trophic interactions, with consequences for autotroph chemistry and the structure of ecological communities (Hunter and Willmer 1989). Chapters 4 and 5 consider in more detail how feedback processes link trophic interactions with variation in the chemistry of primary producers.

1.6 WHICH ECOSYSTEM PROCESSES?

A central focus of ecosystem ecology is the cycling of matter and transfer of energy in biological systems (Coleman 2010). In this book, I focus primarily on links between trophic interactions and nutrient cycling, mediated by variation in autotroph chemistry. In terrestrial systems, primary production generates about 100 gigatons of biomass annually (Gessner et al. 2010), of which about 90 percent enters the detrital pathway without being consumed (Cebrian 2004). The chemistry of that plant tissue is a primary determinant of its decomposition rate (Cornwell et al. 2008) and therefore the rate at which nutrients are recycled in terrestrial ecosystems. There are two important facts to note here. First, the chemistry of tissue that is not consumed by herbivores is nonetheless a legacy of trophic interactions. Natural selection for chemical defense (Thompson 1994) and selective foraging by herbivores (Leibold 1989, Pastor and Naiman 1992) combine to influence the palatability and decomposition rate of plant material, whether it is consumed by herbivores or not (Cornelissen et al. 2004). Second, when plants are consumed by herbivores, their chemistry often changes in ways that influence the decomposition process. This includes induced chemical changes in remaining plant tissues (Findlay et al. 1996) and herbivore-derived inputs such as feces and prematurely abscised leaves (Hunter et al. 2012).

In marine systems, total primary production is about the same as that in the terrestrial environment (Field et al. 1998), although annual rates of consumption of marine plant biomass are about four times higher than those in terrestrial systems (Cebrian 2004). As a result, nutrient turnover rates are higher in marine than in terrestrial systems. In the oceans of the world, an average milliliter of seawater may contain a million bacterial decomposers that mineralize between 35 and 80 percent of annual primary production (Duarte and Cebrian 1996). In turn, marine heterotrophic bacteria are consumed by larger organisms and play key roles in the structure and function of marine food webs (Taylor and Stocker 2012). However, just as in terrestrial systems, rates of organic matter decomposition in oceans and lakes vary with the chemical quality of primary production (chapter 6). For example, only bacteria with the appropriate enzyme systems can decompose certain cyanobacterial toxins (Alamri 2012).

These rates of decomposition and nutrient cycling in terrestrial and aquatic environments are important because nutrient availability feeds back to influence the subsequent chemistry of primary producers (Herms and Mattson 1992,

Anderson et al. 2002) (chapter 7). As a consequence, I focus my attention on how rates of nutrient cycling are influenced by, and in turn influence, the chemical traits of autotrophs. I pay particular attention to rates of decomposition, and to the cycling of nitrogen (N) and phosphorus (P), which are often limiting resources that influence both primary productivity and the chemistry of autotrophs. I will also mention more briefly other ecosystem processes, such as productivity, when their expression is linked to both trophic interactions and variation in autotroph chemistry on the phytochemical landscape.

1.7 WEBS OF GREEN AND BROWN

We are perhaps most familiar with the mechanisms by which trophic interactions change the chemical traits of living autotrophs in the "green food web." For example, within communities of living primary producers, chemical induction (Schultz and Baldwin 1982, Hay 1996), selective foraging (Kirk and Gilbert 1992, Pastor and Naiman 1992), and feces deposition by herbivores (Frost and Hunter 2007) all act to change the chemistry of aquatic and terrestrial autotrophs. In turn, these autotrophs produce organic residues that enter the detrital pathway where true decomposers, those species of bacteria, archaea, and fungi that convert organic molecules into inorganic forms, subsequently act upon them, mediating the recycling of nutrients within ecosystems. Most of this book focuses on chemical mediation of interactions in the green food web.

However, complex trophic interactions within the "brown food web" also influence the chemistry of autotroph residues and the dynamics of nutrient recycling. Bacterivorous and fungivorous organisms, detritivores, and a suite of enemies influence the distribution, activity, and abundance of true decomposers (Coleman et al. 2004, Hunting et al. 2012). Moreover, the chemical nature of decomposing autotroph residues is not static, but changes markedly over time, depending on decomposer activity. Recent work has established that the chemical trajectory of decomposing plant material is influenced by trophic interactions within the brown food web (Wickings et al. 2012) such that even higher-order predators can modify the chemistry of decomposing plant material (Hunter et al. 2003a). The key point here is that figure 1.1 can apply just as readily to brown food webs as to green food webs.

Additionally, because brown and green food webs are inextricably linked together (Bardgett and Wardle 2010, Hines and Gessner 2012), figure 1.1 can serve as a template for understanding interactions between them. For example, the trait-mediated indirect effects that link trophic interactions in green food webs with nutrient cycling can then "cascade through" to influence trophic interactions in linked brown food webs (and vice versa) (chapter 8). Such trait-mediated cascades (Liere and Larsen 2010) represent coupled systems of trait-mediated indirect effects (Utsumi et al. 2010) and are common features of natural systems (Pardee and Philpott 2011). In short, variation in the chemistry of primary producers provides a nexus through which trait-mediated cascades can link together trophic interactions in green and brown food webs and the cycling of nutrients (Reynolds et al. 2003).

In chapter 2, I develop in more detail the concept of the phytochemical landscape, which seeks to describe spatial and temporal variation in the chemistry of primary producers. In subsequent chapters, I describe how this spatial and temporal variation in phytochemistry is both a cause and a consequence of trophic interactions and nutrient dynamics, and acts as a fundamental pathway that links them together.

CHAPTER 2

The Phytochemical Landscape

2.1 DEFINING THE PHYTOCHEMICAL LANDSCAPE

Studying landscapes is like time travel. The history of Earth is written on our physical landscapes, where powerful forces including rivers, glaciers, and plate tectonics have carved their characteristic signatures onto the planet's surface. Importantly, while they help to reveal the past, those same signatures of slope, aspect, and parent material also allow us to gaze into the future; current landscape structure constrains the range of physical and biological processes that may follow later. In short, when we study landscapes, we stand on a knife-edge, where past and future meet.

The same is true of our phytochemical landscapes. Throughout this book, we can envisage phytochemical landscapes as contour maps of complex chemistry—of toxins, nutrients, and structural molecules derived from primary producers—that exist in heterogeneous patches on the landscape (figure 2.1). Such patches of phytochemistry occur when samples of autotrophs taken close to one another are more similar chemically than are samples taken at increasing distance (Covelo and Gallardo 2004). On these phytochemical landscapes, we might study lignin topography generated by grasslands, forests, and deserts. Similarly, we can explore phytochemical seascapes of algal toxins, phytoplankton nutrient content, or seagrass cellulose. As with traditional geographical features, phytochemistry may show complex spatial structure, and may change markedly over time. Mountains of recalcitrant organic molecules may be eroded by decomposers, while valleys of nutrient availability may constrain the activities of consumers.

Phytochemical landscapes are not equivalent to maps of autotroph identity, although they may be related to them. The central theme of this book is that variation in phytochemistry derives in part from variation in nutrient availability and variation in trophic interactions (figure 1.1, chapters 5 and 7). At small spatial scales, pervasive genotype by environment interactions (chapter 3) ensure that autotroph identity is often not an adequate proxy for the phytochemistry expressed by individual primary producers. Local nutrient availability (chapter

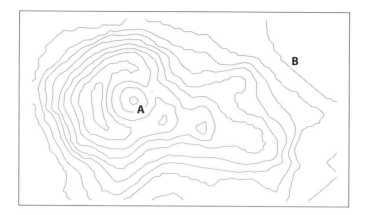

FIGURE 2.1. The phytochemical landscape can be represented as a contour map, where regions of equal phytochemical concentration occur on the same contour line. Across the landscape, we can see regions of relatively high (A) and relatively low (B) phytochemical concentration. Contour maps could be used to represent diverse measures of autotroph chemistry, including nutrient concentrations, structural molecules, and toxins. Contour maps could also illustrate values from multivariate statistics such as principal components analysis that can reduce multiple measures of phytochemical quality into fewer dimensions.

7) and herbivore-induced changes in phytochemistry (chapter 5) both influence the range of chemical phenotypes that any individual autotroph expresses. At larger spatial scales, entire assemblages of autotroph communities can share phytochemical traits because of the environments in which they grow rather than their taxonomic affiliations (Fine et al. 2004). Moreover, taxonomic identity is not an ecological mechanism—it is a proxy for the phenotypes that engage in interactions. And the phytochemical landscape is the expression of vital phenotypic variation that provides a mechanistic link between trophic interactions and nutrient dynamics.

Just as physical landscapes reveal the forces that generated them, so too do phytochemical landscapes; we can time-travel through phytochemical landscapes too. Spatial and temporal variation in phytochemistry result from past ecological and evolutionary processes on timescales ranging from seconds to millennia and on spatial scales from millimeters to biomes (chapter 3). As with physical landscapes, phytochemical landscapes also allow us to predict the future. If we know the current chemistry of primary production in a given location, we can make some powerful predictions about the trophic interactions (chapter 4) and nutrient dynamics (chapter 6) that should follow. Existing on the knife-edge of the present moment, phytochemical landscapes are both the consequences and the causes of ecological interactions.

When we want to understand patterns on physical landscapes, we make maps. In terrestrial systems, we often map the course of rivers, the height and slope of hills, and the location of lakes. In marine systems, we might map the depth of water, major oceanic currents, changes in temperature, and the location of upwelling. Of course, such maps are incredibly useful to biologists, who can superimpose the distribution and abundance of organisms, or ecological and evolutionary processes, onto these physical maps. When we bring together physical maps and biological maps, we can ask and answer important questions about the history (Avise 2000) and the future (Hannah 2011) of life on Earth.

Unfortunately, we have barely begun to map the phytochemical landscape. Many of the tools with which we generate maps of physical features, of climate, and of the distribution of biota could be applied to mapping phytochemistry, and there has been intermittent interest and progress in doing so (Baltensweiler and Fischlin 1988, Hunter 1997, Covelo and Gallardo 2004, Moore et al. 2010, Asner et al. 2014). Some of the best progress in recent years has been made at very large spatial scales in marine systems, wherein nutrient flux between phytoplankton cells and seawater combines with major ocean currents (gyres) to influence patterns of algal chemistry and productivity throughout the oceans of the world (Martiny et al. 2013, Schlosser et al. 2014). There are good examples from terrestrial systems too. For example, studies of European oak forests demonstrate strong spatial correlation in the distribution of foliar phenolics (Covelo and Gallardo 2004) and foliar nutrients (Gallardo and Covelo 2005) on the phytochemical landscape. Critically, such phytochemical patterns can inform us about ecological and evolutionary processes. For example, phytochemical mapping of formylated phloroglucinol compounds and nitrogen in *Eucalyptus* foliage can help to predict the foraging behavior of koalas on Phillip Island, Australia (Moore et al. 2010) (figure 2.2). Moreover, the spatial dependence observed in the phytochemistry of *Eucalyptus* forests may arise in part from spatial genetic structure within tree populations (Andrew et al. 2007). However, it is abundantly clear that the quantitative description of most phytochemical landscapes lags far behind our ability to map other key features of ecosystems, including physical landscape features, soil quality, water quality, and the distribution of species.

Why has there been such slow progress in describing our phytochemical landscapes? Two reasons predominate. First, there remain some methodological difficulties in doing so. It is much easier to count and map the relative abundance of tree species in a forest than it is to map the relative abundance of different phenolic molecules. Likewise, it is probably faster to classify lake algae into taxonomic groups than it is to measure the diversity of their phycotoxins. The equipment and techniques for measuring complex chemistry at landscape scales are not universally available, and methods are not always standardized. However, remote

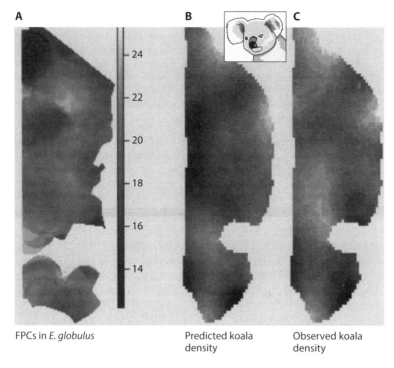

FIGURE 2.2. The spatial distribution of formylated phloroglucinol compounds (FPCs) in *Eucalyptus globulus* trees can be used to generate a phytochemical heat map (A) that predicts the distribution of koala bears in space (B, C). Modified from Moore et al. (2010).

sensing and imaging spectroscopy have the potential to provide detailed maps of primary producer chemistry over very large areas of land and water (Kokaly et al. 2009, Madritch et al. 2014). Similarly, rapid chemical characterization of raw plant samples can be used to explore spatial and temporal variation in autotroph chemistry (Richards et al. 2015). In any case, the fact that analytical chemistry techniques can be challenging cannot serve as an excuse to ignore the important role of the phytochemical landscape in mediating ecological and evolutionary processes. In short, we cannot afford to shy away from this important research area simply because it is sometimes difficult.

Second, there may not have been sufficient emphasis to date on the role that phytochemistry can play in mediating feedback between trophic interactions and nutrient dynamics. A primary goal of this book is to motivate additional research targeted toward better characterization of the causes and consequences of variation in the phytochemical landscapes in which we live. In so doing, we will better integrate population-level and ecosystem-level approaches in ecology. In chapter 1, I described how autotroph chemistry may serve as the nexus that links interactions among organisms with the flux of nutrients in ecological systems (figure 1.1). In the rest of this chapter, I describe why the concept of the phytochemical landscape can serve to clarify the feedback processes operating between trophic interactions and nutrient dynamics. The themes developed here are then considered in much more detail in the remainder of the book, in which I explore in turn each of the pathways of feedback that are outlined below.

2.2 VARIATION ON THE PHYTOCHEMICAL LANDSCAPE

When ecologists study landscapes, they often focus on the relationship between pattern and process (Turner 1989). Patterns of heterogeneity on the phytochemical landscape are both causes and consequences of ecological and evolutionary processes. For example, in terrestrial systems, chemical heterogeneity in plants may map directly onto spatial variation in nutrient availability in soils (Klaper and Hunter 1998) because (a) plant chemistry is in part a function of soil chemistry (chapter 7) and (b) soil chemistry is in part a function of plant chemistry (chapter 6) (figure 2.3). Plants in high-nutrient environments tend to accumulate more nutrients in their foliage than do plants in low-nutrient environments. In turn, senesced litter from nutrient-rich plants is relatively nutrient rich, contributing to the maintenance of high soil fertility (Aber et al. 1990). The presence of strong feedback processes between soil quality and plant chemistry are now well established, and can range from relatively simple patterns of coupled nutrient flux to more complex effects mediated by plant structural and defensive molecules, plant community assembly, and the evolution of plant phytochemistry (Hobbie 1992, Cornelissen et al. 1999, Treseder and Vitousek 2001). Figure 2.3 illustrates the simplest case, in which patches of high N soils on the landscape favor patches of high N plants; in turn, high concentrations of N in plant litter within those patches feed back to maintain high N availability in soils. Exactly this kind of spatial correlation between soil nutrients and foliar nutrients has been reported in research that used phytochemical mapping in stands of oak (Gallardo and Covelo 2005) and pine (Rodriguez et al. 2011).

Figure 2.4 expands upon the simplest case to illustrate how the phytochemical landscape serves as a nexus linking trophic interactions with nutrient dynamics. There exist feedback loops between each layer on the landscape. Phytochemistry is both a cause and a consequence of trophic interactions between herbivores and their enemies (chapters 4 and 5). Likewise, phytochemistry is both a cause and a consequence of spatial variation in nutrient dynamics (chapters 6 and 7). In combination, these feedback loops generate pathways by which nutrient dynamics

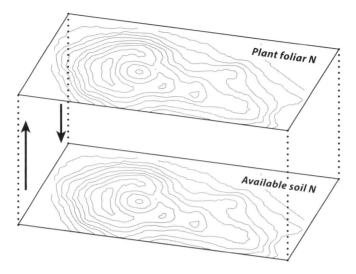

FIGURE 2.3. Spatial variation in autotroph chemistry on the phytochemical landscape is both a cause and a consequence of spatial variation in nutrient availability. In this example, strong feedback between plant and soil nutrient availability generates spatial correlation between plant and soil nutrient concentrations.

influence trophic interactions and vice versa. Population processes and ecosystem processes are combined in the same interaction web and linked spatially on the landscape.

Technically, we could choose to map the densities of herbivores and their natural enemies on the phytochemical landscape. I think that a better approach would be to map variation in the interaction strengths that we typically use in ecological models of trophic interactions (Speight et al. 2008). The strengths of interactions between herbivores and their natural enemies are well known to vary with autotroph chemistry (Price et al. 1980) (figure 2.5). Similarly, nutrient dynamics can be estimated by a variety of techniques at fine spatial and temporal scales (Robertson et al. 1999) and subsequently mapped on the phytochemical landscape (figure 2.4).

Note that the concept is not restricted to any particular ecosystem or any particular measure of phytochemistry. We can consider toxins, structural molecules, and nutrients as metrics of primary producer chemistry, and study them in diverse aquatic and terrestrial ecosystems (chapter 3). Additionally, figure 2.4 represents a snapshot in time; we should expect the spatial patterns illustrated in figure 2.4 to change over ecological and evolutionary timescales (Covelo et al. 2011). The dynamic nature of the phytochemical landscape is an important theme throughout this book.

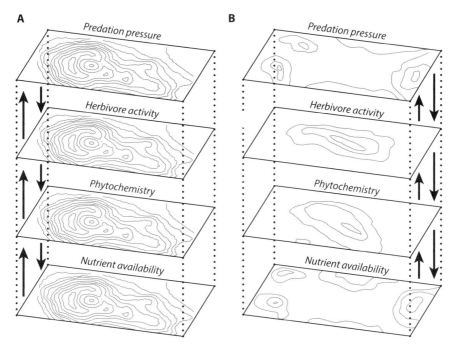

FIGURE 2.4. Feedback loops generate spatial correlation between phytochemistry, trophic inter-actions, and nutrient dynamics. Spatial correlation among layers can be (A) positive or (B) a mixture of positive and negative, depending on the types of interactions taking place. For exam-ple, in (B), negative spatial correlation between predation pressure and herbivore activity might reflect herbivore foraging behavior to avoid predators.

In figure 2.4A, I have illustrated positive spatial correlation among all inter-actions on the phytochemical landscape; areas of high-nutrient availability are associated with hot spots of some aspect of phytochemistry, which are associated positively with both herbivore activity and predation pressure. However, these associations need not be positive; in many cases, for example, we might expect negative spatial correlation between predation pressure and herbivore activity, if herbivores are able to avoid their predators in space (Painter et al. 2014) (figure 2.4B). Likewise, the induction of recalcitrant molecules in primary producers by herbivore activity may reduce the subsequent availability of nutrients in the environment (Findlay et al. 1996, Chapman et al. 2006), resulting in a negative spatial correlation between a particular aspect of phytochemistry and nutrient dynamics (figure 2.4B). In this example, note that spatial variation in predation pressure on the landscape correlates more positively with environmental nutrient availability than it does with herbivore activity, illustrating clearly the role that predators can play in affecting spatial variation in nutrient dynamics (Schmitz et al. 2010).

Whether the spatial associations are positive or negative, the phytochemical landscape still serves to mediate the linkage between trophic interactions and nutrient dynamics. Why focus on landscape-level phytochemistry as opposed to some other level of organization in figure 2.4? Importantly, it is often the same indices of autotroph chemistry (nutrients, structural molecules, and toxins) that influence both trophic interactions and nutrient dynamics; they are mechanistically linked by the common effects of autotroph chemistry on consumers and decomposers (Cornelissen et al. 1999, Cornelissen et al. 2004) (chapter 1). For example, in terrestrial systems, foliar concentrations of cellulose, lignin, polyphenols, and nitrogen are fundamental determinants of plant palatability to diverse herbivore species (Mattson 1980, Bryant et al. 1991, Foley et al. 1999). Those same foliar traits often determine rates of litter decomposition and nutrient cycling (Hobbie 1992, Cornwell et al. 2008), establishing a critical link between herbivore consumers and nutrient cycling in soils (Pastor and Naiman 1992, Chapman et al. 2006, Hunter et al. 2012). In short, it is the premise of this book that predators will exert their greatest impact on nutrient cycling when they cause significant changes in autotroph chemistry on the phytochemical landscape. In complement, nutrient dynamics will exert their greatest influence on trophic interactions, such as predation and disease, when those nutrient dynamics change the chemistry of the phytochemical landscape. Examples of these coupled processes are described in chapter 8.

If it is to be useful, the concept of the phytochemical landscape should help to explain spatial variation in ecological processes and the pervasive environmental contingency that we observe in ecological interactions. As an example, during the 1980s, entomologists were interested in developing a viral pesticide that could be used to control gypsy moth populations in the deciduous forests of eastern North America. Research showed that the virus worked more effectively to kill gypsy moth caterpillars in some forest stands than in others; mortality from virus was much lower on oak (*Quercus*) than it was on aspen (*Populus*) (Keating and Yendol 1987). Additional work has illustrated that high concentrations of polyphenols in tree foliage inhibit the virus (Keating et al. 1988), establishing an effect of the phytochemical landscape on the strength of the virus-caterpillar interaction (figure 2.5); the outcome of the trophic interaction is contingent upon its location on the phytochemical landscape. Additionally, because the prodigious quantities of insect feces deposited during caterpillar outbreaks tend to increase the rate of N cycling in the soils of eastern deciduous forests (Reynolds et al. 2000, Frost and Hunter 2007), the phytochemical (polyphenol) landscape also influences nutrient cycling through its powerful effects on trophic interactions in the forest canopy (Hunter et al. 2012).

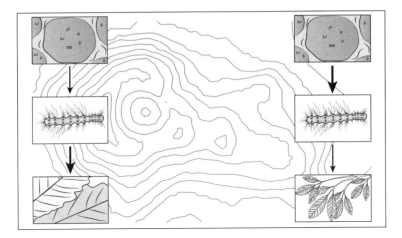

FIGURE 2.5. Trophic interactions are contingent upon their location on the phytochemical land-scape. Here, the contours represents variable concentrations of polyphenolics in the foliage of deciduous trees, while the strength of interactions are indicated by the thickness of the arrows. In areas of low polyphenolic concentration, a viral pathogen exerts strong suppression of a cat-erpillar host, thereby reducing defoliation of trees. In contrast, in areas of high polyphenolic concentration, the virus is inhibited and caterpillars defoliate trees with greater frequency. Based on work described in Keating et al. (1988) and Hunter et al. (1996).

We can simplify figure 2.4 into a condensed form (figure 2.6) that provides an outline and justification for the chapters that follow. In figure 2.6, we see that the phytochemical landscape provides a nexus where trophic interactions and nutrient dynamics meet. Understanding the magnitude of variation that exists on the phytochemical landscape is a prerequisite for all that follows, and describing some of that variation is the subject of chapter 3. In chapter 4 (arrow 4, figure 2.6), I describe in detail how variation on the phytochemical landscape modifies the strength of trophic interactions, including herbivory, predation, parasitism and disease. In turn, in chapter 5 (arrow 5, figure 2.6), I focus on the reciprocal effects of trophic interactions on the phytochemical landscape. In combination, chapters 4 and 5 illustrate that variation on the phytochemical landscape is both a cause and a consequence of variation in trophic interactions; phytochemical pattern and ecological process are intimately linked in a feedback loop. In chapter 6 (arrow 6, figure 2.6), I explore in depth how the phytochemical landscape also mediates variation in nutrient dynamics in ecosystems, focusing on rates of N, P, and C cycling. Chapter 7 (arrow 7, figure 2.6) considers the reciprocal effects of nutrient dynamics on the phytochemical landscape, completing a second feedback loop between phytochemical pattern and ecological process. In combination, chapters 6 and 7 illustrate that variation on the phytochemical landscape is both a cause and

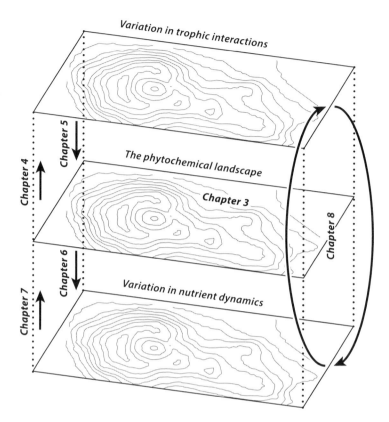

FIGURE 2.6. A roadmap for exploring how the phytochemical landscape links trophic interactions with nutrient dynamics. Spatial and temporal variation in phytochemistry on the landscape (chapter 3) are linked by feedback loops to trophic interactions (chapters 4 and 5) and nutrient dynamics (chapters 6 and 7). In combination, these feedback loops serve to connect (chapter 8) trophic interactions with nutrient dynamics.

a consequence of nutrient dynamics. In chapter 8, I combine the two smaller feedback loops (arrows 4/5 and 6/7) to link trophic interactions with nutrient dynamics (arrow 8, figure 2.6) through the nexus of the phytochemical landscape. Chapter 9 provides a summary and synthesis of what has gone before, and suggests some testable predictions for future studies.

As I noted earlier, there are remarkably few published studies in which the phytochemical landscape has been characterized in a spatially explicit fashion, and then associated with ecological and evolutionary processes. Consequently, a great majority of the examples that I use in this book are drawn from studies that measure the causes and consequences of variation in nutrient dynamics, variation in phytochemistry, and variation in consumer activity without reference to explicit locations in space—most of the examples are not mapped on real landscapes. Given the general paucity of phytochemical mapping, I cite the studies in the following chapters primarily to establish unequivocal support for each of the feed-back loops described in figures 1.1 and 2.4. My motivation and great hope is that, in the years to come, we will increase substantially the frequency with which we study these feedback processes in a spatially explicit fashion (chapter 9).

CHAPTER 3

The Variable Chemistry
of Primary Production

3.1 THE CHALLENGING CHEMISTRY OF AUTOTROPHS

I am lucky enough to live in a green part of the city, and as I look out my window, my view includes a wide variety of trees, shrubs, grasses, and herbs. However, what is notable about this diversity of plant material is that I can eat almost none of it. Even as a polyphagous omnivore, most of the primary production with which I am surrounded is too toxic, too tough, or nutritionally unfavorable to serve me as a source of food. I live within a diverse phytochemical landscape, which is largely unpalatable. And this underscores one of the fundamental features of life on Earth: although primary production provides, directly or indirectly, the energy to power the rest of life on our planet, most primary producers are inedible to most consumers most of the time.

However, one of the other fundamental features of life on Earth is that almost all organisms, including primary producers, are eaten by something. The tough leaves of holly trees are mined by herbivorous flies (Valladares and Lawton 1991), certain *Daphnia magna* genotypes are adapted to the potent protease inhibitors of their phytoplankton prey (Schwarzenberger et al. 2012), while some specialist amphipods protect themselves from predators by feeding on toxic brown algae in the genus *Dictyota* (Hay et al. 1990). During the evolutionary play (Hutchinson 1965), natural selection simultaneously favors mechanisms of autotroph defense against herbivores and mechanisms to overcome such defenses by consumers (Iason et al. 2012). Moreover, the ecological and evolutionary interactions between autotrophs and herbivores have important implications for the ecology of higher trophic levels, including predators (Ginsburg and Paul 2001), parasites (de Roode et al. 2008), and agents of disease (Keating et al. 1990). Simply put, autotroph chemistry and trophic interactions are inextricably linked.

At its heart, life is complex chemistry in a semipermeable bag. The bag is the cell membrane, which allows different chemical conditions to be maintained inside the bag than occur outside. The great trick of life is to use an external source of energy to push locally, vigorously, and temporarily against the second law of thermodynamics to build self-replicating organic molecules, rich in energy. The organic molecules that are built by primary producers in turn become the external sources of energy of Earth's heterotrophs. Trophic interactions, and the food webs we draw to describe them, represent the alternating construction and managed disassembly of organic molecules in bags.

But not all organic molecules are created equal, and some are much, much harder to disassemble than are others (Ding et al. 2012). In addition, many lack the chemical elements (nitrogen, phosphorus, iron, etc.) that are necessary to catalyze replication (Sterner and Elser 2002). Moreover, some organic molecules are toxic, perhaps compromising the cell membranes necessary for life (Zhen et al. 2012) or affecting the enzymatic machinery by which heterotrophs, including microbial decomposers, disassemble organic molecules for their own use (Green and Ryan 1972). Others are sticky, reducing the rate at which herbivores can feed or trapping them in glue-like defenses (Tao and Hunter 2012b). Finally, some organic molecules signal to higher-order heterotrophs that act as plant bodyguards, ameliorating subsequent rates of consumption (Turlings et al. 1990). In other words, organic molecules vary widely in their chemistry, and this variation influences the rates at which they are built and disassembled by the cyclic processes of primary production, trophic interactions, and decomposition (figure 3.1).

3.2 ORIGINS OF VARIATION IN AUTOTROPH CHEMISTRY ON THE PHYTOCHEMICAL LANDSCAPE

From where does Earth's phytochemical diversity arise? The phenotypes of organisms, their genotypes, and the environments in which they live are tightly bound in a "triple helix" of mutual causality (Lewontin 2002). While both genotype and environment influence the expression of autotroph chemical phenotype (Fritz and Simms 1992), autotroph chemistry then feeds back to modify the environment (e.g., nutrient availability) in which primary producers grow and natural selection takes place (Ball et al. 2009b). Within Lewontin's triple helix, the chemical phenotype of autotrophs is both a cause and a consequence of variation in genotype and environment. The phytochemical landscape arises from this mutual causality. This is why the phytochemical landscape is not simply a map of autotroph identity or community structure, important as those may be. It is a map of the chemical heterogeneity, generated in space, by Lewontin's (2002) triple helix.

For example, on land, variation among habitats in soil nutrient availability can shape the genetic architecture of plant populations and the molecular machinery

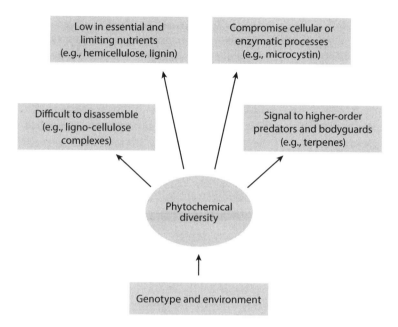

FIGURE 3.1. The challenging nature of autotrophs as food for primary consumers and heterotrophic microbes. Note that some types of organic molecule (e.g., lignin) provide multiple challenges to consumers and may be particularly important mediators of the links between trophic interactions and nutrient dynamics on the phytochemical landscape.

by which plants take up nutrients from soils (Baxter and Dilkes 2012). Using genomic techniques, we can now associate genetic variation in plant populations with the availability of minerals in soil, illustrating how natural selection acts on mechanisms of plant nutrient uptake in soils of different quality (Horton et al. 2012). In other words, the elemental quality of plant tissue for consumers is in part the result of evolutionary adaptations to the soil environment. In turn, because mineral elements are often limiting in terrestrial and aquatic environments (von der Heyden et al. 2012), the mineral nutrient concentration of autotroph cells and tissues has important effects on subsequent levels of consumption and the fitness of consumers (Elser et al. 2010). Finally, as we'll see in chapter 6, plant nutrient concentration feeds back to influence the quality of soils in which plants grow and reproduce.

Natural selection also has promoted considerable plasticity in the elemental nutrient concentrations of some autotroph species, such that nutrient concentrations of individual autotrophs increase or decrease within their lifespan, depending upon environmental conditions (chapter 7). For example, in aquatic systems, turbulence in the water stirs patches of available nutrients into networks of thin filaments. These filaments occur at spatial scales that influence nutrient uptake by a variety of aquatic microorganisms, including phytoplankton (Taylor and Stocker 2012). In turn, the nutrient contents of phytoplankton influence the performance of zooplankton grazers (Elser et al. 2010). In terrestrial systems, the foliar nitrogen concentrations of many land plants increase following nutrient additions, with subsequent increases in herbivore performance or abundance (Throop and Lerdau 2004). Conversely, apparent dilution of plant nitrogen concentration under elevated carbon dioxide concentrations can cause subsequent reductions in herbivore performance (Stiling et al. 2003). In summary, in both terrestrial and aquatic environments, variation in the nutrient concentration of primary producers arises from evolutionary and ecological processes, with subsequent impacts on consumers. In chapter 8, I describe in detail how these trophic interactions can feed back to influence nutrient availability in the environment (Hunter et al. 2012).

What about variation in the secondary metabolites of autotrophs, those chemical compounds that have no obvious primary role in metabolism but can serve to combat environmental stresses, including consumption by herbivores (Fraenkel 1959)? As with elemental nutrient concentrations, ecological and evolutionary processes combine to generate enormous diversity in plant secondary metabolites (PSMs) on the phytochemical landscape (Moreira et al. 2014). Our understanding of the evolution of PSMs is increasing rapidly and the molecular revolution has shown how many arise from the duplication and subsequent evolution of genes from primary metabolic pathways. Gene duplications can release primary metabolic enzymes from constraints on their structure and function (Ortlund et al. 2007), setting the stage for neofunctionalization or the evolution of new enzyme functions. The enzymes that catalyze primary metabolic processes are likely constrained in structure and function to a much greater extent than are the new duplicated genes that may evolve into the specialized metabolic enzymes that produce PSMs.

Once a gene has been duplicated, it may have fewer structural and functional constraints thereby facilitating "catalytic promiscuity" (Weng et al. 2012b). In essence, this means that specialized metabolic enzymes may produce several products from a single enzyme, either by expanded substrate recognition or by catalyzing more than one transformation within the enzyme. Each new enzyme is, itself, potentially available for gene duplication and neofunctionalization, and the result is an ever-increasing evolutionary palette of PSMs (Weng et al. 2012a). In land plants, this evolutionary diversification of specialized metabolic enzymes appears to have begun with the colonization of land, some 500 million years ago, and proceeded in parallel with the diversification of vascular plant lineages (Banks et al. 2011). While diversification is clearly the rule, convergent evolution can result in the synthesis of similar toxins in unrelated taxa. For example, molecular evidence suggests that caffeine production by coffee, tea, and cacao plants is the result of convergent evolution from independent gene duplication events (Denoeud et al. 2014).

Does the evolution of PSMs in aquatic primary producers differ in any fundamental way from that of terrestrial plants? As in the terrestrial environment, aquatic primary producers synthesize a simply dazzling array of secondary compounds that influence the performance of heterotrophic organisms (Cembella 2003). Best known, perhaps, are the phycotoxins produced by cyanobacteria, diatoms, dinoflagellates and brown algae, from which new compounds are discovered almost daily (Fenical 2012). But it seems likely that most aquatic primary producers are defended in some way, and even chemolithoautotrophic bacteria (which gain their energy from inorganic molecules) in deep-sea hydrothermal vents are emerging as producers of potent PSMs (Andrianasolo et al. 2012).

There are few primary producers that can claim the level of importance in the evolution of life on Earth as the cyanobacteria (Bergman et al. 2008), and they may help us to understand the evolution of PSMs in marine environments. Cyanobacteria are a monophyletic group of photosynthetic bacteria that evolved between 2.7 and 3.5 billion years ago, oxygenated Earth's atmosphere, and gave rise to the chloroplasts of plants and algae (Tomitani et al. 2006). Some cyanobacteria produce potent toxins, including cyclic peptides, alkaloids, and polyketides (Sasso et al. 2012) that are active against the enzymes, cells and tissues of many heterotrophs including humans. The cyclic peptides called microcystins are the most prevalent cyanobacterial toxins, and the microcystin synthetase gene cluster that catalyzes their synthesis occurs in distantly related lineages of cyanobacteria (Rantala et al. 2004). The sporadic occurrence of microcystin production within and among genera of cyanobacteria has been used to argue for horizontal gene transfer as a mechanism of toxin spread and evolution in distantly related taxa. Horizontal gene transfer is now known to be an important force in the evolution of bacterial genomes (Francino 2012) and those of some eukaryotes (Schönknecht et al. 2013), and it could therefore provide a very different path for the evolution of PSMs in aquatic bacterial primary producers than the mechanism of gene duplication and neofunctionalization described above for land plants.

However, at least for the microcystins, horizontal gene transfer does not appear to be the primary mechanism of evolutionary spread on the phytochemical landscape. Phylogenetic studies reveal a pattern of coevolution of the microcystin synthetase gene cluster with genes that are unrelated to toxin production for the entire evolutionary history of the toxin (Rantala et al. 2004). This argues for repeated losses of the gene cluster in cyanobacterial strains that do not produce microcystin rather than repeated gains of the gene cluster through horizontal gene transmission. Moreover, the production of the related cyclic peptide nodularin in the cyanobacterial genus *Nodularia* provides further evidence for the evolutionary model of duplication and neofunctionalization. The phylogenetic analyses conducted by Rantala et al. (2004) suggest that the genes encoding nodularin synthetase have been derived more recently from those encoding microcystin synthetase, in a manner identical to that proposed for phytochemical diversification in land plants. In other words, while horizontal gene transmission may provide an additional route for the evolution and spread of secondary metabolites (Moran et al. 2012), particularly during the acquisition of plastids of bacterial origin by eukaryotes (Blunt et al. 2008), gene duplication and neofunctionalization appear to operate in the most common systems we currently know. Such processes appear to have been taking place in the evolution of some cyanobacterial toxins for a staggering two billion years (Murray et al. 2011).

Other aquatic primary producers also have evolved diverse PSMs. For example, dinoflagellates are well known producers of toxins in aquatic environments (Kellmann et al. 2010), including the neurotoxin saxitoxin that can accumulate in shellfish and cause paralysis if eaten. Saxitoxin, by mass, is 1,000 times more potent than cyanide and 50 times stronger than curare (Hackett et al. 2004). Based on molecular phylogenetic analysis, toxin production appears to have evolved early in the radiation of dinoflagellates (Murray et al. 2009). Some brown, red and green algae produce dimethylsulfoniopropionate (DMSP), which is converted to acrylic acid and dimethyl sulfide on damage to the algae (Van Alstyne et al. 2001). Brown algae also produce a diversity of phlorotannins and terpenoids with biological activity against at least some consumers (Hay and Fenical 1988). Also, many aquatic primary producers synthesize or accumulate chemical structures that are difficult to digest or decompose, in the same way that the lignocellulose complexes of terrestrial plants are recalcitrant to digestion. For example, armored dinoflagellates produce a protective theca from cellulose plates. Similarly, diatom cells are surrounded by a frustule of hydrated silicon dioxide (silica), which is generated intracellularly by the polymerization of silicic acid monomers. In this way, the silica of diatoms shows similarities with the silica defenses of some terrestrial plants (Cooke and Leishman 2011). Likewise, the calcification of seaweeds may provide direct defense against marine grazers and synergistic effects with seaweed secondary metabolites (Hay 1996).

As with autotroph nutrient concentrations, the expression of PSMs by primary producers has a strong environmental component in both terrestrial and aquatic habitats (chapter 7). For land plants, factors including soil nutrient availability

(Bryant et al. 1983), drought (Mattson and Haack 1987), light (Hunter and For-kner 1999), temperature (Bauerfeind and Fischer 2013), and atmospheric pol-lutants (Kinney et al. 1997) contribute to spatial and temporal variation in the expression of PSMs. Similarly, in aquatic ecosystems, light (Kaebernick et al. 2000), temperature (Kurmayer 2011), and nutrient limitation (Bar-Yosef et al. 2010) can all influence the expression of PSMs by aquatic primary producers. Nutrient availability also influences the relative abundance of toxin-producing algae and bacteria (Downing et al. 2001) and their per capita toxin production (chapter 7). Moreover, phytoplankton can be transported significant distances in water by physical forces including winds and currents. When phytoplankton transported in this way produce toxins, physical forcing can be a major driver of changes in the local concentration of algal PSMs (Raine et al. 2010) (chapter 9).

3.3 MICROBIAL SYMBIONTS AND VARIATION IN AUTOTROPH CHEMISTRY: WHOSE PHENOTYPE IS IT ANYWAY?

To conclude this discussion of the origins of phytochemical diversity, it's import-ant to note that eukaryotic organisms, including the primary producers, are a mixed bag of cross-kingdom DNA, genetic chimeras containing high proportions of microbial genetic material. Symbiotic associations between microbes and their host species can be highly intimate and obligate, such as the mitochondria within the cells of all eukaryotes or the plastids that facilitate carbon fixation in primary producers. Alternatively, symbioses may be much looser, with the identity, abun-dance, and presence/absence of microbes exhibiting considerable variation in time and space (Smith and Read 2008). This is important because the chemical traits of aquatic and terrestrial primary producers are influenced to a significant degree by their symbiotic associations with bacteria and fungi (Blunt et al. 2008, Vannette and Hunter 2011b). Variation on the phytochemical landscape, then, arises from interactions between eukaryotic and microbial genotypes.

Excellent books have been written on the various symbioses between autotro-phs and microbes (Smith and Read 2008, Cheplick and Faeth 2009, Lindström and Mousavi 2010) and there is not room to consider those symbioses in full detail here. For the purposes of this book, I limit my discussion to a brief con-sideration of three important symbioses (mycorrhizal fungi, N-fixing bacteria, and foliar endophytes) to highlight their role in generating variation in primary producer chemistry. Because of their important role in generating variation on the phytochemical landscape, we should expect these symbioses to participate in the feedback processes that link trophic interactions with nutrient dynamics (figure 3.2).

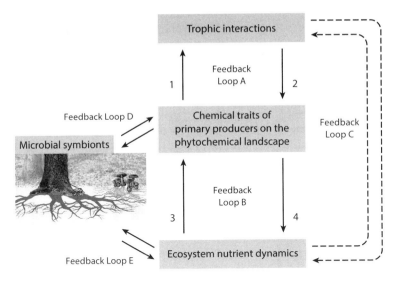

FIGURE 3.2. Microbial symbionts that associate with primary producers influence both the nutritional and defensive chemistry of those producers. In turn, microbial symbionts are affected by the chemistry and nutrition of their hosts (loop D). Similarly, microbial symbionts both take up and return nutrients to the environment (loop E). Microbial symbionts increase significantly the potential pathways for indirect interactions between trophic interactions and nutrient dynamics (compare with Figure 1.1 for further details).

3.3.1 Nutritional Symbioses

Most terrestrial plants, some aquatic macrophytes and certain aquatic protists gain at least some of their mineral nutrients through the action of symbiotic organisms that exchange nutrients in return for photosynthate. For example, about 90 percent of land plants associate with mycorrhizal fungi that increase rates of nutrient uptake (Smith and Read 2008) while N-fixing bacteria such as *Rhizobium*, *Frankia*, and some cyanobacteria colonize the roots of many plant families (Sprent and Parsons 2000, Heath and Tiffin 2007). Additionally, mutualisms between mosses and cyanobacteria play a large role in the N and C dynamics of high-latitude ecosystems (Lindo et al. 2013). In other words, the N and P concentrations of many terrestrial plant species are mediated in part by microbial symbionts, with subsequent effects on trophic interactions (chapter 4) and nutrient cycling (chapter 6). Although we know less about them, some aquatic primary producers also associate with microbes to gain nutrients from the environment. For example, aquatic ferns in the genus *Azolla* associate with the N-fixing cyanobacteria *Anabaena azollae* (van Kempen et al. 2013), while some submerged aquatic plants also host mycorrhizal fungi (Kohout et al. 2012). Likewise, some N-fixing cyanobacteria such as *Richelia intracelluaris* associate as either epibionts (externally attached symbionts) or as endosymbionts of marine diatoms (Kneip et al. 2007). Recently, widespread marine N-fixing cyanobacteria with reduced genomes (incapable of their own C-fixation) have been found in mutualistic associations with photosynthetic picoplankton (Thompson et al. 2012), confirming that nutritional mutualisms are pervasive in aquatic ecosystems.

Because nutrients and carbon represent principal currencies in allocation of resources to autotroph defense (Herms and Mattson 1992), exchanges of nutrients and C between primary producers and symbionts should influence the chemical quality of autotrophs for trophic interactions and nutrient cycling. Moreover, colonization by microbial symbionts can prime the defense signaling pathways of terrestrial plants so that they react more quickly or more strongly to subsequent attack (Jung et al. 2012). Effects of microbial symbionts on herbivore preference and performance are now well known in terrestrial systems (Hartley and Gange 2009) and likely emerge from a combination of changes in plant nutritional quality, plant size, and chemical defense. For example, mycorrhizal colonization increases concentrations of the iridoid glycosides aucubin and catalpol in *Plantago lanceolata* (Gange and West 1994). Similarly, rhizobia in lima bean increase cyanogenic chemical resistance against bean beetles (Thamer et al. 2011). Nutritional mutualists may also influence the impact of the natural enemies of herbivores on rates of parasitism and predation (Moon et al. 2013) (chapter 4), confirming that microbial symbionts participate in the links illustrated in figure 3.2. Moreover, mycorrhizal fungi can carry defense signals that induce resistance between individual plants through common mycelial networks, and so alert plants that their neighbors are under herbivore attack (Babikova et al. 2013).

Do microbial symbionts always confer increasing chemical resistance against herbivores on the phytochemical landscape? For those of us seeking simple patterns, the unfortunate news seems to be that microbial symbionts have quite variable effects on plant resistance and tolerance to herbivores, with the magnitude and direction of effects varying among species of plant, species of microbe, and species of herbivore (Gehring and Bennett 2009, Garrido et al. 2010). Among all this unsettling idiosyncrasy, there is some evidence that specialist herbivores are less affected by arbuscular mycorrhizal fungi (AMF) than are generalist herbivores (Koricheva et al. 2009), again supporting the hypothesis that AMF induce increases in PSMs to which specialists are already adapted (Bennett et al. 2006). But the key point is that plant-mycorrhizal symbioses generate additional variation and complexity on the phytochemical landscape (figure 3.2).

For example, comparing eight species of milkweed in the genus *Asclepias,* we have found substantial variation among congeneric plant species in their chemical responses to AMF (Vannette et al. 2013). One clade within the genus *Asclepias* (including *A. curassavica, A. linearis,* and *A. verticillata*) appears to exhibit reductions in foliar cardenolide concentrations in response to AMF colonization whereas a second clade (including *A. purpurascens, A. syriaca,* and *A. latifolia*) appears largely unaffected. Additionally, milkweed species also exhibit idiosyncratic effects of AMF colonization on foliar cardenolide composition, with some species expressing substantial changes in the relative concentrations of different cardenolide molecules. Strikingly, the presence of AMF in milkweed roots completely eliminates the trade-off between milkweed growth and chemical resistance that is evident in nonmycorrhizal plants. The take home message is that nutritional symbionts make a significant difference to the chemistry of their autotroph hosts, but determining why those differences are so variable among taxa will take much more work.

Beyond leaves and stems, nutritional symbionts also influence the expression of defense in the roots of terrestrial plants. For example, the presence of the AMF *Glomus mosseae* reduces the survival and biomass of black pine weevil on dandelion roots by half (Gange et al. 1994). Similarly, across twelve species of milkweed in the genus *Asclepias*, plants inoculated with mycorrhizal fungi suffer lower levels of attack by root-feeding fungus gnats than do uninoculated plants, although the precise chemical mechanism remains unclear (Vannette and Rasmann 2012). AMF can certainly increase the root cardenolide concentrations of some milkweed species, and cardenolides reduce the performance of some herbivores on milkweed roots (Rasmann and Agrawal 2011). We have noted, however, that the cardenolides in milkweed roots are characterized by more polar (less toxic) forms than are the cardenolides in foliage (Vannette et al. 2013), perhaps to protect the AMF themselves from the toxic effects of nonpolar forms.

Nonetheless, most published studies report increases in root defensive chemistry following inoculation with AMF. For example, barley roots with AMF express high concentrations of hydroxycinnamic acid amides that are implicated in defense (Peipp et al. 1997). Likewise, experimental inoculation with AMF increases iridoid glycoside concentrations in the roots of some *Plantago lanceolata* genotypes (De Deyn et al. 2009). However, plant genotypes that naturally produce high concentrations of defense in roots also support lower natural levels of AMF colonization, suggesting a trade-off between defense production and nutritional mutualism. AMF also increase concentrations of terpenes and phenolics in the roots of some plant species (Grandmaison et al. 1993, Akiyama and Hayashi 2002) while priming others so that their local and systemic chemical defenses respond more rapidly and vigorously to subsequent attack (Jung et al. 2012).

In addition to their associations with mycorrhizal fungi, many terrestrial plant species associate with N-fixing bacteria from a variety of bacterial taxa. N-fixing bacterial symbionts might be expected to increase the quality of plants for herbivores because they generally increase plant N concentrations, and N is often limiting to herbivore fitness (Mattson 1980). However, when nodulated plants also produce N-containing defenses such as alkaloids, N-fixing bacteria may facilitate plant resistance. For example, nodulated lupines produce more alkaloids than do N-limited plants (Johnson et al. 1987), which may reduce their susceptibility to some herbivores.

As our understanding of plant-symbiont interactions has increased, more subtle effects of symbioses on plant chemistry have been detected. For example, inoculation of soybean with rhizobia induces the expression of some pathogen defense-related genes that reduces soybean susceptibility to subsequent pathogen attack (Gao et al. 2012). Interestingly, the effect of rhizobia on soybean defense appears to depend in part on the presence or absence of mycorrhizal fungi. It may be common that the chemical quality of autotrophs depends upon the activity of more than one microbial symbiont. For example, white locoweed, *Oxytropis sericea*, is defended by an alkaloid called swainsonine that can cause livestock poisoning in western North America. *O. sericea* produces higher concentrations of swainsonine when inoculated with *Rhizobium*. In this case, however, the alkaloid is actually synthesized by the fungal endophyte, *Embellisia*; nodulation with *Rhizobium* facilitates toxin production by the endophytic fungus, presumably by increasing N availability to the fungus (Valdez Barillas et al. 2007).

So despite the common currencies involved in the nutritional exchange (C, N, P), autotrophs seem to vary widely in their chemical responses to microbial mutualists. Why should this be? The answer may lie in other life-history traits, including rates of photosynthesis and C gain. Because autotrophs pay their microbial mutualists for the service of nutrient uptake, there are both costs and benefits for mutualistic associations (Holland et al. 2002) that will influence phytochemistry. Primary producers pay for nutrient uptake with labile C, a currency that is also fundamental to their defense budget (Herms and Mattson 1992). As a consequence, there may be trade-offs between the allocation of carbon to mutualisms (nutrient uptake), growth, and defense. At the same time, N and P are required to synthesize or to power the enzymatic machinery that is used by autotrophs to make secondary metabolites, and nutrient uptake by microbial symbionts should influence defense production.

We have proposed the resource exchange model of plant defense (REMPD) to account for the costs and benefits of nutritional mutualisms between autotrophs

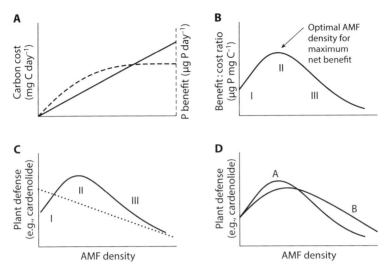

FIGURE 3.3. The resource exchange model of plant defense (REMPD), illustrated here by the interaction between terrestrial plants and their arbuscular mycorrhizal fungi (AMF). (A) Carbon costs to plants (solid line) increase with increasing mutualist density, while benefits (dashed line) saturate. As a result (B) the benefit:cost ratio is related nonlinearly to mutualist density. Zone I represents limited symbiont abundance and nutrient transfer, zone II represents optimal exchange with mutualistic symbionts and maximal nutrient benefits, and zone III represents parasitism, wherein carbon costs exceed nutrient benefits. The benefit:cost ratio translates directly to the (C) expression of plant defenses predicted by REMPD (solid line), in comparison to the C:nutrient balance hypothesis (Bryant et al. 1983) (dotted line). (D) The shapes of the phenotypic response curves to symbiont abundance vary among plant genotypes (A and B). Based on Vannette & Hunter (2011b).

and microbial symbionts (Vannette and Hunter 2011b). The model recognizes that C costs for autotrophs may continue to increase as symbiont density increases even though the benefits to autotrophs saturate when the resource exchanged (e.g., P from mycorrhizal fungi or N from rhizobia or cyanobacteria) no longer limits autotroph performance. The result is an optimal density of symbionts at which autotroph fitness is maximized and neither nutrients nor C limit greatly the production of secondary metabolites. A key prediction of the model is a quadratic relationship between the density of symbionts and the production of chemical defense by autotrophs (figure 3.3).

We have conducted some preliminary tests of the REMPD by manipulating the densities of two mycorrhizal fungi associated with the roots of common milkweed plants (Vannette and Hunter 2011b). As predicted by our model, increasing root colonization by the arbuscular mycorrhizal fungus *Scutellospora pellucida* produces quadratic responses in plant growth, latex exudation, and cardenolide production. In contrast, *Glomus etunicatum* appears to act as a parasite of *A. syriaca*, causing exponential decline in both plant growth and latex exudation. In fact, both results support the REMPD because, under our experimental conditions, *G. etunicatum* provided no net benefit to milkweed, and defenses should therefore decline as its density increases. Such parasitism of plants by certain mycorrhizal fungi may be widespread (Kiers and Denison 2008, Grman 2012).

The problem with the simplest form of the REMPD is that it doesn't take into account natural variation in the availability of nutrients. As I described earlier, variation in the availability of inorganic nutrients in the environment has a significant impact on the expression of autotroph defense (figure 3.2) (chapter 7). However, it also influences the cost-benefit function between autotrophs and their nutritional symbionts. Simply put, why should primary producers pay the C costs of maintaining symbionts if nutrients are already widely available in the environment? Under such circumstances, they are paying costs without receiving benefits. Therefore, we should not always expect simple quadratic relationships between symbiont abundance and autotroph chemical defense, and the relationships can be much more complex (Vannette and Hunter 2013). We have proposed a model that combines aspects of the REMPD with those of the growth-differentiation balance hypothesis (Herms and Mattson 1992) to predict how microbial symbionts and natural variation in nutrient availability combine to determine defense expression in primary producers. In short, the model predicts that autotrophs will be most heavily defended at intermediate symbiont densities and intermediate availability of nutrients from the abiotic environment (figure 3.4) (Vannette and Hunter 2011b). To my knowledge, our combined model has not yet been tested experimentally in either terrestrial or aquatic ecosystems.

3.3.2 Nonnutritional Symbioses—Who Makes the Toxins?

Under the nutritional symbioses described above, autotroph chemistry changes largely because the symbionts alter the resources available to primary producers, or prime their defensive responses; the autotrophs themselves still make any defensive compounds that are affected by the symbioses. However, primary producers also associate with a wide variety of endophytic fungi and bacteria that can influence autotroph chemistry directly. I have already mentioned the production of alkaloids by endophytic fungi in white locoweed (Valdez Barillas et al. 2007), and mycotoxin production by endophytic fungi appears to be common (Clay 1990). Mycotoxins produced in grasses can include alkaloids,

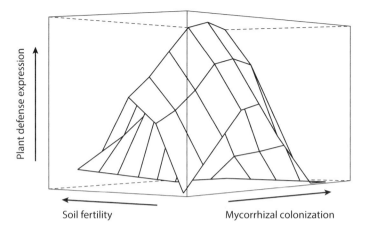

FIGURE 3.4. An integration of the growth differentiation balance hypothesis (Herms & Mattson 1992) with the resource exchange model of plant defense (Vannette & Hunter 2011b). In this terrestrial example, soil fertility alters the benefits associated with nutritional symbionts (e.g., mycorrhizal fungi) and the subsequent effects on defense expression. When soil fertility is very high, mycorrhizal fungi act only as parasites, and increasing mycorrhizal costs result in declines in defense expression. When mycorrhizal colonization is very high, defense expression is insensitive to variation in soil fertility.

lolitrems, lolines, and peramines, with toxic effects on a variety of vertebrate and invertebrate herbivores (Clay and Schardl 2002). Although notably important in grasses (Clay 1991), fungal endophytes appear to be ubiquitous in land plants (Friesen et al. 2011), and may also occur in marine autotrophs (Sakayaroj et al. 2010).

The effects of fungal endophytes on autotroph chemistry vary from substantial to negligible (Faeth and Fagan 2002). There may be some taxonomic bias in the examples that illustrate strong effects of endophytes on herbivores, with much of the work having been conducted on two invasive grass species. However, native grass–endophyte assemblages also appear to deter herbivores from feeding on grasses (Crawford et al. 2010). In comparison, there appear to be no significant defensive effects of endophytes on tree-feeding herbivores (Saikkonen et al. 2010). In some cases, endophytes may actually increase damage levels to plants by having disproportionately greater deleterious effects on natural enemies than on herbivores (Jani et al. 2010, Saari et al. 2010) (chapter 4). However, when the effects of endophytes on autotroph chemistry are large, subsequent effects of trophic interactions and ecosystem processes are also substantial. For example, endophytic fungi in grasses can influence plant community composition, subsequent patterns of succession, and nutrient dynamics in soils (Rudgers et al. 2007).

In addition to fungi, nonnutritional bacterial endophytes can have significant effects on the chemical phenotype of autotrophs. For example, bacteria in the genus *Streptomyces*, ubiquitous as free-living decomposers in soil, are also important symbionts of plants and other eukaryotes (Seipke et al. 2012). *Streptomyces* species produce a great variety of secondary metabolites (Challis and Hopwood 2003), including many that have been exploited by humans as pharmaceutical drugs. For example, the first effective antibiotic used against tuberculosis (a disease caused by another member of the Actinobacteria) was derived from *S. griseus* (Schatz and Waksman 1944). In plants, *Streptomyces* range from parasites to mutualists, the latter of which produce a range of antimicrobial secondary metabolites that confer resistance against plant pathogens (Seipke et al. 2012). *Streptomyces* endophytes can also prime the defense pathways of plants so that they are activated more rapidly after attack by enemies (Conn et al. 2008). Antibiotic production by endophytic *Streptomyces* is a potentially valuable tool in the biological control of plant pests (Franco et al. 2007).

3.4 SUMMARY AND CONCLUSIONS

Aquatic and terrestrial primary producers vary enormously in their nutritional and defensive chemistry. Mutation and recombination generate genetic diversity in the enzyme systems that regulate nutrient acquisition and secondary metabolism. Environmental variation interacts with genetic variation in autotrophs, and in their microbial symbionts, to determine variation on the phytochemical landscape (figure 3.2). Next, I will consider how phytochemical variation then influences trophic interactions (arrow 1, figure 3.2, chapter 4) and in turn is influenced by trophic interactions (arrow 2, figure 3.2, chapter 5). Subsequently, I will illustrate how the phytochemical landscape influences nutrient dynamics in ecosystems (arrow 4, figure 3.2, chapter 6), and in turn is influenced by those nutrient dynamics (arrow 3, figure 3.2, chapter 7). The end result is a dynamic system of feedback (loop C, figure 3.2) that interlinks trophic interactions with nutrient dynamics through the nexus of autotroph chemistry on the phytochemical landscape (chapter 8).

CHAPTER 4

Effects of Primary Producer Chemistry on Trophic Interactions

4.1 HERBIVORES AND HERBIVORY: THE INTERACTIVE EFFECTS OF AUTOTROPH CHEMISTRY AND NATURAL ENEMIES

Many of the patterns that we observe in nature result from interactions among primary producers, their herbivores, and a suite of natural enemies at higher trophic levels. Studying variation in the nutritional and defensive chemistry of autotrophs provides one method of organizing these interactions (Hunter and Price 1992a) as we try to understand their broader implications for the structure and function of ecosystems (figure 4.1). Interactions between autotrophs and herbivores, and between herbivores and their natural enemies, take place on a dynamic phytochemical landscape that is both a cause and a consequence of trophic interactions.

The arrow that is circled in figure 4.1 represents the effects of variation in autotroph chemistry on trophic interactions, and is the subject of this chapter. Studying how phytochemical variation influences herbivore ecology is a useful starting point to explore the impacts of the phytochemical landscape on trophic interactions. However, it is critical to recognize from the outset that nearly every facet of the behavior, performance, population dynamics, and community structure of herbivores reflects the combined and interactive effects of autotroph chemistry and natural enemies. Table 4.1 lists the principal effects that autotroph chemistry and natural enemies can have on herbivore ecology. In no sense does either the chemistry of primary producers or predation pressure by natural enemies exert primacy over the ecology of herbivores.

Traditionally, ecologists have focused on how natural enemies influence herbivore mortality, and thereby reduce herbivore densities. However, natural enemies can have as large an effect on herbivore ecology and evolution by changing the ways that herbivores behave as they do by actually attacking them (Schmitz et al. 2004). By altering how, where, and for how long herbivores forage on the diversity of food sources available to them, natural enemies can influence herbivores by all of the routes listed in table 4.1. For example, the deleterious

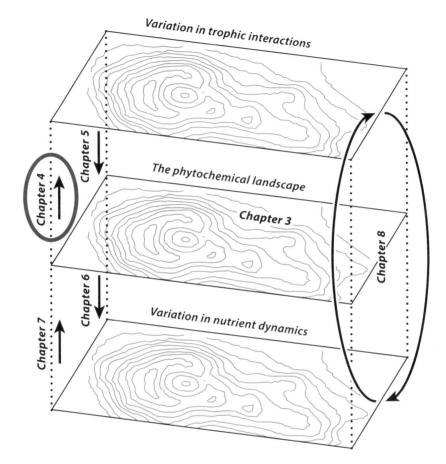

FIGURE 4.1. Variation on the phytochemical landscape is the nexus linking trophic interactions and nutrient dynamics. In this chapter (heavy circle), I focus on the role of phytochemistry in mediating trophic interactions, including herbivory, predation, parasitism, and disease.

effects of secondary metabolites on the growth of herbivores may occur in part because it is too risky for herbivores to forage elsewhere. Effects of enemies on the foraging behavior of herbivores can be flexible, such that behaviors change in the presence or absence of the enemy (Lefèvre et al. 2010), or fixed over evolutionary time to influence the dietary breadth of herbivores (Bernays and Graham 1988).

Selective foraging among primary producers of different chemistry provides a compelling example of how phytochemistry and predation pressure interact to influence herbivore ecology. Selective foraging arises through natural selection because herbivores perform better on some autotrophs (species, individuals,

TABLE 4.1. Effects of variation in autotroph chemistry
and natural enemies on the ecology of herbivores.

Individual Level	Foraging Behavior
	Growth Rate
	Fecundity
	Mortality
	Dispersal
Population Level	Per Capita Rate of Increase
	Abundance
	Genetic Architecture
	Sex Ratio
	Distribution
Community Level	Species Richness and Relative Abundance

tissues) than they do on others. But measures of herbivore performance must include minimizing the deleterious effects of their natural enemies. From early work on multitrophic interactions (Price et al. 1980), through concepts of "enemy-free space" (Jeffries and Lawton 1984) and enemy-driven evolution of specialization (Bernays and Graham 1988, Hay 1991), ecologists have worked to understand how autotroph chemistry and predation pressure interact to influence the performance of herbivores (Hunter and Price 1992b). Given the toxicity and toughness of many primary producers (chapter 3), it is no surprise that the growth rate and fecundity of herbivores are related to the chemistry of what they eat. But those same patterns of consumption can vary with the risk of predation such that natural enemies also influence herbivore foraging behavior, subsequent growth rate, adult fecundity, and patterns of plant damage (Peacor and Werner 2001, Schmitz et al. 2004).

Critically, it is not that natural enemies cause variation in autotroph chemistry to matter less. Rather, natural enemies often increase the importance of phytochemistry in herbivore ecology and evolution by constraining herbivores to a subset of the total phytochemical landscape. Often, that subset of the phytochemical landscape is nutritionally poorer and/or higher in plant toxins. For example, elk may migrate to reduce the risk of predation from wolves but, in doing so, are constrained to feed on plants of lower than average nutritional quality (Hebblewhite and Merrill 2009). Likewise, some marine amphipods choose to graze on nutritionally poorer brown algae because those algae are also toxic to predators and competitors of the amphipods (Lasley-Rasher et al. 2011). In essence, the foraging decisions of all herbivores represent trade-offs between avoiding risks from natural enemies and gaining nutrition from autotrophs (Fryxell and Sinclair 1988). But even a scared herbivore will not eat anything or everything available.

As they forage across "landscapes of fear" (Brown and Kotler 2004), herbivores are simultaneously exploiters and victims of chemical heterogeneity on the phytochemical landscape.

The time is long overdue to abandon false dichotomies between whether autotrophs or predators drive the foraging behavior of herbivores, and any subsequent effects on herbivore ecology. The landscape of fear is also a phytochemical landscape, flavored with tannins, phenolics, lignins, alkaloids, and essential nutrients. This chemical heterogeneity constrains herbivore foraging, yet provides opportunities to exploit the deterrent properties of plant secondary metabolites (PSMs) (Rosenthal and Berenbaum 1991). Fear, or predation risk, is another axis in the multivariate space within which herbivores make their foraging decisions (Simpson and Raubenheimer 1993). In box 4.1 and figure 4.2, I use some simple, well-established theory to illustrate a basic example of how the dynamics of herbivore populations on the landscape of fear depend upon interactions between autotroph chemistry and natural enemies. Because natural enemy dynamics are coupled to those of their herbivore prey, the dynamics of natural enemy populations also vary with autotroph chemistry (see section 4.2).

In the sections that follow, when describing the effects of phytochemical traits on herbivore preference and performance, I make the assumption that both natural enemies and autotroph chemistry have influenced the foraging decisions of herbivores over ecological and evolutionary timescales. In chapter 5, I will consider explicitly how trophic interactions feed back to influence the phytochemical landscape.

4.1.1 Selective Foraging

Selective foraging, when herbivores include a nonrandom subset of available autotroph resources in their diet, is the norm (Crawley 1983). In terrestrial ecosystems, the dietary specialization of many insect herbivores serves as an extreme example of selective foraging (Novotny et al. 2010, Schallhart et al. 2012). We have known for decades that many species of insect herbivore will feed on only a few related species of plants (Forbes 1956) that share similar chemical traits (Ehrlich and Raven 1964). As far as specialists are concerned, plants without those chemical traits might as well be concrete.

Specialization on the chemistry of plants has led to some degree of phylogenetic congruence between insect herbivores and the plants on which they feed (Becerra 2003). However, the evidence for tight coevolution driving reciprocal bouts of speciation between insects and plants is not strong (Percy et al. 2004, Brandle et al. 2005). The evolution and diversification of plant clades may have as

BOX 4.1.

Herbivores, Phytochemistry, and the Landscape of Fear

Risk of attack can cause herbivores to forage on primary producers of lower chemical quality than they do in the absence of enemies (Hebblewhite and Merrill 2009, Lefèvre et al. 2010, Zamzow et al. 2010), causing herbivore fecundity to decline. However, for the change in foraging to be adaptive, the efficiency with which natural enemies attack their prey should also decline, otherwise herbivores would not move. Effects of the trade-off between food quality and risk of attack can be explored by varying herbivore fecundity and natural enemy efficiency in simple models.

Here, I use the well-known Nicholson-Bailey model (Nicholson and Bailey 1935), stabilized by the addition of herbivore self-limitation, to illustrate the consequences of changing herbivore fecundity and enemy efficiency on the dynamics of herbivore and enemy populations. The equations for the herbivore and enemy respectively are:

$$N_{t+1} = N_t \cdot \exp(r - aP_t - bN_t)$$
$$P_{t+1} = cN_t \cdot (1 - \exp(-aP_t))$$

where N and P are the population densities of herbivore and enemy, respectively, t is time, r is herbivore fecundity, a is the efficiency with which enemies attack herbivores, b is the coefficient of self-limitation acting on the herbivore population, and c is the number of new enemies produced from each herbivore that is killed. For the simple illustration below, c is kept constant with a value of 1 and b is kept constant with a value of 0.02; these can of course be varied to explore model behavior in more detail.

Figure 4.2 illustrates how the densities of (a) herbivores and (b) natural enemies change with variation in herbivore fecundity, r, and enemy efficiency, a. Inserts (i), (ii), and (iii) illustrate temporal dynamics for 3 sets of parameter values:

(i)   r = 0.7, a = 0.1
(ii)   r = 0.5, a = 0.1
(iii)   r = 0.5, a = 0.05

If risk from enemies causes herbivores to move from feeding on high quality autotrophs to low quality autotrophs (e.g., from [i] to [ii]), dynamics of herbivores and enemies will change, with average declines in the densities of both. However, if the move by herbivores is adaptive, the efficiency of enemies will also decline (e.g., from [ii] to [iii]), causing yet further changes in the dynamics of both herbivores

(Box 4.1 continued)

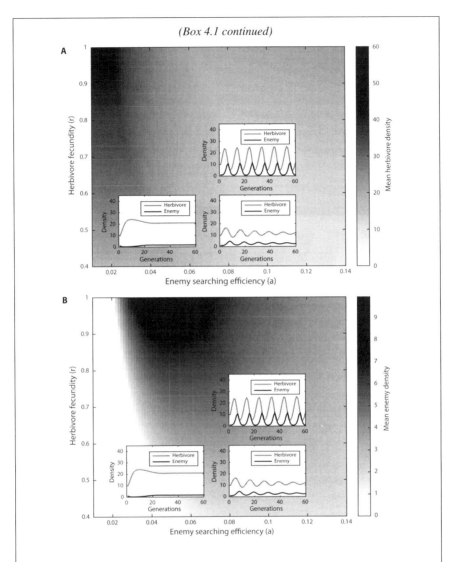

and their enemies. Note that there is a potential area of "enemy free space" (Jeffries and Lawton 1984) to the lower left of figure 4.2B wherein natural enemies cannot persist if herbivores pay a large cost in fecundity, perhaps providing initial incentive for specialization on toxic autotrophs (Bernays and Graham 1988).

The simple point is that the interaction between autotroph chemistry and predation risk influences the dynamics of both herbivore and enemy populations whenever herbivores trade the risk of predation (represented by a) for declining fecundity, r.

Figure 4.2 kindly provided by P. Rohani. □

much to do with patterns of climate (Nyman et al. 2012) as with their interactions with herbivores. Moreover the "catalytic promiscuity" of the enzymes that synthesize PSMs (Weng et al. 2012b) (see chapter 3) contributes to the lack of strict phylogenetic signal in their occurrence across plant species; distantly related plant species may sometimes express similar chemistries. This might help to explain some of the host shifts of certain insect herbivores across plant clades (Janz 2011), with host shifts occurring preferentially—but not exclusively—among closely related plants (Nyman 2010). As we might expect, chemical similarity appears to be more important than phylogenetic relatedness in driving host use by insect herbivores (Becerra and Venable 1999).

Beyond the extreme specialization of many insect herbivores, it is now clear that autotroph chemistry and escape from enemies combine to influence the foraging strategy of most herbivores (Forister et al. 2012). Small herbivores such as amphipods in marine systems, despite being dietary generalists, still show phylogenetic constraint in their choice of seaweed diet (Poore et al. 2008). Similarly, large generalist terrestrial herbivores, as they forage across the landscape of fear, still choose among alternative plant species based in part on plant nutrient contents and secondary metabolites. For example, moose in North America feed preferentially on deciduous tree species compared with conifer species and drive changes in the relative abundance of these groups of trees (Pastor et al. 1993). Domesticated sheep will avoid grass species that are tougher and lower in nutrients, sometimes altering the relative abundance of native and exotic grass species (Yelenik and Levine 2010).

Critically, the spatial scales at which plants vary in chemistry on the phytochemical landscape, relative to the mobility of herbivores, determine how and when herbivores may forage selectively within and among plants. Mobile herbivores select among chemistries at larger spatial scales than do more sedentary herbivores (Barbosa et al. 2009). For example, the greater sage grouse, *Centrocercus urophasianus*, makes foraging decisions on at least three spatial scales based in part on the chemistry of its host plants (Frye et al. 2013). First, it chooses habitats of black sagebrush over big sagebrush because the former has lower concentrations of PSMs than does the latter. Second, within black sagebrush habitat, grouse choose habitat patches within which plants exhibit high nutrient concentrations and low secondary metabolite concentrations. Finally, within patches, grouse choose the most nutrient-rich and defense-poor individuals upon which to feed (figure 4.3). Similarly, koalas forage selectively among patches in which their *Eucalyptus* host trees are high in protein and low in phenolics. Within those habitat patches, they also forage selectively on the most nutritious and least defended individual trees (Moore et al. 2010). Both grouse and koalas

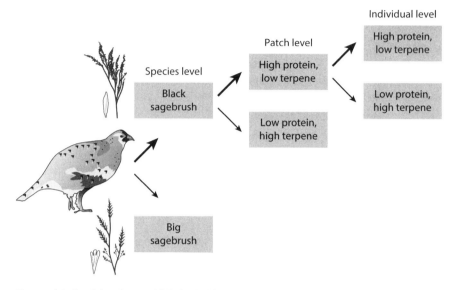

FIGURE 4.3. Spatial scales at which the herbivorous greater sage grouse, *Centrocercus uropha-sianus*, makes foraging decisions among sage plants. Birds preferentially choose to feed on black sagebrush over big sagebrush. Additionally, birds choose to feed in patches of plants, and on individual plants within those patches, that contain high concentrations of protein and low concentrations of terpene. Based on Frye et al. (2013).

are foraging selectively on the phytochemical landscape based on heterogeneity in the nutrients and PSMs of autotrophs.

Of course, some herbivores forage selectively on the phytochemical landscape to increase, not decrease, their consumption of PSMs. For example, wood-rats increase the proportion of juniper in their diet when ambient temperatures fall, because the PSMs in juniper increase woodrat metabolic rates and facilitate thermoregulation (Dearing et al. 2008). The study by Dearing et al. (2008) raises an important issue. Choices made by foraging herbivores, particularly those with some degree of generalist diet, are dynamic and vary with the physiological needs of the foraging animal and its risk of predation. Work by Stephen Simpson and David Raubenheimer revolutionized the way that ecologists view the foraging strategies of consumers, based on what they describe as the geometry of nutritional decisions (Simpson and Raubenheimer 1993). Figure 4.4A illustrates a simplified version of their ideas. At a given moment in time, an herbivore will be in a certain physiological state (S) with respect to its needs for resources such as two nutrients, A and B. It will also have a target, T, representing the ideal balance of the two. However, the herbivore's available sources of food are either low in nutrient A compared to B (plant 1) or low in nutrient B compared

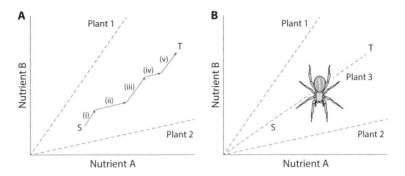

FIGURE 4.4. An herbivore with internal nutrient state S feeds on two primary producers that vary in their ratios of nutrients A and B. The herbivore has a target nutrient state T that can be reached only by alternating feeding bouts (arrows) on plants 1 and 2. The length of the arrows represents the length of the feeding bout, and arrows are parallel to the isoclines that represent the nutrient ratios of the two plant species. In (B), a third plant species with an ideal nutrient ratio for the herbivore's target state is unavailable because of high predation pressure, illustrating the interactive nature of phytochemistry and natural enemies in determining the foraging decisions of herbivores. Sometimes the best things to eat are simply too dangerous.

to A (plant 2). By alternating between plants 1 and 2, the nutritional state of the herbivore can move toward the target, T. The length of each arrow represents the duration of each meal, which moves the internal state of the herbivore on a nutritional trajectory that is parallel to the lines representing the ratios of nutrients in plants 1 and 2. The key point is that foraging decisions made by the herbivore (what to eat and how long to do so) are driven by variation in the chemistry of the diet relative to the current needs of the foraging herbivore. The simple model illustrated in figure 4.4A can be expanded to include the multidimensional and complex competing demands of herbivore diet including elemental stoichiometry (Raubenheimer and Simpson 2004), avoiding deleterious PSMs (Raubenheimer and Simpson 2009), and risks imposed by natural enemies (Sotka et al. 2009) (figure 4.4B).

Aquatic herbivores also forage selectively based on the combined forces of phytochemistry and predation pressure (Hay 2009). In most aquatic systems, unicellular phytoplankton (bacteria, diatoms, dinoflagellates, and others) are responsible for the majority of primary production. As a consequence, many aquatic herbivores are filter feeders. While we may imagine that filter feeding is a rather passive mechanism for eating plant material, at least some aquatic filter feeders have considerable choice in what they actually consume (Kirk and Gilbert 1992). For example, some copepods are able to feed selectively and choose among different species of phytoplankton, or between toxic and nontoxic strains of the same phytoplankton species (Porter 1977, Barreiro et al. 2006). The ability of zooplankton to forage selectively and to avoid toxic algae may increase with prior exposure to toxic species (Tillmanns et al. 2011). As a consequence, naïve copepod populations, with no prior experience of toxic algae, may suffer greater deleterious effects from algal toxins than do experienced copepod populations (Colin and Dam 2003); they have not yet learned to navigate the phytochemical landscape. Omnivorous copepods, which can feed on both autotrophs and heterotrophs, can switch to heterotrophic prey in the presence of toxic phytoplankton (Saage et al. 2009).

There are additional strategies that result in selective foraging by zooplankton. For example, some *Daphnia* species, when presented with filamentous cyanobacteria, can change the depth at which they forage and the morphology of their filtering apparatus (Bednarska and Dawidowicz 2007). Temporary feeding inhibition may provide a short term equivalent to selective foraging, whereby feeding ceases until a better source of food is encountered or until energetic demands require the consumption of low-quality phytoplankton (DeMott 1999). At the other end of the zooplankton size spectrum, unicellular microzooplankton such as the heterotrophic dinoflagellate, *Oxyrrhis marina*, use PSMs released by their phytoplankton prey (chemical gradients in the phycosphere) as cues in selective foraging (Breckels et al. 2011).

Macrophytes in both marine and freshwater systems are also often well defended against herbivores (Taylor et al. 2003). As a result, many aquatic grazers forage selectively on the phytochemical landscape, avoiding some macrophytes while consuming others. Selective foraging and various degrees of specialization have been documented in herbivorous amphipods, fish, mollusks, and sea urchins (Hay et al. 1990, Taylor and Steinberg 2005, Poore et al. 2008). For example, generalist consumers avoid feeding on the highly invasive red alga *Bonnemaisonia hamifera* in its new range in the North Atlantic. The alga is protected by a novel chemical weapon, 1,1,3,3-tetrabromo-2-heptanone, that confers resistance against local generalist herbivores (Enge et al. 2012). The toxicity of the alga may also provide a refuge for some herbivore species from their fish predators (Enge et al. 2013).

The key point here is that variation in the chemistry of primary production combines with predation risk to influence the foraging patterns of herbivores in both aquatic and terrestrial ecosystems. Together, autotroph chemistry and predation pressure affect how herbivores behave, what they eat, and how much they consume in a given meal. These foraging decisions have profound consequences for autotroph community structure and ecosystem processes (chapters 5 and 8). But how do these foraging decisions influence the subsequent performance of herbivores?

4.1.2 Effects of Autotroph Chemistry on Herbivore Performance

4.1.2.1 NUTRIENT LIMITATION AND ECOLOGICAL STOICHIOMETRY

Foraging herbivores, constrained by a combination of biotic and abiotic forces, encounter primary producers that vary markedly in chemistry. Variation on the phytochemical landscape influences multiple indices of herbivore performance, from their attractiveness to potential mates (Geiselhardt et al. 2012) to their differential susceptibility to parasites and predators (Smilanich et al. 2009a). As I noted in chapter 3, autotroph biomass may often be high, but is also generally challenging to use as a source of food. That the chemistry of primary producers limits the growth of individual herbivores is aptly demonstrated by the differences in growth efficiencies (unit of mass gained per unit of mass ingested) of most herbivores (2–38%) compared with those of most predators (38–51%) (Speight et al. 2008). Evidently, primary producers often represent much lower quality food per unit bite than do animals.

The nutritional quality of primary producers is therefore a fundamental determinant of trophic structure on the phytochemical landscape, controlling the distribution of biomass between producers and consumers (Cebrian et al. 2009) and the diversity of those consumers (Olff et al. 2002). First, ecosystems based on producers of high nutritional quality are characterized by high ratios of herbivore to producer biomass, high herbivore productivity, and rapid rates of nutrient cycling (Odum 1957, Cebrian 2004). For example, the higher concentrations of nutrients relative to recalcitrant structural compounds in aquatic primary producers promote higher ratios of herbivore to producer biomass in aquatic than in terrestrial systems (Cebrian et al. 2009); terrestrial herbivores appear to suffer greater nutrient limitation than do aquatic herbivores (figure 4.5A, B). Second, high nutritional quality of primary producers promotes high consumer diversity; patches on the phytochemical landscape must be both productive and nutrient rich to support the highest herbivore diversity (Olff et al. 2002) (figure 4.5C).

Early work on the nutritional quality of primary production stressed the importance of nitrogen (N) as a factor that limits herbivore performance in terrestrial ecosystems and (with iron) in marine ecosystems (Mattson 1980). Phosphorus (P) limitation was historically considered more important in freshwater ecosystems (Boersma and Elser 2006). Certainly, many terrestrial herbivores respond positively to N additions to their plants, and we have seen ample evidence of this from our own work. For example, densities of most insect herbivores on oak trees increase after N fertilization of those trees (Forkner and Hunter 2000). Additionally, experimentally induced N limitation decreases the survival of insects on oak trees and the levels of defoliation that they cause (Stiling et al. 2003, Hall et

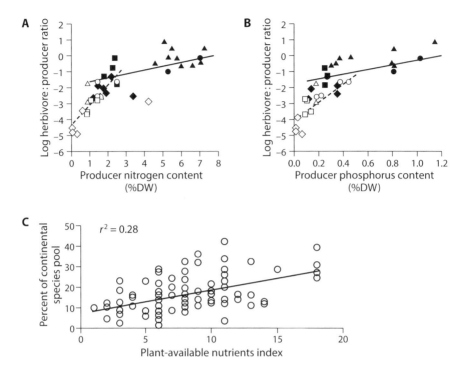

FIGURE 4.5. Relationships among the nutrient content of primary producers, trophic structure (A, B) and herbivore diversity (C). Autotroph nitrogen (A) and phosphorus (B) concentrations predict the ratio of herbivore to producer biomass in both terrestrial (open symbols) and aquatic (closed symbols) ecosystems. The availability of nutrients in autotrophs also predicts the proportion of the available herbivore diversity that will occur in a given habitat (C). In A and B, closed symbols: triangles, pelagic systems (phytoplankton as dominant producer); circles, sediment flats (benthic microalgae as dominant producer); squares, macroalgal beds; diamonds, submerged grass meadows (seagrasses or freshwater macrophytes as dominant producer). Open symbols: triangles, marshlands; circles, grasslands; squares, tundra heathlands; diamonds, shrublands and forests. Modified from Cebrian et al. (2009) and Olff et al. (2002).

al. 2005b). Similarly, on milkweeds, growth rates of monarch caterpillars, aphid population growth rates, and aphid carrying capacities are all positively correlated with foliar N concentrations (Zehnder and Hunter 2008, Tao and Hunter 2012b). The leafhoppers (Cicadellidae: Cicadellinae) provide archetypal examples of adaptations to the chronic low availability of N in xylem. Their enlarged heads hold muscles that can pump up to 100 times their body mass in fluid per hour, and their convoluted guts recirculate xylem to extract nutrients. Moreover, cicadellids excrete ammonia as their primary nitrogenous waste, and concentrations of glutamine and essential amino acids in their diet are good predictors of leafhopper behavior and performance (Brodbeck et al. 2011).

Certainly, in association with lignin concentration, dietary N remains a compelling index of plant quality for many terrestrial herbivores (Loranger et al. 2012). While herbivorous endotherms (e.g., mammal herbivores) may also require substantial quantities of carbon (C) to maintain their high metabolic rates (Klaassen and Nolet 2008), there remains ample evidence that N availability in plant diets limits the performance of many terrestrial mammal herbivores (Dearing et al. 2005). Likewise, the performance of herbivorous marine fishes may be limited in part by the N content of their diets (Goecker et al. 2005).

Again supporting historical views, there are many studies from freshwater ecosystems reporting deleterious effects of low dietary P concentrations on herbivores (Boersma and Elser 2006). Low P availability in algae influences the performance of herbivorous snails (Stelzer and Lamberti 2002), zooplankton (Boersma and Kreutzer 2002), and fish (Borlongan and Satoh 2001). Gradients in the C:P ratios of phytoplankton within and among lakes on the phytochemical landscape are associated with the relative abundance of high-P and low-P zooplankton (Laspoumaderes et al. 2013), illustrating the effects that phytoplankton stoichiometry can have on herbivore community structure. P pollution continues to stimulate record-setting algal blooms in freshwater lakes (Michalak et al. 2013), generating patches of high production on the phytochemical landscape. P limitation may be particularly severe for fast-growing organisms that rely on rapid rates of cell division. Such rapid growth demands substantial allocation of P to ribosomes, and there are very strong relationships across broad taxonomic groups among the growth rate of organisms, their RNA content, and their body P content (Elser et al. 2003).

However, recent advances are challenging the traditional view that freshwater herbivores are always limited by different nutrients than are herbivores in terrestrial and marine systems. The theory of ecological stoichiometry (Sterner and Elser 2002) asserts that the ratios of dietary nutrients (e.g., C:N:P), relative to those of consumers, determine the quality of food for herbivores more than does the absolute concentration of nutrients. The great value of ecological stoichiometry theory is that it makes explicit links between population processes and ecosystem processes (Sterner and Elser 2002) and serves as an excellent model for understanding feedback between trophic interactions and nutrient cycling (Elser et al. 2012). Under this paradigm, a diet is of low quality (for herbivores, predators, or decomposers) when its elemental ratios differ substantially from those of its consumers. In other words, a diet has no intrinsic value, high or low, until its elemental ratios are compared with those of the organism that is trying to eat (or decompose) it. We might therefore expect the foraging decisions of animals to reflect their changing requirements for nutrients relative to their own body tissues (Raubenheimer and Jones 2006).

A stoichiometric perspective has illustrated that it may be hard to make generalizations about which elements limit herbivore growth in which ecosystems. First, recent comprehensive analyses have illustrated that N and P limit primary production about equally across marine, freshwater, and terrestrial environments (Elser et al. 2007, Harpole et al. 2011). As a consequence, herbivore growth may also be limited by both N and P across diverse ecosystems (Elser et al. 2000). Studies from terrestrial systems certainly illustrate the potential of P to limit rates of herbivore population growth (Huberty and Denno 2006, Joern et al. 2012). For example, both densities and larval mass of the winter moth correlate strongly in space with needle P concentrations in Sitka spruce trees (figure 4.6A, B) and not with needle N concentrations. On this coniferous host, P availability on the phytochemical landscape appears to override the more typical factors (host plant phenology and parasitism) that influence the densities and performance of winter moth among its deciduous tree hosts (Hunter et al. 1991). Similarly, low dietary P concentrations can limit the growth rates of *Manduca sexta* caterpillars (Perkins et al. 2004).

Second, ecological stoichiometry provides a theoretical basis to remind us that it is always possible to have too much of a good thing. Stoichiometric theory suggests that, for a given consumer, there are optimal ratios of nutrients in primary producers, and that high concentrations of individual nutrients, even those that are sometimes limiting, can be deleterious. This underlies Bertrand's Rule, which states that there will be costs to overingesting essential nutrients (Mertz 1981). Boersma and Elser (2006) provide examples illustrating fitness costs in aquatic systems from the consumption of P in excess of nutritional demand. Their brief review provides examples from diverse taxa including fish, crayfish, cladocerans, mollusks, and insects. Likewise, an overabundance of N can impose fitness costs upon terrestrial herbivores. For example, fecundity of some grasshopper species is highest at intermediate concentrations of foliar N (Joern and Behmer 1998). In our experiments with *Aphis nerii* on milkweed, we have observed increases in both aphid per capita growth rate and carrying capacity as foliar N concentrations increase (Zehnder and Hunter 2008). However, when foliar N concentration rises very high, the performance of *A. nerii* on milkweed declines, suggesting a metabolic cost to eliminating N that is in excess of demand (Zehnder and Hunter 2009). As a consequence, aphid per capita rates of increase are actually highest at intermediate concentrations of foliar nitrogen (figure 4.6C). Interestingly, P availability has no direct effect on population growth by *Aphis nerii*, but has a strong indirect effect; high soil P increases the uptake of N from soil, exacerbating the deleterious effects of excessive N availability on aphid per capita growth rates.

Stoichiometric theory provides some testable predictions of how the nutritional quality for herbivores of primary producers might vary with nutrient

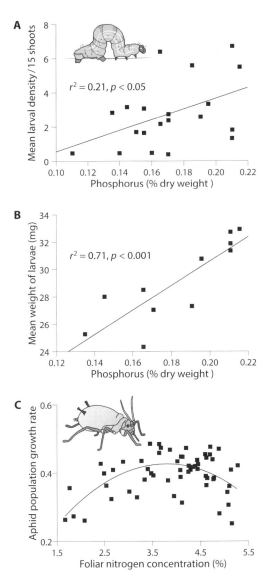

FIGURE 4.6. Plant N concentrations are not necessarily limiting for all terrestrial herbivores. For winter moth, *Operophtera brumata*, feeding on Sitka spruce trees in Scottish moorland, needle P concentration, not N, limits (A) the density and (B) the mass of winter moth larvae. For *Aphis nerii* feeding on common milkweed (C), very high foliar N concentrations reduce rates of population growth. Data from Hunter et al. (1991) and Zehnder and Hunter (2009).

pollution. Fertilizer production for agriculture, and fossil fuel consumption, have resulted in substantial N deposition in ecosystems worldwide (Vitousek et al. 1997), with the potential to induce P limitation in systems that were previously N limited (Vitousek et al. 2010). For example, increasing anthropogenic N deposition in Sweden is expected to increase phytoplankton biomass overall, and impose increasing P limitation (Bergström et al. 2013). Already, oligotrophic lakes in Sweden are becoming more P limited in the south of the country, where N deposition rates are greater. N:P ratios of algae are increasing in these lakes, generating novel heterogeneity on the phytochemical landscape and causing an apparent increase in P limitation of grazing invertebrates (Liess et al. 2009). However, P limitation in herbivores is not an inevitable consequence of anthropogenic N deposition. We have explored the effects of N deposition on the potential for P limitation in the milkweed aphid, *Aphis asclepiadis*. We expected to observe increases in aphid population growth rates under P fertilization, particularly at high levels of N deposition. In stark contrast, population growth rates of *A. asclepiadis* actually declined with increasing P fertilization of milkweed because P fertilization exacerbated the nutritional mismatch between the high N:P ratios of the aphids and the lower N:P ratios of the milkweed. In other words, *A. asclepiadis* is always more strongly limited by N than by P (Tao and Hunter 2012b).

While they are clearly important, the relative ratios of nutrients in resources and consumers on the phytochemical landscape provide a partial picture only of how autotroph chemistry affects trophic interactions (this chapter) and nutrient cycling (chapter 6). Multiple chemical traits of primary producers, and interactions among those traits, mediate the links between trophic interactions and nutrient dynamics. First, and perhaps most simply, not all nutrients present in autotrophs are readily available for herbivores (Foley et al. 1999). For example, N that is associated with fiber or bound by tannins may be unavailable for assimilation, especially by browsing mammals (Wallis et al. 2012). Plant tannins reduce the availability of foliar N for brushtail possums feeding on *Eucalyptus* in Australia, so reducing their reproductive success and the growth rates of their young (Degabriel et al. 2009). Likewise, foliar tannins reduce the availability of N for moose in Alaska, reducing their fecundity (McArt et al. 2009). In marine systems, grazing by herbivores within temperate and tropical seagrass meadows does not correlate closely with seagrass nutrient content in part because most nutrients are bound to indigestible fiber (Cebrian and Duarte 1998).

Even when nutrients in primary producers are not bound in complexes with fiber and tannin, they may be in forms that are difficult for herbivores to process. The concentrations of limiting essential amino acids may be more important to herbivore performance than is total N concentration (Barbehenn et al. 2013), and imbalances between the amino acid and ureide composition of autotrophs and herbivores can make some dietary N of low value (Anderson et al. 2004). For example, populations of the soybean aphid rise and fall in concert with the ureide-N concentration of soybean plants. In contrast, total N is a poor predictor of aphid population size on soybean, illustrating that the form of N matters to herbivore performance (Riedell et al. 2013). More broadly, the organic molecules in which limiting elements reside have profound effects on the palatability of primary producers. For example, plant N that is in the form of a toxic alkaloid cannot alleviate C:N imbalance in a generalist herbivore in the same manner as can the N available in the (nontoxic) enzyme RuBisCO that catalyzes an early step of carbon fixation. In other words, the elements essential for life are often trapped within recalcitrant or toxic PSMs (Dyer et al. 2003).

Additionally, while N and P availability in autotrophs often impose fundamental limitations on herbivore performance on the phytochemical landscape, this is not always the case. For example, the performance of *Oedaleus asiaticus*, a dominant locust in north Asian grasslands, actually declines as the N content of its diet increases. In this species, livestock grazing decreases the N content of grassland plants, improves the performance of the locust, and appears to initiate locust outbreaks (Cease et al. 2012). In reality, diverse mineral nutrients are associated with the distribution, abundance, and performance of grasshoppers (Joern et al. 2012). Elements, such as sodium, may be more limiting for herbivorous ants (Kaspari et al. 2008b, Chavarria Pizarro et al. 2012) whereas other elements and their oxides (e.g., silica) can serve as potent defenses against herbivores (Cooke and Leishman 2011, Reynolds et al. 2012).

Beyond measuring simple elements and their ratios, multiple characteristics of phytochemistry, and interactions among those characteristics, can influence the performance of herbivores (Loranger et al. 2012). For example, polyunsaturated fatty acids (PUFAs) are synthesized almost exclusively by primary producers and can limit the growth, survival, and reproductive rates of marine and freshwater consumers (Brett and Muller-Navarra 1997). Sterols and PUFAs interact to colimit the growth of *Daphnia magna* in freshwater (Sperfeld et al. 2012). Moreover, food webs may become increasingly lipid limited at higher trophic levels (Wilder et al. 2013). PUFA availability constrains the metamorphosis of omnivorous mussels, which depend upon algal-derived PUFAs rather than zooplankton fatty acids to initiate larval settlement (Toupoint et al. 2012). Moreover, subsequent mussel performance is limited more by the quality than the quantity of dietary lipids, with higher cohort survival when algal lipids are available. In short, lipid availability on the phytochemical landscape is an important determinant of trophic interactions in aquatic ecosystems.

More fundamentally, effects of autotroph nutrients and their ratios on herbivore performance cannot adequately be assessed without reference to the myriad

of PSMs that occur on the phytochemical landscape (chapter 3), and their consequences for herbivore fitness. Additionally, the impacts of nutrient stoichiometry and PSMs on herbivore performance are not independent of one another; the
toxicity of PSMs can vary with the elemental mismatch between autotroph and
herbivore. Examples of these two processes are described below.

4.1.2.2 INDEPENDENT EFFECTS OF PLANT SECONDARY
METABOLITES (PSMs) ON HERBIVORE PERFORMANCE

Terrestrial PSMs. As discussed above, nutrient ratios in primary producers can
impose a fundamental constraint on herbivore performance. However, nutrients
alone do not determine the quality of food for herbivores. For example, the chemical elements required for life often come packaged in toxic or indigestible forms.
Limiting nutrients such as N may be incorporated into alkaloids that are toxic to
a majority of generalist herbivores (Wink 1994). Even some (nonprotein) amino
acids can be extremely toxic to herbivores (Rosenthal 1977). Carbon, too, can
occur in forms that are toxic and deterrent (Barbehenn et al. 2005), in forms that
bind with other nutrients (Foley et al. 1999, Wallis et al. 2012), or in molecules
such as lignin that are enzymatically challenging to catabolize (Ding et al. 2012).
Assuming that an herbivore is able to include a particular toxic autotroph in its
diet at all, the amount of primary producer biomass that the herbivore is able
to consume may be constrained by the rate at which it can detoxify the chemical defenses (Marsh et al. 2005), which may select for diet mixing in herbivores
as different as mammals and bees (Dearing et al. 2005, Eckhardt et al. 2014).
Excellent volumes have been written on the many effects of PSMs on herbivores
(Rosenthal and Berenbaum 1991, Iason et al. 2012), and there is no doubt that
autotroph defenses influence multiple aspects of herbivore ecology and evolution
on the phytochemical landscape (Wittstock and Gershenzon 2002). Because of
space limitations, I provide just a few examples below to emphasize that autotroph
quality is not simply a function of nutrient stoichiometry and that PSMs are a fundamental component of the phytochemical landscape upon which herbivores feed.

The magnitude of spatial and temporal variation in phytochemistry is often
astonishingly high. Earlier, I noted that primary producers vary significantly in
the concentrations of nutrients (e.g., N and P) in their tissues. However, concentrations of PSMs generally vary much more than do those of elemental nutrients.
For example, among three species of milkweed in the genus *Asclepias*, foliar N
concentrations vary from about 1.62 to 3.68 percent of dry weight (2.27-fold)
while foliar P concentrations vary from about 0.19 to 0.28 percent of dry weight
(1.47-fold). Among the same milkweed species, foliar cardenolide concentrations
vary over 50-fold (Tao et al. 2014), generating direct negative effects on aphid
(de Roode et al. 2011) and caterpillar (Sternberg et al. 2012) performance. The

key point is that variation in food quality on the phytochemical landscape may be driven much more by variation in PSMs than by variation in primary nutrient availability.

Within autotroph species, levels of variation in chemical defense among individuals are often surprisingly high. On a hillside in Pennsylvania, we have observed foliar tannin concentrations varying over 13-fold among individual oak trees and foliar astringency varying over 22-fold. Although trees were of a similar age, growing on similar soils, and on a slope of constant aspect, this phytochemical landscape exhibited substantial spatial variation. Notably, protein concentration varied only 1.5-fold among the same individual trees. Moreover, the phytochemical variation mattered to diverse members of the herbivore community on those trees. Densities of gall-forming and leaf-mining insects increased with foliar condensed tannin concentrations whereas densities of leaf-chewing insects decreased with foliar gallotannin concentrations (Hunter 1997). Phenolics and tannins have diverse modes of action against herbivores, depending in part of the gut physiology of the herbivore and the structure of the phenolics in question (Barbehenn et al. 2009). The end result is that spatial and temporal variation in phenolics and tannins on the phytochemical landscape have pervasive effects on vertebrate and invertebrate herbivores (Foley et al. 1999, Barbehenn and Constabel 2011).

What happens when autotroph chemical defenses are combined? Most primary producers contain diverse mixtures of PSMs, yet we know remarkably little about how they function synergistically to influence the preference and performance of herbivores. In the tropical shrub *Piper cenocladum* three individual foliar amides have negative effects on both generalist and specialist insect herbivores. Importantly, mixtures of the amides have more deleterious effects on herbivore performance than would be predicted from the effects of individual amides alone. In other words, amide mixtures act synergistically to defend *Piper cenocladum* from a wide range of herbivores (Dyer et al. 2003) (figure 4.7). Because the rate at which herbivores can detoxify PSMs mediates their effects on performance (Marsh et al. 2005), mixtures of defense chemicals may interfere with the ability of herbivores to detoxify the individual constituents (Berenbaum and Neal 1985). Moreover, synergistic defenses may prevent herbivores from compensating for low dietary quality by simply eating more autotroph biomass (Steppuhn and Baldwin 2007). Additionally, synergistic chemical defenses may increase herbivore susceptibility to their natural enemies (Richards et al. 2010) (below). Understanding the interactive effects of PSM mixtures on trophic interactions should be a high priority for future research.

PSMs also affect the performance of herbivores through their impact on nutritional mutualists. That herbivores associate with beneficial microbes to aid their nutrition is well known (Buchner 1965, Douglas 1998), but the degree to which

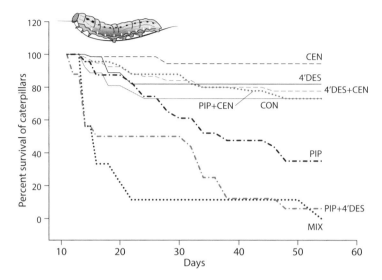

FIGURE 4.7. Synergistic effects of amides in the foliage of *Piper cenocladum* on the survival of the generalist herbivore, *Spodoptera frugiperda*. Caterpillars were reared on artificial diet containing 3 amides (CEN, 4'DES, PIP) in all possible combinations, including a 3-compound mixture (MIX). Control diets (CON) contained only solvent. Data from Dyer et al. (2003).

PSMs mediate interactions between herbivores and their nutritional mutualists remains poorly understood (Janson et al. 2008). However, evidence is accumulating rapidly that plant PSMs influence herbivore performance through their indirect effects on nutritional symbionts. For example, the terpene profiles of ponderosa pine trees influence the growth of the fungus *Entomocorticum*, which is a critical nutritional mutualist of the bark beetle, *Dendroctonus brevicomis* (Davis and Hofstetter 2012). In other words, the effects of pine PSMs on bark beetles are indirect, mediated by their effects on fungal symbionts.

It is also becoming apparent that microbial communities within the guts of herbivores participate in the detoxification or metabolism of PSMs. Metagenomic approaches illustrate that the gut microbiomes of grasshoppers, cutworms, and termites reflect their capacity to degrade and utilize different sources of phytochemistry (Shi et al. 2013). Herbivores with simple diets, like fruit flies that feed primarily on sugar-rich sources of food, have impoverished gut microbial communities (Wong et al. 2011). Moreover, prior exposure to a particular suite of PSMs may select for certain microbial communities in herbivore guts, which may serve to decrease the impact of toxic PSMs on herbivore performance. In woodrats, for example, prior exposure to the toxins in creosote bush influences how gut microbes respond to experimental addition of creosote bush toxins to diet; gut microbial diversity increases in pre-exposed woodrats, but declines in those individuals without prior experience of creosote bush toxins. Essentially, the foreguts of woodrats may serve as detoxification chambers, supporting microbial communities appropriate to the PSMs of their diet (Kohl and Dearing 2012) and enabling woodrats to exploit chemically defended plants across the phytochemical landscape that they might otherwise avoid (Kohl et al. 2014).

Aquatic PSMs. As I noted in chapter 3, aquatic primary producers also synthesize an extraordinary diversity of secondary metabolites. As in terrestrial systems, the phytochemicals produced by aquatic autotrophs influence a wide variety of ecological processes, with defense against herbivore attack only one among many potential roles played by aquatic PSMs (Cembella 2003). Whether they have evolved as defenses or not, the key point for our current purpose is that variation in the expression of aquatic PSMs on the phytochemical landscape influences profoundly the preference and performance of aquatic herbivores.

Secondary metabolites produced by algae are often referred to as phycotoxins (from *phycology*, the scientific study of algae). Perhaps the best-known phycotoxins are those released during harmful algal blooms, which can cause the death of organisms and the disruption of food webs from the tropics to the poles (Pohnert et al. 2007). For example, the paralytic shellfish poisons (PSPs) produced by some dinoflagellates block sodium channels in cell membranes, causing rapid swelling and lysis of cells. Because ion channels occur throughout the metazoa, PSPs are toxic to a broad range of organisms including zooplankton, fish, and mammals (Hackett et al. 2004). Likewise, the amnesic shellfish poison (ASP) domoic acid acts as a potent neurotoxin in mammalian, avian, insect, and crustacean nervous systems (Lewitus et al. 2012). Domoic acid is a secondary amino acid that acts as a glutamate agonist, so causing extensive neuronal depolarization in areas rich in glutamine receptors. As a consequence, domoic acid can poison species at many levels in aquatic food chains, with ecosystem-level consequences (Cembella 2003). In short, spatial and temporal variation in phycotoxin concentration on the phytochemical landscape can have a dramatic influence on trophic interactions and food web structure by killing diverse metazoa.

Of course, not all effects of phycotoxins on trophic interactions are as dramatic as those that arise from the most toxic algal blooms. The (rather beautiful) diatom *Thalassiosira rotula* produces unsaturated aldehydes that induce programmed cell death (apoptosis) in the embryos of both copepods and sea urchins. As a result, diatom toxicity results in reproductive failure of both herbivore groups (Romano et al. 2003), reducing herbivore recruitment at the population level (Miralto et al. 2003). Other phycotoxins have more variable effects on the preference and performance of consumers. For example, a wide variety of phytoplankton produce dimethylsulfoniopropionate (DMSP) and its cleavage enzyme DMSP-lyase (Cembella 2003). While the evolutionary advantage of DMSP production remains unclear, it appears to act as a feeding deterrent against some heterotrophic dinoflagellates and ciliates but as a feeding stimulant to many other organisms (Seymour et al. 2010).

Clearly, phycotoxins are not equally deleterious to all herbivores and may play no significant defensive function in some systems (Flynn and Irigoien 2009). This had led to some controversy surrounding the role of phycotoxins as evolved defense mechanisms against herbivores (Sieg et al. 2011). Their ecological effects are certainly highly variable. For example, the effects of the cyanobacterial hepatotoxin, microcystin, on zooplankton appear to range from strongly negative to completely neutral (Hansson et al. 2007). Microcystin can reduce rates of feeding, movement, and growth in some zooplankton species (Lampert 1987) while having no apparent effects on others (Wilson et al. 2006). Moreover, the toxicity of *Microcystis* to *Daphnia* appears to depend upon genotype-by-genotype interactions, establishing the potential for local adaptation between grazer and cyanobacterial populations (Lemaire et al. 2012). *Daphnia* populations may also adapt locally to the protease inhibitor defenses of their cyanobacterial food (Schwarzenberger et al. 2012). Given that the enzymatic machinery to produce microcystins appears to predate the evolution of metazoa (chapter 3), current effects of microcystins on grazers must be unrelated to their original evolutionary value.

In comparison to larger species of cladocerans, small-sized cladocerans and copepods appear to be less affected by toxin-producing cyanobacteria (Kirk and Gilbert 1992, Guo and Xie 2006), becoming relatively more abundant during cyanobacterial blooms (Edmondson and Litt 1982). A key point here is that phycotoxins such as microcystin, by affecting differentially the performance of zooplankton species or genotypes, can change herbivore community structure on the phytochemical landscape of aquatic ecosystems. For example, extensive sampling across six lakes in southern Sweden, in combination with mechanistic laboratory experiments, has shown that microcystin can profoundly change zooplankton community structure (Hansson et al. 2007). Specifically, high microcystin concentrations reduce the densities of *Daphnia* and calanoid copepods, thereby facilitating increases in the densities of smaller and less efficient phytoplankton feeders such as cyclopoid copepods, *Bosmina*, and rotifers (figure 4.8). In short,

FIGURE 4.8. In Swedish lakes, declines in total zooplankton biomass (A) caused by high microcystin concentrations derive from differential effects of microcystin on zooplankton classes (B), orders, and genera (C). Consequently, microcystin alters zooplankton community structure across the phytochemical landscape. Data from Hansson et al. (2007).

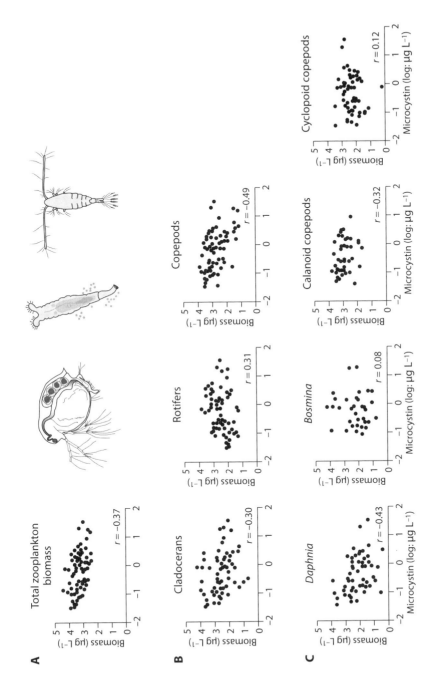

A

Total zooplankton biomass

Biomass (µg L⁻¹)

$r = -0.37$

B

Rotifers

$r = 0.31$

Copepods

$r = -0.49$

Cladocerans

$r = -0.30$

Microcystin (log: µg L⁻¹)

C

Cyclopoid copepods

$r = 0.12$

Microcystin (log: µg L⁻¹)

Calanoid copepods

$r = -0.32$

Microcystin (log: µg L⁻¹)

Bosmina

$r = 0.08$

Daphnia

$r = -0.43$

Microcystin (log: µg L⁻¹)

Biomass (µg L⁻¹)

the effects of phycotoxins on individual herbivore taxa can generate substantial changes in herbivore community structure across the phytochemical landscape of Swedish lakes.

As with phytoplankton, the toxicity of marine multicellular algae (the seaweeds) varies substantially among algal species, among populations within species, and among geographic regions of the ocean (Bolser and Hay 1996, Taylor et al. 2003). In other words, the seaweeds generate substantial heterogeneity on the phytochemical landscape. For example, tropical reef seaweeds (e.g., bryopsidalean green algae and dictyotalean brown algae) more commonly produce lipophilic secondary metabolites than do temperate reef seaweeds (e.g., laminarian and fucalean brown algae, which favor water-soluble phlorotannins). Because they lack significant evolutionary history with lipophilic defenses, sea urchin herbivores from temperate regions are more susceptible to these secondary metabolites than are sea urchins from tropical regions (Craft et al. 2013). Additionally, closely related sea urchin species exhibit more similar responses to lipophilic secondary metabolites than do distantly related species, illustrating the important role of herbivore phylogeny in influencing their responses to variation on the phytochemical landscape. Similarly, the phylogeny of marine amphipod herbivores predicts in part their responses to algal chemical defense (Poore et al. 2008). Not surprisingly, seaweed phylogeny also plays a role in determining resistance to grazers, presumably because closely related taxa share algal defense traits (Poore et al. 2012). In combination, these studies emphasize the importance of evolutionary history in the interactions between marine herbivores and their chemically defended algal resources.

The chemical defenses of freshwater macrophytes also influence the preference and performance of herbivores. For example, crayfish tend to avoid the freshwater angiosperm *Saururus cernuus*, despite its high nutritional value and soft texture (Bolser et al. 1998). Crayfish are deterred from eating *S. cernuus* because it contains a complex mixture of PSMs, including seven lipid-soluble lignans (Kubanek et al. 2001). Similarly, the high N foliage of watercress is protected from generalist herbivores by the sulfur-containing compound 2-phenylethyl isothiocyanate (Newman et al. 1996). When forced to feed on fresh watercress leaves, generalist herbivores grow slowly or actually lose mass. Finally, the aquatic orchid *Habenaria repens* deters crayfish because it contains the hydroxybenzyl-succinate derivative, habenariol (Wilson et al. 1999).

Root PSMs. One fundamental difference between the algae and cyanobacteria that dominate primary production in aquatic systems, and the land plants that contribute most to production in terrestrial systems, is that the latter have roots. In terrestrial ecosystems, the plant tissue that we see aboveground represents only a fraction of that available to herbivores (Hunter 2008) and plants defend their roots from herbivores with a wide variety of PSMs (Van Der Putten 2003). In short, root chemistry contributes to the phytochemical landscape of terrestrial ecosystems.

To put things in perspective, the area of fine roots in terrestrial ecosystems often exceeds that of leaf area aboveground (Jackson et al. 1997). Commonly, over 50 percent of net primary production is allocated belowground in forests (Hendrick and Pregitzer 1992) and grasslands (Sims and Singh 1978), establishing an enormous potential resource for herbivores that can overcome the challenges of feeding belowground (Hunter 2001c). Even after accounting for the C that is allocated to rhizodeposition (secretion of C around the root) and directly to root symbionts, the biomass of fine root tissue belowground can be enormous, exceeding 8,000 kg ha^{-1} in forests (Hendrick and Pregitzer 1993) and 6,000 kg ha^{-1} in grasslands (Sims and Singh 1978). With such an abundance of fine root tissue, it is no surprise that there are herbivores able and willing to attack roots, and their densities can reach staggering levels. Some reported densities of invertebrate herbivores include 3.75 million cicadas ha^{-1} (Dybas and Davis 1962), 0.2 million chrysomelid beetles ha^{-1} (Pokon et al. 2005), and 15 million nematodes m^{-2} (Sohlenius 1980). In other words, if we want to understand how plant chemistry mediates the links between trophic interactions and nutrient dynamics, we cannot ignore the chemistry of roots (van Dam 2009).

Roots vary substantially in their structure and function and, as a result, in their chemistry. Perennial plants necessarily have perennial roots, and the large diameter perennial roots of trees and shrubs are principally engaged in structural support, storage (often starch), and xylem and phloem transport. The woody portions of roots are built primarily of cellulose and hemicellulose fibers, bound together with lignin. Only herbivores that host anaerobic bacteria or protozoa can usefully catabolize woody root tissue, the most notable examples being termites (Fensham and Bowman 1992). In contrast, starch is a glucose-based polymer that can provide an energy rich source of food for herbivores. However, without additional sources of N and P, woody root tissue presents substantial challenges as a food source for most herbivores. Interestingly, some termites may receive supplemental N from spirochete bacteria in their hindguts that are able to fix atmospheric N (Lilburn et al. 2001), so substantially supplementing the quality of their woody diet.

In contrast to large woody roots, fine roots are the principal sites of exchange between plants and the soil environment, including the symbiotic microbes that live in, on, or surrounding fine roots (chapter 3). Because they are highly active metabolically, transporting nutrients and water into plants while transporting organic molecules to the rhizosphere, fine roots are richer in N- and P-containing enzymes than are large perennial roots. For example, fine root N concentrations of temperate forest trees can vary from 0.85–3.09 percent (Pregitzer et al. 2002). These fine root N concentrations are slightly lower than, but overlapping with, values of foliar N concentrations (Speight et al. 2008). The youngest fine roots are also nonwoody and lack the layer of suberin of older roots. Suberin is a waxy, irregular polymer that is highly hydrophobic and controls the progress of water and solutes in the root systems of plants (Bernards 2002). Suberin is also a major component of cork and, like lignin, is a poor food source for most of the world's herbivores.

There is considerable variation in the nutritional and defensive chemistry of fine roots for herbivores foraging on the phytochemical landscape. At least some of that variation appears based on environmental conditions, with soil nitrate availability increasing the N concentrations of fine roots and decreasing the relative concentrations of recalcitrant carbon compounds such as lignin (Hendricks et al. 2000); root chemistry is in part a product of nutrient dynamics (figure 4.1, chapter 7). There are also taxonomic differences in fine root chemistry, with gymnosperm tree roots exhibiting lower nutritional quality than those of angiosperm trees (Pregitzer et al. 2002). At least for trees, the concentrations of recalcitrant organic molecules in fine roots appear to be consistently high. It is not unusual for lignin, suberin, and tannin-protein complexes to make up 50 percent of the dry mass of young fine roots, suggesting that they are better defended than older roots (Hendricks et al. 2000). Lignin concentrations alone in young fine roots can sometimes approach 50 percent of root mass (Muller et al. 1989), imposing a significant challenge for most herbivores.

In addition to recalcitrant carbon compounds, plant roots are well defended with toxic PSMs, including alkaloids (Baldwin 1989), toxic steroids (Rasmann et al. 2009), iridoid glycosides (Jamieson et al. 2012), glucosinolates (van Dam et al. 2004), terpenes (Rasmann et al. 2005), furanocoumarins (Zangerl and Rutledge 1996), and astringent polyphenolics (Muller et al. 1989). As with defensive chemistry aboveground, the expression of PSMs on the phytochemical landscape belowground varies with plant genotype and environment. For example, there is heritable genetic variation in pyrrolizidine alkaloid production in the roots of *Senecio jacobaea* (Hol et al. 2004). Similarly, there is genetic variation in the cardenolide concentration of milkweed fine roots (Vannette and Hunter 2014) and in the rhizome alkaloid concentration of bloodroot, *Sanguinaria canadensis* (Salmore and Hunter 2001a). However, the expression of bloodroot alkaloids declines with increasing light and increasing nutrient availability (Salmore and Hunter 2001b), illustrating the important role that environmental variation plays in the expression of root PSMs at landscape scales. As a particularly powerful example, supplementary N availability increases iridoid glycoside concentrations in the roots of *Linaria dalmatica* by an astonishing 400 percent while having no influence on foliar iridoid glycoside concentrations of the same plant species (Jamieson et al. 2012).

It is worth noting that root chemistry is not expressed entirely independently of foliar chemistry in some systems. Genetic correlations between foliar and root chemical defenses suggest that selection on one trait can influence the expression of the other, thereby linking the phytochemical landscapes aboveground and belowground. This is illustrated by milkweed, in which foliar and root cardenolides are correlated among species (Rasmann et al. 2009), and by *Plantago lanceolata*, in which the expression of iridoid glycosides in roots and leaves are correlated among genotypes (De Deyn et al. 2009). A volatile chemical defense in maize, (E)-β-caryophyllene, is also correlated genetically aboveground and belowground (Erb et al. 2011). In short, this means that selection by herbivores aboveground can influence root chemistry belowground and vice versa. It is both a sobering and fascinating thought that some of the chemical variation that we see driving trophic interactions and ecosystem processes aboveground may in fact emerge from selection by root herbivores imposed on the phytochemical landscape belowground.

4.1.2.3 INTERACTIONS BETWEEN PSMs AND ELEMENTAL STOICHIOMETRY

We make an unfortunate oversimplification when we treat the nutrient and secondary metabolite components of phytochemistry as independent of one another in determining the quality of primary production as food for herbivores. In reality, autotroph quality depends upon interactions between elemental nutrients and PSMs (Behmer et al. 2002). Such interactions have been documented for a long time, and I have mentioned one familiar type already; in vertebrate herbivores with acidic guts, tannins appear to have their primary effect on food quality by reducing the digestibility of plant protein (Dearing et al. 2005). Phlorotannins in brown algae can serve a similar function in marine ecosystems (Targett and Arnold 2001). While tannin-protein complexes are no longer viewed as an effective defense mechanism against insect herbivores (Barbehenn and Constabel 2011), there is little doubt that they can serve as a potent defense against vertebrate browsers and some marine grazers (Degabriel et al. 2009, Wallis et al. 2012).

However, there are more subtle interactions between elemental nutrients and PSMs that influence the quality of the phytochemical landscape for herbivores. For example, in work with monarch caterpillars, we have shown that high concentrations of foliar N have negative impacts on larval growth, but only when foliar cardenolide (toxin) concentrations are also high (figure 4.9A) (Tao et al. 2014). While autotroph nutrient ratios that differ substantially from those of consumers impose metabolic costs associated with excretion of the overabundant elements (Boersma and Elser 2006), our data suggest that elemental nutrient imbalance may be particularly deleterious when PSM concentrations are also high. A complementary way of stating the same result is that the per mg toxicity of cardenolides

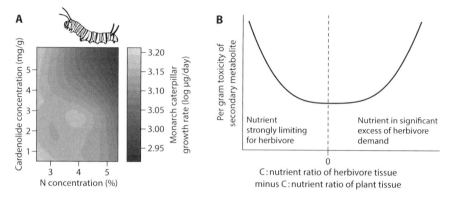

FIGURE 4.9. Interactive effects of autotroph nutrient and PSM concentrations on herbivore performance. (A) Combinations of high foliar N and cardenolide concentrations in milkweed foliage result in the lowest rates of monarch caterpillar growth (darkest shading) (data from Tao et al. 2014). (B) In general, we hypothesize that the per unit toxicity of PSMs increases when C:nutrient ratios in autotrophs are either much higher or much lower than are those in herbivores.

to monarch caterpillars increases as foliar N concentrations rise above the optimum for larval growth (figure 4.9A). In this example, it is really a matter of perspective whether we consider the elemental nutrient or the steroid as the toxin because the action of each is contingent on the concentration on the other. As David Raubenheimer and Stephen Simpson have pointed out, the dividing lines between toxins, nutrients, and medicines are "thin, vague, and heavily contingent" (Raubenheimer and Simpson 2009). For monarch caterpillars, the cost of accommodating one type of nutritional stress (high cardenolide) presumably compounds that of accommodating a second type (excess N).

At the other end of the nutritional spectrum, very low autotroph nutrient concentrations relative to herbivore requirements might also increase the per mg toxicity of PSMs. For example, the toxicity of cardenolides for monarch caterpillars declines as the C:P ratio of the foliage also declines (L. Tao, unpublished data), suggesting that alleviating P limitation can reduce the toxicity of at least some secondary metabolites. Taken together, our experiments with monarchs suggest that either very low or very high concentrations of nutrients in autotroph tissue, relative to the demand of herbivores, can increase the effective toxicity of PSMs (figure 4.9B). Similar effects have been reported for grasshoppers, in which the toxicity of condensed tannins is greatest at either very high or very low ratios of protein to carbohydrate in the diet (Simpson and Raubenheimer 2001).

Of course, interactions between nutrients and toxins are not limited to terrestrial herbivores. In an elegant study, investigators manipulated the toxin and nutrient content of the diets of six species of marine crustacean (Cruz-Rivera and

Hay 2003). Concentrations of diterpene alcohols from the brown alga *Dictyota menstrualis* were added at natural concentrations to artificial diets varying in their nutrient concentrations. For three of the herbivore species, the algal secondary metabolites decreased feeding to a greater extent in the low-nutrient diets than in the high-nutrient diets. In other words, for fully half of the species tested, the impacts of PSMs on herbivore performance were contingent upon dietary nutrient concentrations. If interactions between dietary nutrients and PSMs matter to mammals, monarch butterflies, and marine crustaceans, they are likely important in most ecosystems and merit much more detailed study. Importantly, nutrient-PSM interactions may explain variation in herbivore performance on the phytochemical landscape, and therefore variation in the links between trophic interactions and nutrient dynamics (figure 4.1)

4.1.2.4 Temporal Variation on the Phytochemical Landscape

In the preceding sections, I described how the individual and interactive effects of elemental nutrients and PSMs influence the preference and performance of herbivores. So far, we have seen how autotroph genotype and environment combine and interact to influence phytochemistry, with a focus on chemical variation in space. However, in describing the effects of the phytochemical landscape on herbivore performance, it's important to note that the phytochemical landscape is dynamic in time, with important consequences for the ecology of herbivores (below) and their natural enemies (section 4.2). Here, I provide just a few brief examples of temporal variation in autotroph chemistry and the effects of that variation on herbivore performance.

It is difficult to overestimate the importance of seasonal phenology in driving changes in the chemistry of autotrophs and subsequent effects on herbivores, trophic interactions, and nutrient dynamics. In terrestrial ecosystems, ecologists have recognized for decades that changes in plant nutrient concentration, chemical defense, water content, and toughness vary seasonally, with profound implications for the preference, performance, and population dynamics of herbivores (Feeny 1970). On oak trees, the timing of spring budburst and autumnal leaf fall determine in large part variation among individual trees in the densities of defoliating and leaf-mining insects (Stiling et al. 1991, Hunter et al. 1997). For defoliators, trees that burst bud too late in the season are unavailable for spring colonization whereas those that burst bud too early become poor quality food at a rapid rate (Hunter 1990). Within oak forests in general, the phenology of spring budburst and autumnal leaf fall generates two peaks of defoliator activity and species diversity, a large spring peak and a moderate fall peak (Feeny 1970). These same peaks then drive temporal patterns of defoliation (Hunter 1987) which feed back to influence the chemistry of leaves (chapter 5). For example, on the English oak, *Quercus robur*, 90 percent of annual defoliation occurs during a 6-week period between mid April and early June (Hunter and Willmer 1989). Strong seasonality in oak foliage chemistry therefore drives the timing and chemistry of herbivore-derived materials (feces, abscised green leaves) that fall from the canopy to the forest floor and influence nutrient cycles in soils and streams across the phytochemical landscape (chapter 8). Through their effects on plant chemistry, spring defoliators influence the entire structure of herbivore communities on oak trees (Hunter 1992a) and nutrient dynamics at the watershed scale (Hunter et al. 2012). Insect herbivores in tropical forests are perhaps even more tightly linked to the budburst phenology of their host plants (Dajoz 2000). Most tropical trees accumulate a majority of their leaf damage very early during expansion and appear well defended thereafter.

Seasonal variation in phytochemistry can supersede other biotic and abiotic forces in driving changes in food web structure. For example, temporal variation in the chemistry of striped maple, *Acer pensylvanicum*, foliage has a larger effect on herbivore community structure than does spatial variation in biotic and abiotic forces across a 600-m gradient in elevation in the southern Appalachian mountains (Zehnder et al. 2009). Foliar nitrogen concentrations decline and foliar lignin concentrations increase as the growing season progresses, causing changes in the relative abundance of herbivore groups and their arthropod natural enemies on the phytochemical landscape.

Likewise, in lakes and marine systems, phenological changes in PSM production have important effects on trophic interactions. The relative abundance of eukaryotic microalgae and of cyanobacteria vary seasonally, with concomitant changes in the expression of aquatic toxins (Hansson et al. 2007). For example, the toxic dinoflagellate *Alexandrium* is well known for generating red tides and secreting paralytic shellfish neurotoxins. In reality, *Alexandrium* spends much of its lifecycle in a resting (cyst) stage in marine sediments and is only active as a pelagic primary producer for a few weeks each year. Consequently, there is seasonal variation in the concentration and identity of neurotoxins in both the sediment and the water column of the phytochemical landscape (Persson et al. 2012). Similarly, concentrations of the cyanobacterial toxin, microcystin, vary seasonally in Swedish lakes. Peak concentrations occur in early summer and again in fall, and are associated negatively with zooplankton biomass (Hansson et al. 2007). In Ford Lake in Michigan, blooms of toxin-producing *Microcystis aeruginosa* occur in late summer, as nitrate levels rise in the lake and *M. aeruginosa* outcompetes the N-fixing cyanobacteria, *Aphanizomenon flos-aquae* (McDonald and Lehman 2013). In all of these cases, phenological variation in phycotoxins on the phytochemical landscape influences subsequent trophic interactions.

Perhaps most dramatically, seasonal variation in autotroph chemistry facilitates the evolution of some of the most spectacular animal migrations in the natural world. Migratory consumers can take advantage of spatially separated resource pulses and reach higher overall densities than can nonmigratory species (Fryxell et al. 1988, Middleton et al. 2013). The iconic movement patterns of wildebeest, elk, and mule deer all serve to track the seasonal greening of their food plants. However, as I noted earlier, the foraging behavior of herbivores has multiple ecological drivers, and phenological variation in the chemistry of plants combines with variation in predation pressure and climate to influence the migratory behavior of many herbivores (Fryxell and Sinclair 1988, Hebblewhite and Merrill 2009). For example, seasonal movement patterns of winged planthoppers in salt marshes appear to balance the competing demands of finding high-quality food while minimizing predation pressure (Huberty and Denno 2006). Likewise, seasonal changes in forage quality combine with body size to determine altitudinal migration of Galápagos tortoises (Blake et al. 2013). Moreover, as a changing environment influences the distribution of high-quality plants and the foraging behavior of predators, we should expect subsequent effects on migratory animals. Migratory elk populations in Yellowstone National Park in the United States appear to be suffering declines based on both increasing predation pressure and decreasing availability of high-quality plants across their migration routes (Middleton et al. 2013).

4.1.2.5 PRIMARY PRODUCER CHEMISTRY AND PARENTAL EFFECTS

As illustrated above, spatial and temporal variation on the phytochemical landscape affect the preference and performance of herbivores. Importantly, effects of variation in phytochemistry can persist across generations through non-Mendelian parental effects, wherein the dietary chemistry experienced by the parent influences the phenotype of the offspring (Rossiter 1996). Parental effects are widespread in nature, and there has been a recent surge in interest in epigenetic effects of the current environment on the phenotype of humans and other animals (Liu et al. 2013b). Among natural populations, parental effects may represent adaptations to fine-tune the phenotype of offspring to better fit the environment experienced by parents (Mousseau and Fox 1998). The effects of parental diet chemistry can even be strong enough to decouple herbivore populations from density-dependent processes, with the potential to drive outbreak dynamics (Rossiter 1994, Benton et al. 2005).

Perhaps the simplest form of parental effect occurs when parents invest differentially in their offspring based on current dietary conditions. For example, egg-laying herbivores provide a packed lunch for their offspring in the egg, and the quality of that lunch can be directly related to the chemistry of autotrophs upon which parents feed (Rossiter et al. 1993). Mothers therefore have the opportunity to adjust their reproductive strategy based upon their perception of diet quality. For example, some *Daphnia* mothers produce a few large offspring when phytoplankton resources are poor and many smaller offspring when phytoplankton resources are good (Gliwicz and Guisande 1992), establishing a strong link between parental diet and offspring phenotype. Moreover, individual *Daphnia magna* can develop induced resistance to the toxic cyanobacterium *Microcystis* and transfer that toxin resistance to their offspring via maternal effects (Gustafsson et al. 2005). Back in the terrestrial environment, *Pieris rapae* butterflies modify the size of their eggs based on the protein content of the diet. As a consequence, mothers that feed on very high or very low protein diets produce offspring that grow best under those same conditions (Rotem et al. 2003). Similarly, parental diet chemistry in the gypsy moth, *Lymantria dispar*, influences the growth rate and fecundity of offspring (Rossiter 1991) with consequences for population dynamics (Rossiter 1996).

In monarch butterflies, we have found important effects of parental dietary chemistry on offspring susceptibility to disease. Specifically, parents that feed on high-cardenolide milkweed species contribute those cardenolides to their eggs; both mothers and fathers contribute cardenolides to eggs, although the contributions of mothers is greater than that of fathers (figure 4.10A). Subsequently, monarch offspring from parents reared on high-cardenolide milkweeds are more resistant to a protozoan parasite than are offspring from parents reared on low-cardenolide plants (figure 4.10B), in part because egg cardenolides appear to serve as medicinal compounds against the parasite (Sternberg et al. 2015) (see section 4.2, below). Interestingly, although fathers contribute less cardenolide to eggs than do mothers, their contribution appears to be more pharmaceutically active against the parasites than is that of monarch mothers.

Of course, not all maternal effects are adaptive. Poor nutrition in parents, from humans to *Daphnia*, can simply produce offspring of low fitness. In a striking example, the quality of forage experienced by both parents and grandparents in Minnesota deer populations influences the susceptibility of offspring (and grand-offspring) to wolf predation (Mech et al. 1991). The important point is that variation in defensive and nutritional chemistry on the phytochemical landscape influences herbivore phenotypes across generations, with consequences for trophic interactions and population dynamics (Hunter 2002b).

In summary, spatial and temporal variation in the chemistry of primary producers influence the preference and performance of herbivores in terrestrial and aquatic systems. In the next section, I explore the consequences for herbivore population dynamics of variation on the phytochemical landscape.

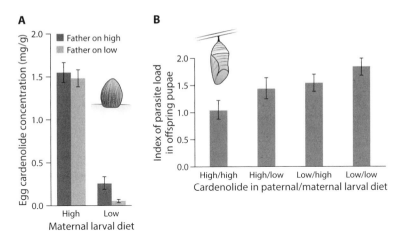

FIGURE 4.10. Effects of parental diet chemistry on resistance of offspring to a protozoan parasite. (A) Monarch butterfly parents reared on high-cardenolide milkweeds ("high") contribute significantly more cardenolide to their eggs than do parents reared on low-cardenolide milkweeds ("low"). (B) Offspring from parents reared on high-cardenolide diets suffer lower parasite loads in the pupal stage than do offspring from parents reared on low-cardenolide diets. Data from Sternberg et al. (2015).

4.1.3 Effects of Autotroph Chemistry on Herbivore Population Dynamics

Studying population dynamics should be easy. The dynamics of all populations are influenced by only four vital rates: rates of birth and immigration, which add individuals to populations, and rates of death and emigration, which remove individuals (Hunter 2002a). Accordingly, the extent to which the phytochemical landscape influences these four vital rates is the extent to which it can influence population dynamics (Hunter 1997). However, phytochemistry influences the vital rates of herbivores by multiple direct and indirect mechanisms (Price et al. 1980), adding considerable complexity to the study of population dynamics. In some ways, it is almost easier to define the indirect effects—whenever the effects of phytochemistry on the vital rates of herbivores are imposed by a third ecological factor (climate, mutualists, enemies, etc.), then the effects are indirect. Examples might include autotroph-mediated variation in the susceptibility of herbivores to disease (Keating and Yendol 1987) or to cold temperatures (McLister et al. 2004). In cases such as these, natural enemies or climate are the proximate agents of population change, but their strengths are determined in part by the chemistry of the herbivores' diet. By simple contrast, direct effects of phytochemistry on herbivore vital rates occur when no other ecological factors mediate their effects on birth, death, or movement (Awmack and Leather 2002). Of course, single direct or indirect effects rarely dominate the population dynamics of species. Rather, overall dynamics often result from subtle combinations of multiple direct and indirect effects (Hunter 1998). It is this interactive nature of population dynamics that make it both a challenging and fascinating area to study.

Box 4.2 illustrates a simplified example of how the spatial and temporal dynamics of herbivore populations on the phytochemical landscape generally arise from interactions between phytochemistry and predation pressure. A significant proportion of the confusion that surrounds this topic has arisen primarily because of poor or inconsistent use of terminology—many of us have tended to use largely meaningless phrases such as "the system is driven primarily from the bottom up" in which the system is some unspecified response in vital rates, driven refers to some unspecified ecological process, and bottom up refers to some unspecified combination of spatial and temporal variation in the quantity and quality of resources (Hunter 2001b). Box 4.2 describes how different components of the same system can be driven by entirely different processes. For example, figure 4.11 in box 4.2 illustrates three herbivore populations, distributed across the phytochemical landscape, in which equilibrium densities (E1, E2, E3) derive from the combination of phytochemistry, predation pressure, and resource limitation. In this simplified example, substantial temporal variation (e.g., V_{T1}) derives from delayed density-dependent predation, whereas spatial variation (V_{S1} and V_{S2}) derive from variation in phytochemistry. To understand how the chemistry of primary production and predation pressure combine to influence population dynamics, we need to be explicit about the emergent response variables that we are trying to understand (e.g., equilibrium, spatial variance, temporal variance, regulation), the vital rates that generate them (birth, death, immigration, emigration), and the ecological factors that impinge upon those vital rates (resource chemistry, resource abundance, predation pressure, climate, pH, etc.) (Hunter 2001b).

There is abundant empirical evidence that phytochemical variation influences the population dynamics of herbivores. For example, in our own work, population growth rates of the milkweed-oleander aphid, *Aphis nerii,* are lower on *Asclepias curassavica* than on *A. incarnata,* apparently because the former expresses much higher concentrations of cardenolides than does the latter (de Roode et al. 2011). What is particularly interesting about aphids on milkweed is that plant chemistry influences both per capita rates of increase and the strength of density-dependent processes (Agrawal 2004), those factors that generate equilibria around which herbivore populations may fluctuate (box 4.2). For example, density-dependent reductions in aphid fecundity, longevity, and adult mass are much greater on *Asclepias viridis* than on other the other milkweed species that we have studied

BOX 4.2.

PHYTOCHEMISTRY, PREDATION,

AND HERBIVORE POPULATION DYNAMICS

The chemistry of primary production interacts with other ecological factors to influence herbivore population dynamics through effects on birth, death, immigration, and emigration. In this simplified example, herbivore population dynamics vary among three patches (P1–P3 in figure 4.11A and B, below) on the phytochemical landscape. The contour lines in figure 4.11 represent points of equal phytotoxin concentration, so that patches P1 and P2 are in a poor-quality (high-toxin) region of the phytochemical landscape whereas P3 is in a high-quality (low-toxin) region of the phytochemical landscape. For simplicity, I ignore rates of movement here, although they can be incorporated easily into such models (Helms and Hunter 2005). Birth rates vary in space on the phytochemical landscape among patches of different chemistry (patches P1, P2, and P3 with birth rates 1, 2, and 3). Birth rates also decline at high population density. In all patches, death rate is initially density dependent at low herbivore density, caused by predators that respond numerically or behaviorally to herbivore abundance. At intermediate densities, predation becomes saturated, and death rate is inversely density dependent. At high herbivore density, competition for food causes density-dependent death rates once again (figure 4.11A).

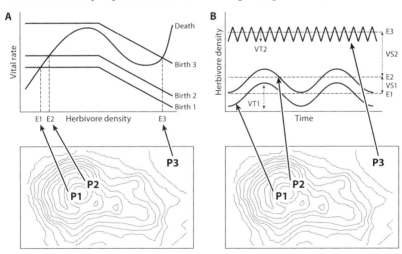

FIGURE 4.11.

In high-toxin patches (P1, P2) on the phytochemical landscape, herbivore birth rates are low and predators regulate herbivores around low-density equilibria.

(Box 4.2 continued)

Critically, the equilibria E1 and E2 exist because of the **combined effects** of low food quality and density-dependent predation on the landscape. Similarly, the equilibrium E3 exists because of the combined effects of a low-toxin diet and density-dependent competition. A key observation is that autotroph chemistry contributes to the generation of population equilibria even when predators regulate herbivore dynamics.

We can illustrate the spatial and temporal dynamics that might result on the phytochemical landscape from these combinations of birth and death rates (figure 4.11B). I have assumed that predation acts as a delayed-density dependent factor, and so can generate cyclic dynamics around E1 and E2 (Turchin 1990, Royama 1992). Competition causes more rapid feedback, and cycles are of shorter period (E3). Here, temporal variation (V_{T1}) around equilibria E1 or E2 is caused by delayed density-dependent predation. In contrast, temporal variation V_{T2} around E3 is caused by competition for resources. Note that spatial variation in herbivore abundance among sites (e.g., V_{S1} and V_{S2}) is caused by spatial variation in autotroph chemistry on the phytochemical landscape.

Overall, variation on the phytochemical landscape mediates the spatial and temporal dynamics of predators and their prey. It is simply not possible to describe the population dynamics of herbivores in even simple systems like this without reference to the combined forces of phytochemistry and predation pressure. Importantly, we should distinguish clearly among response variables (e.g., equilibrium, spatial variance, temporal variance) as we assess the interactive effects of ecological forces on dynamics (Hunter 2001b). □

(Zehnder and Hunter 2007a). Notably, *A. viridis* contains higher concentrations of cardenolide than do the other milkweed species in this study.

A combination of life table analysis, modeling, and experimental work often provides the best understanding of the interactive processes that generate herbivore population change (Hunter 2001b). For example, temporal variance in the long-term population dynamics of the gall-forming sawfly, *Euura lasiolepis*, on the arroyo willow, *Salix lasiolepis*, is determined in large part by the availability of long shoots on willow trees (Price and Hunter 2005). Long shoots are more nutritious and less well defended than are short shoots, resulting in less gall abortion and lower levels of early larval mortality (Price and Hunter 2005). Consequently, long shoots are the preferred oviposition sites for adult female sawfly (Craig et al. 1989). In this desert ecosystem, the availability of long willow shoots, and their average length, varies profoundly in space, increasing with water availability (Price and Clancy 1986). Local patterns of precipitation therefore determine

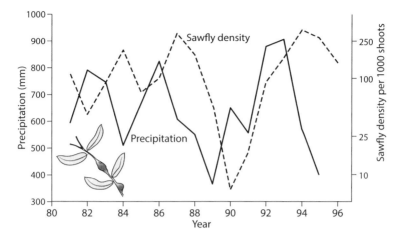

FIGURE 4.12. Impact of precipitation on the dynamics of sawfly populations on willow trees. High levels of precipitation increase the relative abundance of long willow shoots that are high in nutrients and low in chemical defense. As a result, sawfly densities track the rise and fall of precipitation-driven changes in plant chemistry on the phytochemical landscape. Data from Hunter and Price (1998).

long-term temporal variance in population dynamics by generating variation in willow shoot size and chemistry (figure 4.12). Time-series analysis of 22 years of data implicates winter precipitation (snowfall) and the Pearson Drought Severity Index (PDSI) as primary drivers of sawfly temporal variance, explaining from 67 to 73 percent of variation in rates of population change (Price and Hunter 2005). Population models built on these 22 years of data were highly effective at predicting the subsequent 10 years of population change (Price and Hunter 2015). Because sawfly densities have little impact on the availability of long shoots in the following year (and cannot influence snowfall), there is no feedback between sawfly density and food quality, and only very weak density dependence (Hunter and Price 1998, 2000). Similarly, effects of parasitism and predation on sawfly population dynamics are negligible (Hunter and Price 1998).

Some of the most compelling examples of the effects of autotroph chemistry on herbivore population dynamics come from systems in which the dynamics of radically different herbivore species are synchronized on the same primary producers. For example, in southern Norway, the dynamics of moth populations that feed on bilberry leaves are synchronized with those of voles that feed on bilberry twigs. The nutritional and defensive chemistry of bilberry varies among years, depending in part on how plants are allocating their limited resources among growth, reproduction, and defense. Years in which tissues are nutritious for foliar feeders coincide with years in which tissues are nutritious for twig feeders, thereby synchronizing the dynamics of voles and moths. In the same ecosystem, populations of two moss-feeding moths are synchronized with those of moss-feeding lemmings, again because temporal variation in plant chemistry synchronizes the population cycles of taxonomically disparate herbivore species (Selås et al. 2013).

While phytochemistry contributes to the intrinsic rate of increase of all herbivore species (see box 4.2), it is important to recognize that phytochemistry plays a negligible role in explaining temporal variance in some herbivore populations. In some systems, the quantity rather than the chemistry of plant resources may determine the temporal dynamics of herbivores (Abbott and Dwyer 2007, Karban et al. 2012), while in others, abiotic factors such as temperature determine temporal variation (Nelson et al. 2013). In yet other systems, temporal variance in herbivore populations appears linked primarily to the impacts of natural enemies. For example, in a 35-year time series of the gall-forming midge, *Taxomyia taxi*, on yew trees, we find no signal of temporal variation in the chemistry of trees on gall midge dynamics (Redfern and Hunter 2005). Rather, regular 14-year cycles of the hemivoltine midge result from delayed density dependence in a coupled parasitoid-host interaction with *Torymus nigritarsus*. Interestingly, the dynamics of a second parasitoid in the same system, *Mesopolobus diffinis*, provide a compelling example of donor control, with parasitoid densities depending purely on the availability of gall midge hosts. The yew-gall midge-parasitoid system is therefore best described as one in which (a) plant chemistry, by its impact on intrinsic rate of increase, combines with parasitism to generate an equilibrium density for gall midges, but (b) temporal variance of the gall midge and its parasitoid *Torymus nigritarsus* around this equilibrium results from a coupled parasite-host interaction that (c) determines the gall densities available for parasitism by *Mesopolobus diffinis*.

Similarly, temporal variance in the population dynamics of larch budmoth in the Swiss Alps appears to be determined in large part by budmoth-parasitoid interactions. A simple model of interactions between parasitoids and budmoth explains about 90 percent of the variance in budmoth density over time, whereas effects of plant chemistry are much weaker (Turchin et al. 2003). What is notable about this example is that temporal variance in larch budmoth dynamics was thought previously to be determined largely by delayed responses of plant chemistry to budmoth feeding (Baltensweiler and Fischlin 1988). While it can sometimes be challenging to untangle the proximate causes of population change in organisms, high-quality data and strong analytical approaches are vital ingredients (Hunter 2001b).

Despite examples where either phytochemistry (Price and Hunter 2005) or predation pressure (Redfern and Hunter 2005) dominate temporal variance in herbivore populations, it is probably more common that they interact to influence the spatial and temporal dynamics of herbivore populations. Interactions across three trophic levels, or "tri-trophic interactions," are commonly reported in the literature and have had a significant influence on the development of ecological and evolutionary theory (Price et al. 1980, Abrahamson and Weis 1997). A common type of tri-trophic interaction occurs when the impact of a natural enemy on its herbivore prey varies with the chemistry of primary production that the prey consumes. For example, in saltmarsh ecosystems, the abundance of cordgrass herbivores such as planthoppers is determined primarily by spatial variation on the phytochemical landscape; patches of high N plants with low phenolics support the highest planthopper densities (Marczak et al. 2013). However, spiders are better able to constrain the population growth of planthopper prey when those planthoppers are feeding on plants of low nutritional quality (Denno et al. 2002). The result is a stronger effect of spider predation on planthopper population dynamics in high-marsh habitats, where plant quality is lower, than in low-marsh habitats, where plant quality is higher; predator-prey dynamics in saltmarsh phytochemical landscapes behave just like those illustrated in box 4.2. Similarly, rates of parasitism of *Aphis nerii* by the parasitic wasp, *Lysiphlebus testaceipes*, vary spatially from less than 10 percent to over 30 percent among milkweed species (Helms et al. 2004), reducing rates of per capita population growth of aphids (figure 4.13A). Notably, both the strength and form of density dependence vary among the milkweed species (figure 4.13B). However, parasitism rates are not able to prevent exponential population growth of aphids on any of the milkweed species (figure 4.13C), emphasizing that density dependence is necessary, but not sufficient, for population regulation by natural enemies (box 4.2, figure 4.11A).

On oaks, we have studied, experimentally and analytically, how direct effects of phytochemistry on herbivore fecundity combine with indirect effects mediated by natural enemies to influence temporal dynamics of herbivore populations. First, defoliation of oak trees by gypsy moth caterpillars induces increases in foliar tannin concentrations (Hunter and Schultz 1995). High foliar tannin concentrations reduce the growth rate and fecundity of the gypsy moth (Rossiter et al. 1988), representing a direct negative effect of tannins on population growth rate. However, foliar tannins also decrease the susceptibility of gypsy moth caterpillars to a baculovirus (Keating et al. 1988, Hunter and Schultz 1993), an indirect positive effect on gypsy moth population growth rate. We combined these direct and indirect effects into an analytical model that included relationships between foliar tannin concentrations and (a) gypsy moth fecundity and (b) gypsy moth mortality from virus (Foster et al. 1992). In other words, plant chemistry was included explicitly in the population model, and we used it to assess

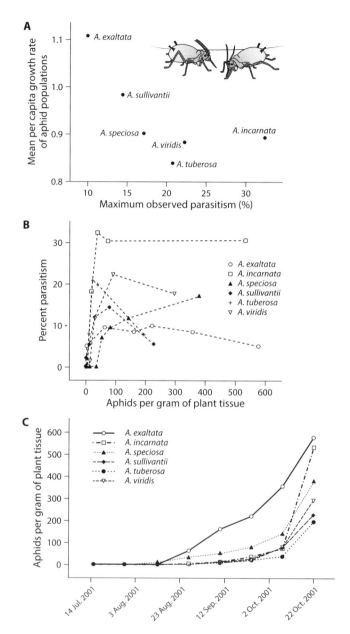

FIGURE 4.13. Autotroph quality and parasitism combine to influence population growth of *Aphis nerii* on milkweed. Aphid per capita growth rates decline with increasing parasitism, which varies among milkweed species (A). Although parasitism rates are density dependent, the form of that density dependence also varies among milkweed species (B), and is insufficient to prevent exponential growth in aphid populations (C). Modified from Helms et al. (2004).

the dynamics that might result from opposing direct and indirect effects of that chemistry. Our results suggested that increases in foliar tannin concentrations should initially increase the amplitude of fluctuations in gypsy moth populations because it takes longer for the virus to initiate population crashes. However, at high foliar tannin concentrations, negative effects on gypsy moth fecundity should come to dominate dynamics, and the amplitude of outbreaks would decrease again. A more recent analysis illustrates that wound-induced changes in oak chemistry, and the resultant chemical interactions with the gypsy moth virus, can drive the alternating high- and low-density outbreaks of gypsy moth in forests that are dominated by oak trees (Elderd et al. 2013). In other words, spatial variation on the phytochemical landscape influences the form of temporal dynamics exhibited by herbivores. Perhaps an additional insight to emerge from our work is the relative ease with which we can incorporate phytochemistry directly and explicitly into analytical population models, yet attempts to do so remain depressingly rare (Elser et al. 2012).

More generally, studies of insects on oak trees serve to illustrate clearly how variation in phytochemistry combines with other ecological factors to determine patterns of herbivore population dynamics. Long-term population monitoring combined with experimental work and modeling has provided detailed information on the combination of ecological factors that determine both spatial and temporal variation in herbivore abundance. In the United Kingdom, both native oak species, *Quercus robur* and *Q. petraea*, can suffer high levels of spring defoliation generated by two species of moth, *Operophtera brumata* (the winter moth) and *Tortrix viridana* (the green oak tortrix). In some years, combined defoliation by these two insects completely removes the first flush of leaves from oak trees, and annual levels of defoliation average 40 percent at some sites on the phytochemical landscape (Hunter and Willmer 1989). *O. brumata* and *T. viridana* have proven particularly interesting to study because they show both marked similarities and some marked differences in their population dynamics.

First, spatial variation in the density of both herbivore species among trees derives primarily from phenological variation in the chemistry of oak leaves. The timing of budburst in spring varies markedly among individual trees and among patches of forest (Hunter 1987, 1992b), and oak leaves decline rapidly in quality for herbivores as they age, losing both N and water, while gaining condensed tannins and becoming tougher (Feeny 1970, Tikkanen and Julkunen-Tiitto 2003). Consequently, *O. brumata* and *T. viridana* time their emergence from eggs to take advantage of oak foliage of the best possible chemistry. Because some larvae hatch too early and either disperse or starve (Hunter 1990), trees that burst bud early are hot spots of high nutritional quality on the phytochemical landscape— such trees accumulate the highest densities of caterpillars (Hunter et al. 1997)

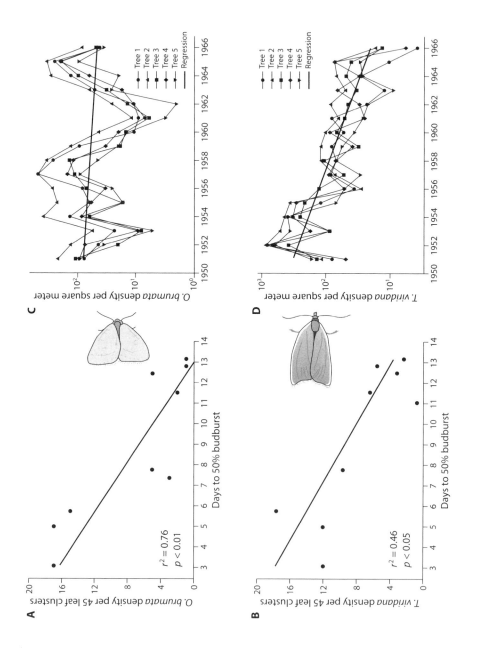

A

O. brumata density per 45 leaf clusters

Days to 50% budburst

$r^2 = 0.76$
$p < 0.01$

B

T. viridana density per 45 leaf clusters

Days to 50% budburst

$r^2 = 0.46$
$p < 0.05$

C

O. brumata density per square meter

Tree 1
Tree 2
Tree 3
Tree 4
Tree 5
Regression

D

T. viridana density per square meter

Tree 1
Tree 2
Tree 3
Tree 4
Tree 5
Regression

(figure 4.14 A, B). For both herbivore species, variation in plant chemistry on the phytochemical landscape explains much of the spatial variation in caterpillar density within and among forests.

However, the two species of herbivore differ substantially in the ecological factors that determine temporal variation in population density. Pupal predators drive much of the temporal variance in *O. brumata* populations, generating cycles in *O. brumata* abundance (figure 4.14C) (Hunter et al. 1997). In contrast, temporal variance in *T. viridana* derives largely from the combination of intraspecific competition among *T. viridana* larvae and some unknown ecological factor that has caused systematic declines in the density of *T. viridana* in our study populations (figure 4.14D). Declines in a number of British moth species have been linked to environmental change (Fox 2013), and *T. viridana* may belong to the group of species at risk. In summary, studying the dynamics of these two moth populations has illustrated the fundamental importance of phytochemistry in generating spatial variation in density, while temporal variance arises from a combination of interacting ecological factors. As is likely the case with most herbivore species, no single factor dominates all aspects of their population dynamics.

At larger spatial scales on the phytochemical landscape, differences in autotroph chemistry among populations of primary producers interact with other habitat-specific factors to generate spatial variation herbivore dynamics. For example, some grasshopper populations in the rangelands of the United States exhibit unpredictable eruptive dynamics based in part on variation in primary producer chemistry. When plant nutritional quality is low for grasshoppers, predation pressure from birds maintains grasshopper populations around a low-density equilibrium. In contrast, areas and/or periods of high rainfall can improve plant nutritional chemistry to the point where grasshopper birth rates rise, death rates decline, and their populations escape regulation by birds and erupt (Belovsky and Joern 1995). A key point to emerge from this study is that landscape level variation in population dynamics emerges from landscape level variation in phytochemistry (Hunter 2002a).

A central theme of this book is that spatial variation on the phytochemical landscape matters profoundly to trophic interactions and nutrient dynamics. Focusing on herbivore population dynamics, the chemistry of a few individuals within a

---

FIGURE 4.14. Spatial and temporal variance in the densities of the winter moth (A, C) and the green oak tortrix (B, D) on oaks in the United Kingdom. For both moth species, spatial variance in density (A, B) derives primarily from the effects of oak budburst date on the chemistry of oak leaves. In contrast, temporal variance in density (C, D) derives from species-specific combinations of ecological factors. Modified from Hunter et al. (1997).

population of autotrophs can be sufficiently important to determine patterns of temporal variance of herbivores across the entire autotroph population (Helms and Hunter 2005). Scaling up, variation in autotroph chemistry among patches of primary production can influence the dynamics of consumer metapopulations, with some patches determining the dynamics across many others (Pulliam 1988). Using average values of autotroph (or patch) chemistry in models to predict herbivore population dynamics can cause substantial problems.

We have illustrated this effect in a simple patch model of aphid populations on individual plants that vary in nutritional quality across the phytochemical landscape. Assuming that aphids can move among plants at a moderate rate, then a single high-quality plant that supports exponential aphid population growth can induce apparent exponential growth on low-quality plants in the same population (Helms and Hunter 2005). Insect population growth on low-quality plants actually arises from immigration of individuals from high-quality plants. Essentially, plants (or plant patches) of good chemistry on the phytochemical landscape are sources of herbivores whereas plants (or plant patches) of bad chemistry are sinks (Pulliam 1988). Our model simulations illustrated an important outcome; increasing the variance in nutritional chemistry on the phytochemical landscape increases the likelihood that the chemistry of a small subset of patches will dominate temporal variance in the dynamics of herbivores across all patches within the phytochemical landscape (Helms and Hunter 2005).

We then used Bayesian parameter estimates from a field population of *Aphis nerii* on milkweed to simulate the effects of spatial variation in plant chemistry on aphid population dynamics. We found that using average values of plant nutritional quality grossly underestimated aphid population growth in over 93 percent of simulations because herbivore dynamics on a majority of plants were determined by dynamics on a subset of high-quality plants on the phytochemical landscape (Helms and Hunter 2005). Our results emphasize that the population dynamics of herbivores cannot simply be predicted by averaging the chemistry of patches on the phytochemical landscape; the spatial heterogeneity in chemistry is central to the dynamical processes. Similar conclusions have been reached by Nora Underwood, who has shown that both variance in plant nutritional quality and skew in plant quality around the mean cause herbivore population dynamics to deviate from those expected under the average quality of plants (Underwood 2004). Such results are important because they illustrate the potential nonadditive effects on herbivore population dynamics that may arise from genetic variation in plant chemistry on the phytochemical landscape (Underwood 2009).

Moving beyond just herbivores, we have explored recently how spatial variation in phytochemistry interacts with predation pressure to generate spatial patterns of herbivore and enemy population dynamics. Coupled parasite-host

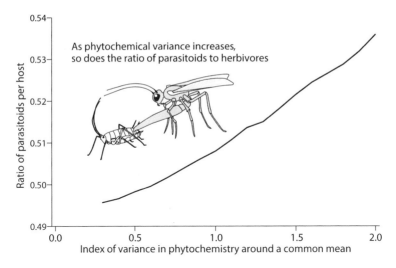

FIGURE 4.15. Ratios of parasitoids to herbivores that result from mathematical models of phytochemical landscapes. Landscapes have the identical average chemical quality, but differ in the variance in chemistry around the common mean. Importantly, the ratio of parasitoids to herbivores increases as variation in autotroph nutritional chemistry increases, despite a common average phytochemical quality. Modified from Riolo et al. (2015).

interactions can generate fascinating dynamical patterns in space, including spirals, traveling waves, and spatial chaos (Comins et al. 1992, Rohani et al. 1997), even on homogenous landscapes. However, as I have emphasized above, landscapes are not chemically homogenous, and the phytochemical landscape provides a profoundly variable template upon which predator-prey interactions take place. Using spatially explicit models, we have generated landscapes upon which autotrophs vary in their nutritional chemistry at different spatial scales, and then superimposed herbivore-enemy interactions upon those landscapes. We have found that increasing variation in nutritional phytochemistry (while maintaining the same average chemistry) generates increases in the mean and variance of both herbivore and enemy population size. Critically, fine-scale variation in phytochemistry generates disproportionate increases in enemy populations, and hence increases the ratios of enemies to herbivores (Riolo et al. 2015). One important conclusion from this work is that maintaining variation in phytochemistry in natural populations may serve to increase herbivore suppression by natural enemies (figure 4.15).

In short, as far as phytochemistry is concerned, means are probably meaningless, and herbivore population models that are based on average autotroph qualities may be woefully inadequate. This disconnect between the average autotroph nutritional quality used in most population models and the true impact of variation in phytochemistry on herbivore and predator population dynamics has impeded the development of a balanced population ecology that gives appropriate weight to the combined effects of phytochemistry, natural enemies, and abiotic factors on herbivore population dynamics (Hunter 1997, Hunter et al. 2000).

As I noted in chapter 3, a wide variety of ecological factors influence spatial and temporal variation on the phytochemical landscape, and therefore variation in herbivore population dynamics. There is not space to describe all the possible contributing factors here. They include effects of plant stress (White 1984) and plant vigor (Price 1991) on herbivore population dynamics, as well as the role of herbivore-induced changes in plant chemistry on subsequent population dynamics (Edelstein-Keshet and Rausher 1989). While studies of plant stress and vigor may be quite useful for understanding interactions between individual herbivore species and their food resources, I no longer find them a compelling framework for synthesis in ecological systems broadly writ. The dynamical consequences of induced resistance are discussed in detail in chapter 5.

4.1.4. Effects of Autotroph Chemistry on Herbivore Diversity

The diversity of consumers in ecosystems depends on interacting ecological and evolutionary forces, including predation pressure, disturbance, and the diversity of food resources (Paine 1966, Murdoch et al. 1972, Bishop 2002). To what extent do phytochemical traits contribute to variation in the diversity of herbivores? Over evolutionary time, it seems likely that phytochemical diversity has influenced the diversification of herbivores, and vice versa (Ehrlich and Raven 1964, Hay et al. 1987). Such coevolutionary interactions are a well-reviewed mainstay of the literature on plant-herbivore interactions, and I will consider them only briefly in chapter 5. Here, I focus on the responses of herbivore diversity in ecological time to variation on the phytochemical landscape.

In chapter 1, I distinguished between effects of trait variation and trait diversity on ecological processes, and it is important to revisit these ideas here. First, there are particular phytochemical traits that are linked to the diversity of herbivores. For example, high nutrient availability in primary producers facilitates high mammalian herbivore diversity at regional and global scales (Olff et al. 2002) (figure 4.5C). In other words, variation in specific traits of autotrophs can be linked to variation in the diversity of herbivores that live in particular patches on the phytochemical landscape. Second, there are effects of trait diversity per se (Tilman et al. 1996, Cardinale et al. 2006) on herbivore diversity. For example, while not all herbivores in tropical wet forests are specialists (Novotny et al. 2002), it seems clear that the diversity of chemical traits expressed within communities of tropical forest trees promotes diversity in the assemblage of herbivores that feed on those trees (Becerra 2003, Richards et al. 2015). Below, I consider briefly some effects of plant trait variation and trait diversity on herbivore diversity.

4.1.4.1. TRAIT VARIATION

Why should variation in particular chemical traits on the phytochemical landscape influence herbivore diversity? The positive association in terrestrial ecosystems between plant nutrient content and mammalian herbivore diversity (Olff et al. 2002) (figure 4.5) provides an interesting example. Regions of moderate rainfall and high soil nutrient availability support the highest diversity of mammalian herbivores because they yield plants of high nutritional quality (supporting smaller herbivore species that lack the ability to ruminate) and high productivity (supporting larger herbivore species that must consume substantial quantities of food). In other words, as nutrient availability declines, smaller species drop out of mammalian herbivore communities. A key result of the work by Olff and colleagues is that patches of low nutritional quality on the phytochemical landscape may support only a subset of the consumer species that persist in patches of higher nutritional quality. Do such effects hold in communities of other kinds of herbivore?

Studies of gall-forming insects on oak trees provide at least partial support for this idea. Species that form galls on plants are master manipulators of phytochemistry (Nabity et al. 2013), and their ability to manipulate plant tissues might make them much less responsive to variation in plant chemistry than are other types of herbivore. Yet gall diversity also declines with rather broad indices of plant nutritional chemistry. We conducted a survey of gall-forming cynipid wasps on six oak species in a Florida scrub ecosystem. Overall, we recorded 88 cynipid gall species, 23 of which were new to Florida and 17 of which were new to science (Price et al. 2004). Remarkably, 81 percent of the variation in cynipid species richness among tree species was associated with variation in foliar hemicellulose concentration, with gall species richness declining steeply as foliar hemicellulose concentrations increased (figure 4.16). Although this correlation does not demonstrate any cause-and-effect relationship, it does suggest some important link between variation in phytochemistry and the species richness of galling herbivores on oak. Hemicellulose generally combines with lignin to form cross-links with cellulose in plant cell walls, a combination that reduces the digestibility of plant tissues (Scheller and Ulvskov 2010) (chapter 6). It is possible that hemicellulose interferes with the ability of at least some gall-forming species to make galls on potential oak hosts. In addition to simple measures of species richness, we have associated the composition of cynipid gall wasp communities on oaks with the chemistry of oak foliage. Foliar lignin, cellulose, hemicellulose, phenolics, and N are all strongly

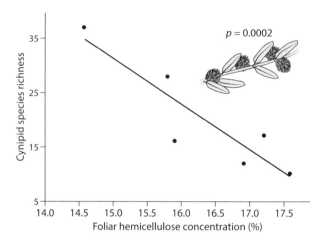

FIGURE 4.16. The species richness of cynipid gall wasps declines as foliar hemicellulose concentration increases among six species of oak in Florida. Data illustrate how variation in a phytochemical trait may influence the diversity of herbivore communities. Data from Price et al. (2004).

correlated with the composition of gall communities on these Florida oak trees, indicating a fundamental role for phytochemistry in structuring herbivore communities (Abrahamson et al. 2003).

4.1.4.2. TRAIT DIVERSITY

The loss of biodiversity on Earth has stimulated much recent research on the effects of species diversity on ecological processes (Knops et al. 1999). Of course, it is the diversity of phenotypes, rather than the diversity of nomenclature, that influences ecological interactions; in this context, species diversity is a proxy for trait diversity (Tilman et al. 1997). Because we now recognize that trait diversity at multiple levels, from genotypes to broad functional groups, influences how species interact with their environment (Madritch and Hunter 2002), emphasis is shifting from a species-based to a trait-based approach in studying diversity effects (Luck et al. 2012, Edwards et al. 2013). One component of trait diversity within ecological communities is the diversity of chemical traits that influence the quality of primary producers as food for herbivores (Asner et al. 2014). We should therefore expect to see links between autotroph trait diversity (or its proxy, species diversity) and the diversity of herbivores in communities. However, it is rarely possible to associate trait diversity effects on herbivores with the chemical diversity of primary producers, perhaps in part because measuring the diversity of chemical forms in autotroph communities remains logistically challenging. Additionally, specialization by herbivores on phytochemical traits is only one pathway among several that can link autotroph and herbivore diversity (below).

Nonetheless, primary producer diversity is consistently related to the diversity of herbivores in communities. In a recent meta-analysis of terrestrial systems, correlations between plant diversity and animal diversity were observed for arthropods, herps, birds, and mammals, regardless of the habitat in question (Castagneyrol and Jactel 2012). The authors even suggest that plant diversity may serve as a surrogate for measuring animal diversity when rapid biodiversity assessments are necessary. In addition to species diversity, high phylogenetic diversity among plants is also associated with diverse communities of herbivores and predators (Dinnage et al. 2012). Consequently, losses of autotroph diversity have effects that cascade up in food webs, reducing the complexity, connectance, interaction diversity, and interaction strength within those webs (Rzanny and Voigt 2012). In turn, associated losses of natural enemies can cascade down again to reduce rates of predation or parasitism on herbivores (Fenoglio et al. 2012) (chapter 5).

What mechanisms drive the positive relationships between autotroph and animal diversity? Specifically, how important is chemical variation among primary producers? When herbivores are specialists on their host species, or when resource partitioning is strong, we might expect simple and direct relationships between the diversity of autotrophs, the diversity of chemical traits, and the diversity of herbivores (Hunter and Price 1992b, Haddad et al. 2009). Underlying such relationships is the premise that the identity of autotroph species (and their chemical phenotype) is related to the identity of herbivores that they support, which should generate associations between autotroph and herbivore community composition. As expected, across 47 grassland and heathland sites in the Netherlands, the best predictor of arthropod (herbivore, pollinator, and predator) community composition is the species composition of plants in those communities (Schaffers et al. 2008) (figure 4.17A). Similarly, experimental increases in plant diversity in European grasslands generate increases in herbivore diversity both above- and belowground, with somewhat weaker effects at higher trophic levels (Scherber et al. 2010). In both of these European grassland studies, there is substantial evidence of bottom-up trophic (Hunter and Price 1992b) and diversity (Dyer and Letourneau 2003) cascades from plant species through herbivores to higher trophic levels. Moreover, evidence is growing that plant diversity effects, mediated by herbivore host choice rather than root biomass, affect preference and performance of herbivores belowground (Schallhart et al. 2012).

However, when herbivores have more generalized diets, or resource partitioning is weaker, links between autotroph diversity and herbivore diversity may result from a range of indirect mechanisms. For example, high autotroph diversity can increase the productivity of primary producer communities (Tilman et al.

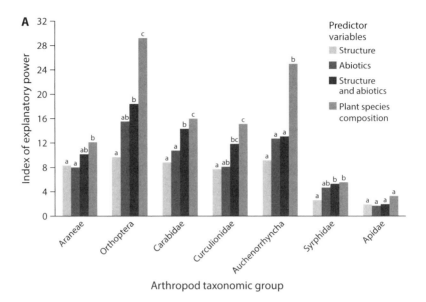

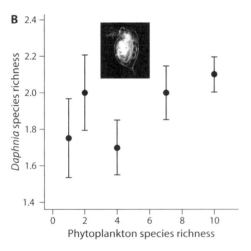

FIGURE 4.17. Relationships among autotroph and consumer diversity in terrestrial and aquatic ecosystems. (A) Among 47 grassland and heathland sites in the Netherlands, plant community composition is a better predictor of arthropod community structure (herbivores, predators, and pollinators) than is vegetation structure or the abiotic environment. (B) Experimental increases in phytoplankton diversity generate increases in the diversity of *Daphnia* consumers. Modified from (A) Schaffers et al. (2008) and (B) Striebel et al. (2012).

1996). In turn, high autotroph productivity begets high consumer productivity, which is positively associated with herbivore diversity. This indirect link, mediated by autotroph and consumer productivity, appears to be important in determining the diversity of herbivore species in Midwestern grasslands in the United States (Borer et al. 2012). In contrast, specialization by herbivores on chemically distinct autotroph hosts may drive direct relationships between autotroph diversity and herbivore diversity in other ecosystems such as species-rich forests (Becerra 1997, Wilson et al. 2012), alpine meadows (Pellissier et al. 2013), and marine algal communities (Poore et al. 2008). Additionally, autotroph species diversity may affect herbivore diversity indirectly by influencing the density and diversity of natural enemies that prey upon herbivore species, or the abundance of herbivores on individual host species, and all of these mechanisms may act simultaneously within communities (Siemann 1998, Haddad et al. 2009).

To date, there have been far fewer studies in aquatic systems that explore relationships between the diversity of primary producers and the diversity of consumers. This is an unfortunate bias, given that about 50 percent of Earth's primary production occurs in aquatic environments (chapter 1). In the few studies completed so far, patterns in aquatic systems appear similar to those in terrestrial systems. For example, experimental increases in the diversity of freshwater phytoplankton species result in both greater abundance and diversity of zooplankton grazers, even when controlling for phytoplankton biomass (Striebel et al. 2012) (figure 4.17B). Similarly, in networks of freshwater ponds, plant species richness is associated positively with the diversity of macroinvertebrates among ponds (Florencio et al. 2013). However, in neither of these studies is it possible to determine the precise mechanisms that link the diversity of primary producers with the diversity of consumers.

Importantly, trait diversity is not just manifest among different autotroph species; genotypic diversity in chemical traits within species of primary producers may also mediate patterns of consumer diversity (Hughes et al. 2008). As I noted in chapter 3, there is substantial genetic variation in the chemistry of autotrophs throughout the world's biomes. Does the presence of such diversity in autotroph communities influence the diversity of consumers? Among stands of cottonwood trees in the western United States, levels of genetic diversity are correlated with arthropod species richness, explaining nearly 60 percent of the variation in arthropod species diversity among sites (Wimp et al. 2004). This powerful relationship arises at least in part because different cottonwood genotypes express different chemical phenotypes, with differential effects on arthropod preference and performance (Whitham et al. 2003). In short, genetic variation contributes to variation on the phytochemical landscape (figure 3.1), with important consequences for the diversity of consumers.

Similarly, genetic diversity influences consumer diversity in populations of goldenrod, *Solidago altissima*, a dominant plant species in old fields in eastern North America. Arthropod species richness in goldenrod plots increases with the genotypic diversity of goldenrods in those plots, establishing a fundamental link between plant genetic diversity and consumer species richness (Crutsinger et al. 2006). In this system, the diversity of both herbivores and predators increases with plant genotypic diversity, illustrating that bottom-up diversity cascades (Dyer and Letourneau 2003) from plant traits through herbivores to secondary consumers can be initiated by plant genotypic diversity. The effects are nonadditive because the number of consumer species inhabiting high-diversity plots is greater than would be expected by summing the number of species that inhabit each of the lower diversity plots. In goldenrod, higher productivity in high diversity plots (food quantity) combines with facilitation of specialized consumers (phytochemical quality) to generate increases in consumer species richness; effects of genotypic diversity are as large as those of plant species diversity (Crutsinger et al. 2006). Similarly, within sand dune ecosystems of the Great Lakes, increasing genetic diversity within populations of the dominant plant species, *Ammophila breviligulata*, has a greater impact on arthropod species richness than does increasing plant species diversity (Crawford and Rudgers 2013).

As with effects of diversity at the species level, the ecological effects of genotypic diversity among aquatic primary producers are poorly known in comparison to their terrestrial counterparts. In the best-studied system to date, genotypic diversity within populations of seagrass, *Zostera marina*, appears to be particularly important during periods of abiotic or biotic stress (Hughes and Stachowicz 2004, Reusch et al. 2005), and may influence interactions among seagrass, epiphytic algae, and grazers. One principal effect of seagrass genotypic diversity appears to be increases in the impact of specific grazer species, such as sea hares, which remove epiphytic algae from seagrass shoots and so improve seagrass performance (Hughes et al. 2010). However, we are a long way from being able to generalize on potential effects of plant genotypic diversity on herbivore diversity in aquatic ecosystems.

As I noted earlier, associations between the diversity of primary producers and the diversity of consumers are driven by a wide variety of ecological processes. The degree to which phytochemistry is central to such associations remains unclear. In my view, there is a pressing need for techniques and studies that will permit us to estimate chemical diversity at multiple spatial and temporal scales to better understand how ecological communities are assembled (chapter 9). Statistical models that relate the traits of primary producers to herbivore diversity are only as good as the trait data that they contain. Currently, such models contain autotroph traits that are easier to measure (often C, N, or P levels; photosynthetic rates; fiber concentrations; growth form; leaf lifespan; etc.) and generally exclude the complex chemical traits that are harder to measure. Clearly, we cannot estimate the relative importance of chemical traits that remain unmeasured. Likewise, while relating the assembly of herbivore communities to the phylogenetic structure of plant communities may provide important clues to mechanism (Agrawal 2007), it is ultimately unsatisfactory without an understanding of the traits (nutrients, defenses, and others) that are phylogenetically conserved (Whitfeld et al. 2012). Transcriptomics and metabolomics are starting to provide important information on how the diverse chemicals produced by autotrophs influence herbivore ecology (Kersten et al. 2013), and these approaches may help us to move past oversimplified associations between the traits of primary producers and herbivore communities. Developing methodologies and conducting studies that assess chemical diversity within and among ecological communities must be a priority (Richards et al. 2015).

4.2 EFFECTS OF THE PHYTOCHEMICAL LANDSCAPE ON NATURAL ENEMIES

In preceding sections of this chapter, I have described how variation on the phytochemical landscape influences many aspects of the behavior, performance, population dynamics, and community structure of herbivores. However, variation in autotroph chemistry has pervasive effects at higher trophic levels too, influencing many aspects of the ecology and evolution of predators and parasites. There are diverse mechanisms by which variation in plant chemistry cascades up to influence natural enemies of herbivores (Hunter 2003) (figure 4.18). The nutritional and defensive chemistry of autotrophs influences the abundance, biomass, and diversity of herbivores that are available for natural enemies to exploit (Hunter and Price 1992b). Phytochemistry also influences the growth rate and size of herbivores, which in turn influences their susceptibility and relative value to predators (Werren 1980, Benrey and Denno 1997). The nutritional quality of herbivore tissues can also vary with the chemistry of the primary producers upon which they feed, with subsequent effects on predator performance and diet breadth (Fagan et al. 2002). Autotrophs use volatile chemical signals to actively recruit the predators and parasites of herbivores (Turlings et al. 1990), providing some level of protection to the autotrophs and a food source for the bodyguards that respond (chapter 5). In contrast, some herbivores may use autotroph toxins as sequestered defenses (Opitz and Muller 2009), while other toxins may simply prove deleterious to predators and parasites when consumed (Campbell and Duffey 1979). I provide a few

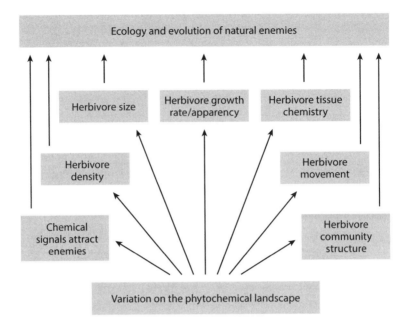

FIGURE 4.18. Major pathways by which variation on the phytochemical landscape influences the ecology and evolution of natural enemies from the bottom up. Corresponding top-down effects of natural enemies on the phytochemical landscape are described in chapter 5.

examples of these mechanisms below to establish the fundamental link between the phytochemical landscape and the ecology of predators and parasites.

4.2.1 Effects of Herbivore Biomass, Growth Rate, Size, and Tissue Chemistry on Predators

Perhaps most simply, autotroph chemistry influences herbivore production and standing stock, the resources upon which natural enemies depend. Ecosystems that are based on primary producers of high nutritional quality are characterized by higher ratios of herbivore to producer biomass (Cebrian et al. 2009) (figure 4.5A, B). In turn, the greater efficiency of trophic transfer engendered by nutrient-rich autotrophs increases the length of food chains and supports greater production of higher-order predators (Kersch-Becker and Lewinsohn 2012). For example, rainfall in the Serengeti improves the nutritional chemistry of primary production available for herbivores; rainfall-mediated increases in wildebeest and rodent populations cascade up to increase consumption by lions and densities of kites, respectively (Sinclair et al. 2013). In many systems, the effects of predation then cascade back down through ecosystems, further influencing ratios of herbivore to primary producer biomass (Carpenter et al. 1985, Moreira et al. 2012). Under this scenario, it is the chemistry, not simply the production, of autotrophs in ecosystems that determines the strength of trophic cascades that those ecosystems support (Hunter and Price 1992b, Cebrian et al. 2009). In turn, because autotroph chemistry has a strong phylogenetic signal, the strength of trophic cascades reflects evolutionary patterns of autotroph nutritional and defense traits (Mooney et al. 2010).

Beyond effects on primary and secondary production, the chemistry of autotrophs influences predator population dynamics. In section 4.1.3, I illustrated how the dynamics of herbivore populations depend in part on the chemistry of autotrophs that herbivores consume. Even when predation pressure maintains herbivore populations well below their carrying capacities, those equilibrium densities depend critically upon autotroph nutritional and defensive chemistry as fundamental determinants of herbivore birth rate (box 4.2). If densities of herbivores depend on autotroph chemistry, so must those of natural enemies. Whatever formulation of predator-prey models one might prefer, effects of phytochemistry influence the temporal population dynamics of predators by their influence on herbivore demography (Rosenzweig 1971, Arditi and Ginzburg 2012). Furthermore, spatial population models confirm that the dynamics of predator populations are linked tightly to variation on the phytochemical landscape (Helms and Hunter 2005, Riolo et al. 2015). As a result, in natural populations, we should observe significant spatial correlation (positive or negative) among autotroph chemistry, herbivore density, and the densities of natural enemies (figure 2.4). Such patterns emerge among oak trees in Pennsylvania, where variation in herbivore density depends in large part upon variation in foliar tannin and nutrient chemistry. In turn, densities of invertebrate predators track herbivore densities among individual trees (Hunter 1997). Moreover, fertilization of individual oak trees with N and P results in combined increases in both herbivore and predator population densities. One result is that the top-down effects of birds on insects are greater on oak trees with high-nutrient than with low-nutrient foliage (Forkner and Hunter 2000). This effect appears to be general among forest trees whereby nutrient-rich trees support higher densities of foliage-feeding caterpillars that, in turn, suffer proportionately higher rates of density-dependent mortality from birds (Singer et al. 2012). In short, predator-prey dynamics are in part a function of variation on the phytochemical landscape.

In addition to the effects on predator dynamics mediated by herbivore density, autotroph chemistry influences the growth rate and final body size of herbivores, with important consequences for the ecology and evolution of natural enemies. When herbivores feed on low-quality primary production, they often grow more slowly (Hunter and McNeil 1997), which can increase their window of vulnerability to natural enemies (Werner et al. 1983). The basic premise is that small or juvenile organisms are more susceptible to certain natural enemies than are large or adult organisms. Growing rapidly into the adult stage or size-class should therefore reduce rates of mortality. In other words, predators may impose a disproportionately large risk for herbivores that feed in low-quality regions of the phytochemical landscape. In addition to growing more slowly, herbivores that feed on poor quality autotrophs may be more apparent to their natural enemies because they have to spend more time feeding. In an interesting example of this effect, low-quality foliage increases the mortality of leaf-mining insects to their natural enemies. On oak trees, when foliar nutrient concentrations are reduced experimentally, leaf-mining larvae eat more to compensate for their poor quality diet. Their leaf mines are therefore larger and provide stronger cues for searching parasitoids, which attack and kill herbivores more frequently (Stiling et al. 2003).

However, the slow growth–high mortality hypothesis is better supported in some systems than in others. For example, variation in the growth rates of insect herbivores among terrestrial plant species is generally not a good predictor of parasitism rates (Clancy and Price 1987, Benrey and Denno 1997), perhaps because species-specific plant traits (including volatile chemicals, chapter 5) have a stronger influence on foraging parasitoids than does the growth rate of herbivore prey (Farkas and Singer 2013). Additionally, those chemical traits that engender low rates of herbivore growth may also give rise to low herbivore densities. Consequently, herbivores in low-density patches on the phytochemical landscape may be at lower risk from predators that forage in a density-dependent fashion. Overall, we might predict (a) fewer predators aggregating in patches of low autotroph quality, but (b) greater risk for herbivores **per individual predator** in such low-quality patches. These opposing effects of the phytochemical landscape on predation pressure are rarely separated in empirical studies.

In addition to influencing herbivore growth rate, autotroph chemistry often influences the size at which herbivores reach maturity (Hunter 1987). As noted above, large herbivores may be less vulnerable to predators than are small herbivores (Adams 2003). But there are also more subtle effects of herbivore size on predators and parasites. For example, parasites tend to produce more female offspring in large herbivore hosts than in small herbivore hosts (Werren 1980). Moreover, large herbivore hosts support more overall parasites with greater longevity and greater fecundity (Mayhew and Godfray 1997, Bernal et al. 1999). In other words, a large or high-quality herbivore may be a particularly attractive host for parasitic organisms.

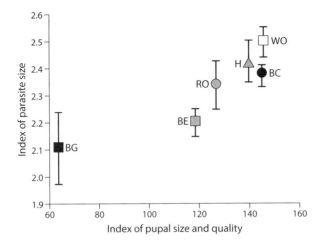

FIGURE 4.19. The size of parasitoids emerging from cocoons of *Euclea delphinii* caterpillars correlates positively with herbivore size and survival among six species of forest trees (BC, black cherry; BE, American beech; BG, black gum; H, pignut hickory; RO, red oak; WO, white oak). Modified from Stoepler et al. (2011).

Therefore, one of the basic mechanisms by which phytochemistry influences natural enemies is by cascading effects of high (or low) food quality (Hunter and Price 1992b, Kagata and Ohgushi 2006). In this simple scenario, high-quality autotrophs produce high-quality herbivores that best support the nutritional needs of natural enemies (Hunter 2003). There are many examples illustrating that well-fed herbivores can support more enemies, each with greater fitness. For example, vigorously growing *Typha latifolia* plants are high-quality food for caterpillars that pupate at larger mass on such plants. In turn, large pupae support the development of larger individual parasitoids (Teder and Tammaru 2002). Likewise, the host plant species on which limacodid caterpillars grow best are also those that produce the largest parasites (Stoepler et al. 2011), suggesting that parasite performance is constrained by the chemistry of autotrophs among which herbivores forage on the phytochemical landscape (figure 4.19). Such relationships may help to explain the equivocal relationship between herbivore growth rate and parasitism that I described earlier; slow-growing hosts may also represent poor-quality hosts for parasites. For example, caterpillars reared experimentally on low-nutrient artificial diets are poorer-quality hosts for parasites than are caterpillars reared on high-nutrient diets (Harvey et al. 1995). In combination, these results suggest that autotroph chemistry may have an impact on the fitness of natural enemies through the quality of herbivore tissues. Equivalent effects have been shown in fish populations, in which rapidly growing individuals support the fittest parasites (Barber 2005).

Are food webs simply aggregations of victims of sequential nutrient limitation? If nutrient-poor autotrophs support nutrient-poor herbivores that limit the performance of nutrient-limited predators, can consumers use any strategies to overcome this apparent cascade of nutrient deprivation? One answer is that low nutrient availability may favor at least two pathways toward the evolution of omnivory (Hunter 2009). First, because primary producers are low in N and P relative to animal tissue, herbivores that include some animal tissue in their diets may gain the combined benefits of abundant plant resources of lower chemical quality and rarer animal resources of higher chemical quality. It turns out that omnivores are much more common in ecosystems than we once thought (Arim and Marquet 2004), suggesting strong selection pressure for including organisms from two or more trophic levels in the diet.

Second, if herbivore tissues have lower N and P concentrations than those of their natural enemies, those enemies may engage in intraguild predation (Polis et al. 1989). For example, the nitrogen content of insect predators is generally around 15 percent higher than that of their insect herbivore prey (Fagan et al. 2002). In response, higher-order predators may include both herbivores and other predators in their diets (Denno and Fagan 2003). In addition to reducing N and P limitation, including both herbivores and predators in the diet may also ease the increasing levels of lipid limitation experienced by predators at higher trophic levels (Wilder et al. 2013). Examples of intraguild predation are now widespread (Hunter 2009) and influence important ecological processes, including the stability of trophic interactions and levels of damage to primary producers (Rosenheim 2007, Vance-Chalcraft et al. 2007). For example, omnivory may dampen the effects of trophic cascades (Strong 1992), reducing herbivore suppression by natural enemies and increasing levels of damage to primary producers (Finke and Denno 2005).

The apparent prevalence of omnivory in terrestrial systems is at least matched by that in aquatic systems. Oceans and lakes are filled with a multitude of mixotrophic bacteria that combine, or switch between, photoautotrophy and chemoheterotrophy, or use light as a source of energy but organic C as a carbon source (Moore 2013, Swan et al. 2013). For example, the cynaobacterium *Prochlorococcus* is the smallest and most abundant oxygenic phototroph in the oceans and one of the main primary producers on Earth (Partensky et al. 1999). *Prochlorococcus* can take up glucose directly from the environment using a sugar transport system (Muñoz-Marín et al. 2013), thereby supplementing the inorganic C that it fixes with organic C. Mixotrophs that sometimes fix inorganic C themselves and sometimes take up organic C previously fixed by others contribute to the trophic tangles in ecological systems, blurring the lines between autotrophs, consumers, and decomposers. At higher trophic levels, omnivory is also a central feature of aquatic ecosystems. Omnivorous fish, urchins, crabs, and mollusks exploit the energy (and nutrient) channels of green and brown food webs, increasing the rates at which nutrients cycle in marine and freshwater ecosystems (chapter 8). In summary, omnivory may be a common strategy by which herbivores and predators ameliorate limitations imposed by the nutritional chemistry of autotrophs. Currently, the role of secondary metabolites in the ecology and evolution of omnivory remains largely unexplored.

4.2.2 Chemical Mediation of Tri-Trophic Interactions

Many of the effects of phytochemistry on natural enemies are more complex than those arising from chemically mediated variation in herbivore density, nutrient concentration, or vigor. For example, toxic or distasteful compounds expressed by autotrophs on the phytochemical landscape impose significant challenges to the foraging behavior and fitness of many natural enemies. As a result, variation in autotroph secondary metabolism can mediate complex tri-trophic interactions (Price et al. 1980), whereby suppression of herbivores by their natural enemies varies markedly with phytochemistry.

Some of the best examples of interactions of this type come from marine systems, in which specialized mesograzers, such as amphipods and polychaetes, gain protection from consumers by associating with particularly toxic macroalgae on the phytochemical landscape. Larger generalist grazers such as urchins and fish remove substantial quantities of macroalgae from marine systems on a daily basis (Carpenter 1986). As a consequence, small mesograzers (which live on the plants that they consume) would likely be eaten rapidly if they foraged on those plants preferred by urchins and fish (Hay 1996). This has led to the hypothesis that marine mesograzers should specialize on toxic seaweeds to avoid consumption by larger generalist herbivores (Hay et al. 1987). There is now good evidence that many mesograzers in marine systems are affected less by algal PSMs than are generalist fish and urchins (Poore et al. 2008), and can therefore exploit enemy-free space on the phytochemical landscape that toxic algae provide. For example, certain amphipods and polychaetes specialize on the toxic brown alga *Dictyota menstrualis* gaining protection from the diterpene alcohols that deter generalist fish herbivores (Duffy and Hay 1994). Similarly, the amphipod *Pseudamphithoides incurvaria* builds a safe house from the toxic brown alga *Dictyota bartayresii* using the same diterpene alcohol that is deterrent to fish as the stimulus for building its retreat (Hay et al. 1990). Some amphipods will even switch their foraging locations on the phytochemical landscape from palatable to toxic algae when they detect cues from foraging predators (Zamzow et al. 2010). Exploitation of enemy-free space also occurs in freshwater systems, in which geese and crayfish prefer to consume less defended vascular plants over toxic aquatic mosses, leaving the latter as refuges for herbivorous isopods and amphipods (Parker et al. 2007).

Back on land, the gypsy moth, *Lymantria dispar*, provides a fascinating example of how phytochemistry interacts with the third trophic level to influence herbivore ecology and evolution. The gypsy moth is a polyphagous forest pest in eastern North America, feeding on a wide variety of plant species (Rossiter 1987a). Historically, outbreaks of the gypsy moth in the United States were terminated by epizootics of a baculovirus, called LdNPV (Doane 1970). Field studies suggested that host use by the gypsy moth and movements among plants by caterpillars under outbreak conditions were strategies to minimize infection by LdNPV (Rossiter 1980, Rossiter 1987b). Subsequent laboratory studies confirmed that the quality of foliage entering the gypsy moth gut with the virus inoculum influenced greatly the probability of viral infection (Keating and Yendol 1987). When larvae were dosed with LdNPV by use of an artificial diet, it took only 800 virus particles to kill 50 percent of experimental larvae. In contrast, it required closer to 60,000 virus particles to kill 50 percent of the larvae on plant leaves, a difference of nearly two orders of magnitude (Keating et al. 1990). Chemical analyses of foliage implicated phenolic compounds in leaves as those potentially active against the virus (Keating et al. 1988), perhaps through interactions between phenolic molecules and the protein coat of the virus (Keating et al. 1990) (figure 4.20).

While the original work by Steve Keating and colleagues focused on chemical variation among tree species on the phytochemical landscape, there is often substantial chemical variation among individuals from the same tree species. For example, among 60 red oak trees on a single Pennsylvania hillside, there exists over 13-fold variation in condensed tannin concentration and over 22-fold variation in foliar astringency (Hunter 1997). This level of chemical variation is sufficient to engender 8-fold variation in susceptibility of gypsy moth larvae to LdNPV on the phytochemical landscape. Intraspecific and interspecific variation in foliar chemistry provides a complex spatial and temporal chemical mosaic upon which gypsy moths and their virus interact. For a constant dose of 50,000 virus particles, we estimate that larval mortality should vary from 13 to 99 percent, depending on local phenolic chemistry on the phytochemical landscape (Hunter et al. 1996).

Although the spread of a fungal pathogen has now complicated the relationship between gypsy moths and LdNPV (Tobin and Hajek 2012), these earlier

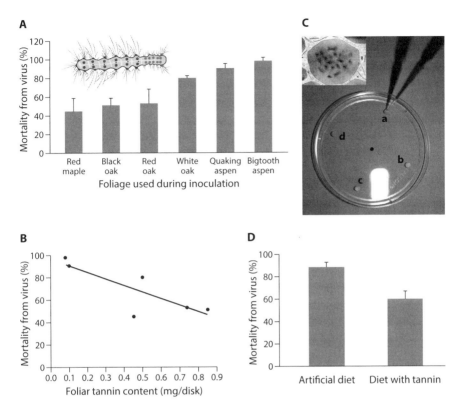

FIGURE 4.20. The effect of a single viral dose of 60,000 PIBS on the mortality of gypsy moth larvae varies markedly among plants species (A), with mortality declining as foliar tannin content increases (B). When added experimentally to artificial diet (C), a viral dose of only 800 PIBS is sufficient to cause 90% larval mortality, unless tannin is added to the artificial diet (D). Data are from Keating et al. (1988) and Keating et al. (1990).

studies illustrate a key theme of this book: The spatial and temporal dynamics of species generally depend upon interactions between phytochemistry and natural enemies. It would be fruitless to ask whether the trees or the virus were more important in determining gypsy moth population dynamics because they interact in such a fundamental way. While the virus is certainly the proximate agent of mortality, its ability to kill caterpillars depends intimately on the chemistry of caterpillar diet (figure 4.20). Such interdependence is the hallmark of tri-trophic interactions (Price et al. 1980) and a cornerstone of the trait-mediated indirect effects that influence so much of the ecology and evolution of species (Ohgushi et al. 2012).

4.2.3 Sequestration of Chemical Defenses

In the previous section, I provided a few examples in which the efficacy of natural enemies attacking their herbivorous prey varied with variation on the phytochemical landscape. The sequestration of PSMs by herbivores provides a specialized case of such tri-trophic interactions, in which the ability of enemies to attack their prey depends upon phytochemicals co-opted by herbivores for their own defense (Ode 2006). It has become a paradigm in evolutionary ecology that some herbivores protect themselves from natural enemies by sequestering toxins from primary producers (Brower 1958, Rothschild 1964). Sequestration is an active process and, in some cases, herbivores sequester phytochemicals at concentrations substantially higher than those in the autotrophs upon which they feed (Wink and Witte 1991), which can impose significant metabolic and storage costs (Opitz and Muller 2009) and compromise their ability to defend themselves from parasites (Smilanich et al. 2009a). Consequently, sequestration must generally confer substantial fitness advantages in protection from predators to overcome the associated metabolic and ecological costs (Opitz et al. 2010). The dynamic balance between the costs and benefits of sequestration is clearly illustrated by the diversity of mimetic species, grading from toxic to palatable, that associate with toxic species (Ritland and Brower 1993); palatable mimetic species, when they occur at low frequency relative to their toxic models, may gain certain benefits of apparent sequestration without paying certain costs. Notably, herbivores that specialize on toxic autotrophs may sequester higher concentrations of toxins than do generalist herbivores feeding on the same autotroph species (Lampert and Bowers 2010), suggesting that specialization may help to reduce some of the costs of sequestration.

While most studies have focused on sequestration of plant toxins by insect herbivores, a wide variety of other herbivores in terrestrial and aquatic environments reduce predation pressure by utilizing chemical defenses in their food. For example, sibling voles, *Microtus levis*, suffer lower rates of predation from least weasels, *Mustela nivalis nivalis*, when the voles feed on ryegrass that is infected by the fungal endophyte *Neotyphodium*. Alkaloids produced by the grass endophytes appear to deter the weasel predators from their vole prey (Saari et al. 2010), supporting the idea that endophyte toxins may have deleterious effects at the third trophic level (Jani et al. 2010). In a remarkable example of co-opting a plant defense, African crested rats, *Lophiomys imhausi*, chew the bark and roots of *Acokanthera schimperi* trees. After masticating the plant tissue, rats apply saliva that is now rich in toxic cardenolides to specialized hairs on their backs that wick up the plant toxins. When threatened, crested rats present the hairs to predators in a characteristic display of their black and white fur. The cardenolide toxins in the specialized hairs of the African crested rat are potent enough to kill or disable large carnivores (Kingdon et al. 2012).

Many aquatic herbivores sequester secondary metabolites from primary producers to gain protection from their natural enemies (Sotka et al. 2009). Perhaps best known are the herbivorous sea slugs and sea hares that sequester diverse toxins from algae and cyanobacteria that help to protect them from predation (Hay 2009). Because these Opisthobranch mollusks have lost their shells, sequestration of PSMs, and specialized glands to secrete them, should represent particularly important adaptations for defense against predators (Pennings and Paul 1993). For example, the sea hare *Aplysia parvula* sequesters the toxins apakaochtedene A and B from the red alga *Portieria hornemannii* and is thereafter unpalatable to fish predators. In contrast, when sea hares feed instead on the nontoxic alga *Acanthophora spicifera*, they are readily consumed by fish (Ginsburg and Paul 2001). However, in some systems, the identity of the fish predators is as important as is the host alga in determining the susceptibility of sea hares to fish predation (Pennings et al. 2001). This illustrates the dynamic ecological and evolutionary nature of tri-trophic interactions on the phytochemical landscape, including those mediated by toxin sequestration.

An interesting alternative to sequestration of chemical defense is simply to wear toxic algae as a form of clothing (Hay 2009). For example, juveniles of the Atlantic decorator crab, *Libinia dubia,* cultivate the toxic brown alga *Dictyota menstrualis* on their carapace, gaining protection from predators that are deterred by the algal toxins. The diterpene alcohol that stimulates the clothing behavior is also the toxin that is most deterrent to local fish populations (Stachowicz and Hay 1999). Likewise, some tropical reef isopods cultivate cyanobacteria on their dorsal surface, using the cyanobacteria both as a source of food and as a defense against predators that are deterred by their phycotoxins (Lindquist et al. 2005). Being clothed in plant defenses, rather than sequestering them, should essentially eliminate metabolic and storage costs, while allowing herbivores to carry the most toxic parts of the phytochemical landscape around with them.

How high up trophic chains can PSMs be sequestered? Can PSMs ever protect natural enemies from their own, higher order enemies? There is certainly evidence that phytotoxins can accumulate in the tissues of natural enemies at the third trophic level (Bowers 2003, Ode 2006), although the ecological impacts of such sequestration for the fourth trophic level are not clear. For example, the nymphalid butterfly *Melitaea cinxia* sequesters iridoid glycosides from its host plants. In turn, the parasitic wasp *Cotesia melitaearum* sequesters iridoid glycosides from its butterfly host. However, the upward cascade of sequestered plant toxins does not appear to protect *C. melitaearum* from its own (hyperparasitoid) enemies, which either catabolize or sequester the iridoid glycosides (van Nouhuys et al.

2012). While the iridoid glycosides are detectable in tissues at the fourth trophic level (i.e. the hyperparasitoids), their concentrations are low and their ecological effects appear to have attenuated to negligible levels. In other words, while sequestration of PSMs by herbivores can reduce predation pressure on those herbivores, evidence for effects at higher trophic levels remains equivocal.

4.2.4 Self-Medication and Pharm-Ecology

Sequestration of cardenolides by monarch butterflies is a classic example wherein sequestered plant toxins protect herbivores from their predators (Brower 1958, Brower et al. 1975). However, recent work has established that cardenolides also provide monarchs with protection against disease. Monarch populations worldwide are attacked by a protozoan parasite, *Ophryocystis elektroscirrha*, which reduces monarch survival, fecundity, longevity, and ability to fly (Bradley and Altizer 2005, de Roode et al. 2007). It turns out that the susceptibility of monarchs to their protozoan parasites varies among species of milkweed (de Roode et al. 2008), with high-cardenolide milkweeds providing monarchs with greater resistance and tolerance to their parasites than low-cardenolide milkweeds (Sternberg et al. 2012) (figure 4.21A, B). In this context, resistance means that monarchs are less likely to get infected, or will suffer lower parasite burdens, when milkweed cardenolides are high. Additionally, tolerance means that, for a given burden of parasites, cardenolides reduce the fitness costs of infection. In an interesting twist to this story, we had observed in previous work that aphids feeding on milkweed could reduce the concentrations of cardenolides in milkweed leaves (Zehnder and Hunter 2007b). Could aphids therefore increase the susceptibility of monarchs to disease by reducing the foliar concentrations of their medicinal cardenolides? As predicted, monarch caterpillars become more susceptible to their protozoan parasites if milkweed aphids have already fed on the plants because the most potent lipophilic cardenolides are expressed in lower concentration after aphid feeding (de Roode et al. 2011). This study illustrates that parasite-host interactions are contingent upon (a) variation in cardenolide chemistry on the phytochemical landscape, and (b) the community of interacting species within which the parasite-host interaction is embedded (Forbey and Hunter 2012).

So should monarch butterflies always select the most toxic milkweeds available? Or should they only choose the most toxic plants when the risk of parasite infection is high? As I noted above, even specialist herbivores pay metabolic and storage costs for consuming toxins, and monarchs pay a fitness cost for consuming high-cardenolide plants (Zalucki et al. 2001, Tao et al. 2014). Consequently, when the risk of parasitism is low, the costs of feeding on high-cardenolide milkweeds

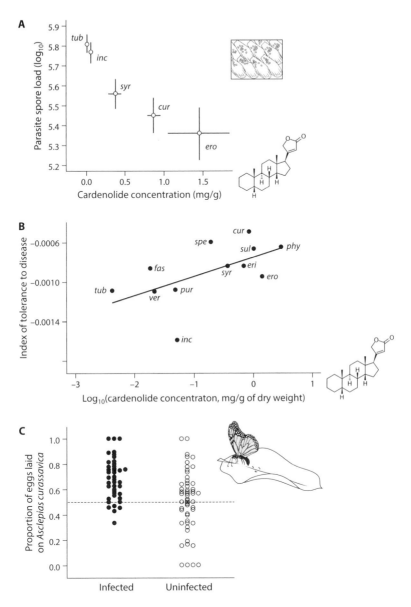

FIGURE 4.21. High-cardenolide milkweeds increase the resistance (A) and tolerance (B) of monarch butterflies to a protozoan parasite. As a consequence, infected monarch mothers choose to lay a greater proportion of their eggs on the high-cardenolide *Asclepias curassavica* than on the low-cardenolide *Asclepias incarnata* (C), whereas uninfected females are not choosy. Milkweed species are *A. tuberosa* (tub), *A. fascicularis* (fas), *A. verticillata* (ver), *A. purpurascens* (pur), *A. incarnata* (inc), *A. speciosa* (spec), *A. syriaca* (syr), *A. eriocarpa* (eri), *A. sullivantii* (sul), *A. curassavica* (cur), *A. erosa* (ero), and *A. physocarpa* (phy). Modified from Lefèvre et al. (2010) and Sternberg et al. (2012).

may outweigh their benefits (Sternberg et al. 2012). As a result, monarchs engage in therapeutic medication behavior in which infected female monarchs lay their eggs on high-cardenolide milkweed species that protect their offspring from subsequent parasite infection (Lefèvre et al. 2010). Uninfected females do not preferentially choose toxic milkweeds (figure 4.21C), and therefore do not impose the costs of high-cardenolide plants on all of their offspring. This transgenerational medication behavior is limited to egg-laying females, presumably because flying females have the greatest opportunity to choose among milkweeds of varying chemistry on the phytochemical landscape (Lefèvre et al. 2012). Additionally, both monarch mothers and fathers transfer cardenolides into their eggs (figure 4.10), which may help to medicate their offspring against disease as they first emerge into the world (Sternberg et al. 2015).

How common is self-medication in animals? Ecologists have known for many years that primates exploit the phytochemical landscape for medicines. For example, primates will chew the bitter bark of trees and swallow leaves whole to self-medicate against gut parasites (Huffman 2003) and use the leaves and fruit of *Balanites* in areas with a high risk of schistosomiasis. However, recent work indicates that medication behaviors in which consumers exploit the pharmaceutical properties of PSMs go far beyond primates (de Roode et al. 2013). Flies, moths, ants, and bees are among the many invertebrates that exploit phytochemicals as medicines. For example, ants (Castella et al. 2008) and bees (Simone-Finstrom and Spivak 2010) protect colony members from pathogen attack by depositing antimicrobial plant resins in their nests, while pollen products in honey upregulate the immunity genes of bees (Mao et al. 2013). Similarly, fruit flies that are exposed to foraging female parasitoids lay their eggs in diets containing high concentrations of alcohol (Kacsoh et al. 2013). This behavior is adaptive because parasitized fruit flies can survive the parasitism when the alcohol concentration of food is high; alcohol has a greater adverse effect on the parasitoids than it does on the flies (Milan et al. 2012). Likewise, woolly bear caterpillars increase their consumption of alkaloid-containing plants when they have been parasitized (Singer et al. 2009) because pyrrolizidine alkaloids act against the parasitoids in caterpillars (Singer et al. 2004a). Self-medication by caterpillars may also include altering the ratios of macronutrients in the diet, rather than the intake of PSMs (Povey et al. 2013); studies that integrate the effects of primary and secondary metabolites on self-medication are sorely needed.

The list of vertebrate species that self-medicate is increasing also. Some urban bird species line their nests with the cellulose from smoked cigarette butts, a concentrated source of nicotine that reduces the number of mites in their nests (Suárez-Rodríguez et al. 2013). Lambs with gastrointestinal parasite infections self-medicate by increasing the concentrations of condensed tannin in their diets

(Lisonbee et al. 2009, Villalba et al. 2010). It turns out that the foraging behavior of many animals on the phytochemical landscape may often represent visits to the pharmacy instead of the grocery store in attempts to medicate themselves or their kin. The apparent ubiquity of medication behavior has significant implications for disease epidemiology and parasite virulence, evolution of the host immune system, and local adaptation between parasites and hosts (de Roode et al. 2013). Whether they act directly on the agent of disease itself, or through priming the host immune system to better resist enemy attack (Smilanich et al. 2011), autotrophs provide a dazzling array of secondary metabolites for use as medicines (chapter 3). PSMs still provide nearly 50 percent of the new drugs registered each year for human use (De Luca et al. 2012), and there are growing concerns over the loss of potential pharmaceuticals that will result from plant extinctions (Ibrahim et al. 2013).

The clear links between human use of plant compounds, and their use by other animals, has led to the emergence of "pharm-ecology" as a growing area of study (Sorensen et al. 2006, Forbey and Foley 2009). The field of pharm-ecology applies the traditional tools and concepts of pharmacology to ecological questions, and can help us to understand how, and why, animals use PSMs as they forage on the phytochemical landscape. In the same way that humans receive a dose-specific prescription from the pharmacist to treat specific diseases, so too must each "herbivore's prescription" take account of the dose-response relationships of PSMs (Forbey and Hunter 2012) and subsequent effects on predators, parasites, and agents of disease.

4.2.5 Additional Effects of PSMs on Predators and Parasites

Based on the ability of some herbivores to sequester PSMs, and to self-medicate when challenged with parasites, we might assume that PSMs are always deleterious for the natural enemies of herbivores. The "nasty host" hypothesis (Gauld et al. 1992), which was originally conceived to explain the relatively depauperate species richness of some Hymenopteran parasitoids in the tropics, argues that the diverse and toxic PSMs of autotrophs limit the ability of some parasites to attack potential hosts. Certainly, some parasites appear to suffer when they grow within herbivores that have consumed toxic autotrophs. For example, an alkaloid of tomatoes, alpha tomatine, decreases the survival and pupal weight of the parasite *Hyposoter exiguae* when attacking the caterpillar *Heliothis zea* (Campbell and Duffey 1979). Similarly, high nicotine concentrations reduce the survival of *Cotesia congregata* parasitoids that attack *Manduca sexta* on tobacco plants (Barbosa et al. 1991). For parsnip webworm feeding within and among plant species in the family Apiaceae, high concentrations of furanocoumarins in plant hosts reduce the probability that caterpillars are parasitized, subsequent parasitoid survival, and parasitoid clutch size (Ode et al. 2004). As we might expect, the deleterious effects of furanocoumarins on the parasitoids of herbivores appear to be lower for specialist than for generalist parasitoids (Lampert et al. 2011). In some cases, extractions of plant-derived toxins from chemically defended caterpillar cuticle can render a previously acceptable caterpillar host unacceptable to its parasitoids (Sime 2002). Finally, diet mixing by some arctiid caterpillars, which include alkaloid-rich plants in their diets, appears to reduce caterpillar susceptibility to parasitoids (Singer et al. 2004a, Singer et al. 2004b).

However, the effects of PSMs on parasite fitness may be more complicated. As I noted earlier, many herbivores gain protection from vertebrate and invertebrate predators by sequestering PSMs (Opitz and Muller 2009). If some endoparasites are generally immune from those sequestered defenses, they may actually gain associational protection in the safe havens of chemically defended hosts. For example, in the lowland wet forest of Costa Rica, caterpillars defended by chemicals that are deterrents to ant predators suffer higher levels of parasitism than do caterpillars lacking chemical deterrents (Gentry and Dyer 2002). In other words, being chemically defended from predators appears to engender higher risk as a safe haven for parasitoids. Similarly, the caterpillar *Ceratomia catalpae* sequesters high concentrations of iridoid glycosides from its catalpa host plants. These iridoid glycosides have minimal direct impact on the fitness of the parasitoid *Cotesia congregata*. Rather, the endoparasite may benefit from developing inside the safe haven of a well-defended caterpillar host that is distasteful to most predators (Lampert et al. 2010).

As I noted earlier, exploiting toxic autotrophs may impose significant costs on herbivores (Dearing et al. 2005), especially when multiple PSMs act synergistically to compromise herbivore fitness (Dyer et al. 2003). Toxin stress can reduce the ability of some herbivores to fight off parasite attack, so increasing the performance of parasites in hosts that feed on toxic autotrophs (Richards et al. 2010). For example, the specialist caterpillar, *Junonia coenia*, sequesters iridoid glycosides from its host plants. While the sequestered toxins provide caterpillars with protection from their predators, the toxins also compromise the caterpillar immune system, rendering them more poorly defended from their parasitoids (Smilanich et al. 2009a) (figure 4.22). Immune responses are central to the ability of herbivores to fight off parasite attack (Smilanich et al. 2009b). As a consequence, any effects of PSMs on the immune responses of herbivores may play an important role in the evolution of herbivore dietary breadth (Richards et al. 2010). It seems likely that the relative selective pressures imposed by predators, parasitoids, and agents of disease, combined with the direct toxicity

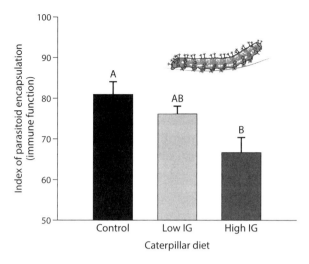

FIGURE 4.22. Diets with increasing concentrations of iridoid glycosides (IGs) decrease immune function in caterpillars, reducing their ability to encapsulate the eggs of attacking parasitoids. Modified from Smilanich et al. (2009a).

of PSMs on herbivores, determine whether the net overall impact of a given diet chemistry is defensive, medicinal, or toxic for herbivores (Smilanich et al. 2011, Sternberg et al. 2012).

In aquatic systems, effects of phycotoxins on predator populations go far beyond those imposed by herbivore sequestration. Because algal toxins are released into the environment when algal cells die, and are also consumed by herbivores and omnivores in food webs, there are multiple mechanisms by which algal toxins influence predator performance. Perhaps most dramatically, harmful algal blooms can change the structure of aquatic food webs by killing aquatic mammals, fish, and seabirds. For example, blooms of the dinoflagellate *Karenia brevis* can kill tens of millions of marine fish from the release of polycyclic ether compounds called brevetoxins (Van Dolah 2000). At the same time, phycotoxins produced by phytoplankton can move through food webs via trophic transfer, and so accumulate at higher trophic levels (Sieg et al. 2011). For example, the phycotoxin nodularin can accumulate in copepods, either via direct consumption of the cyanobacterium that produces it or, more significantly, from microzooplankton that feed first on the cyanobacteria (Sopanen et al. 2009). Likewise, domoic acid, a potent neurotoxin best known for causing paralytic shellfish poisoning in humans, also accumulates in higher trophic levels (Costa et al. 2010). Domoic acid appears to have little impact on the fitness of copepods that consume the diatoms that produce it. However, copepods may then serve as vectors that can transmit the toxin to large invertebrates, fish, birds, seals, and whales, where it can accumulate to toxic levels (Leandro et al. 2010, Sieg et al. 2011).

In summary, variation in nutrients and toxins on the phytochemical landscape influences trophic interactions by a wide variety of direct and indirect mechanisms. By influencing how herbivores forage and subsequent herbivore performance, autotroph chemistry combines with predation pressure to influence the dynamics of herbivore populations. In turn, herbivore dynamics are coupled with those of their predators, resulting in dynamical consequences of autotroph chemistry across multiple trophic levels. Additionally, nutrients and PSMs in primary producers have cascading effects on the fitness of natural enemies, generating both positive and negative effects on the performance of predators, parasites, and agents of disease.

But this is only one quarter of the story. Trophic interactions feed back to influence the chemistry of primary producers and variation on the phytochemical landscape (chapter 5). Moreover, autotroph chemistry influences (chapter 6), and is influenced by (chapter 7), nutrient dynamics at the ecosystem level. In combination, these four processes generate pathways of feedback between trophic interactions and nutrient dynamics (figures 1.1 and 2.6) (chapter 8). In the next chapter, I complete the first feedback loop between primary producer chemistry and trophic interactions by considering how the actions of consumers help to generate variation in the chemistry of autotrophs on the phytochemical landscape.

CHAPTER 5

Effects of Trophic Interactions on the Chemistry of Primary Producers

5.1 BACKGROUND

In the preceding chapter, I described how variation in the chemistry of primary producers has a pervasive influence on the population and community ecology of consumers and the interactions among them. But trophic interactions in turn influence the chemistry of primary producers (figure 5.1) and complete a feedback loop between the phytochemical landscape and trophic interactions (loop A, figure 1.1). In other words, variation on the phytochemical landscape is both a cause and a consequence of complex trophic interactions.

Exactly how do trophic interactions modify the chemistry of primary producers in ecosystems? Four mechanisms dominate, operating over broad spatial and temporal scales (figure 5.2). First, selective foraging by herbivores in ecosystems determines which autotroph cells, tissues, individuals, and species are most abundant in the environment. Note that selective foraging is often prompted by predation pressure, which influences both the number of herbivores available to forage, and the autotrophs on which it is safest to feed (Pastor and Naiman 1992). Although the literature on trophic cascades tends to focus on the standing stock or biomass of primary producers rather than their chemistry (Carpenter et al. 1985), both are affected by top-down forces (Hillebrand et al. 2007). Whatever herbivores persist in the presence of such cascades will inevitably forage selectively on some subset of the potential autotroph community (below), generating an assemblage of primary producers of a certain chemical character (Hunter 1992a). In this way, higher trophic levels generate variation on the phytochemical landscape within and among ecosystems.

Second (and intimately related), trophic interactions influence patterns of succession. Whether it is the turnover of plant species as old fields return to forests or the seasonal turnover of phytoplankton in lakes, consumers play an important role in determining the species composition of primary producers during the successional process (Brown and Gange 1989a, Leibold 1989). By

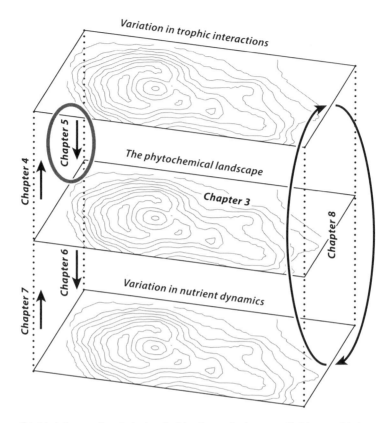

FIGURE 5.1. Variation on the phytochemical landscape is the nexus linking trophic interactions and nutrient dynamics. In this chapter (heavy circle), I focus on the role of trophic interactions (herbivory, predation, parasitism, disease) in generating spatial and temporal variation on the phytochemical landscape.

so doing, consumers participate in the long-term dynamics of phytochemistry at landscape scales. Third, primary producers respond to herbivore attack with a complex suite of chemical changes in their cells and tissues (Karban and Baldwin 1997). These include changes in the elemental stoichiometry of tissues, the induction of PSMs, and the release of volatiles that mobilize the enemies of herbivores as agents of indirect defense. Finally, herbivores act as powerful agents of natural selection on autotroph physiology and defense (Agrawal and Fishbein 2006). Consequently, the forces of predation, parasitism, and disease, which act to structure patterns of herbivory, also act as agents of selection on phytochemistry. Predation broadly writ is a powerful force in the evolution of autotroph chemical traits and their subsequent ecological effects on trophic interactions and ecosystem processes.

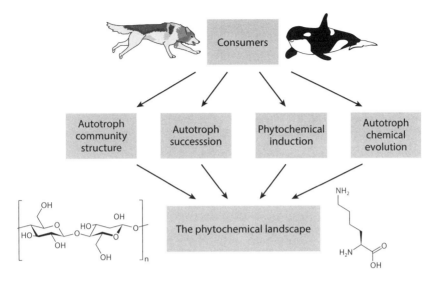

FIGURE 5.2. Four dominant mechanisms by which consumers influence the chemistry of primary producers and thereby spatial and temporal variation on the phytochemical landscape.

In this chapter, I describe briefly the first three of these, illustrating major ecological mechanisms by which trophic interactions feed back to influence the chemistry of primary production in ecosystems. Because many excellent books have been written about the effects of consumers on the evolution of plant defense, I leave that topic for some brief comments at the end of chapter 9.

5.2 EFFECTS OF HERBIVORES AND PREDATORS ON AUTOTROPH COMMUNITY STRUCTURE

5.2.1 Herbivores

The most generalist of herbivore species is still a selective forager. Even a mechanical lawnmower, perhaps the least selective forager we might envisage, removes plant tissues selectively—it culls a greater proportion of the tissues of tall plants than of short plants, so changing their relative representation in the community (Crawley 1983). Real herbivores, like lawnmowers, remove autotroph cells and tissues out of proportion with their representation in communities. Therefore, whenever significant herbivory takes place, selective foraging results, with consequences for spatial and temporal variation in the chemistry of primary producers on the phytochemical landscape (Hunter 1992a). By the same logic, whenever significant predation pressure prevents herbivores from removing substantial quantities of autotroph biomass, or alters the relative consumption of autotroph species by herbivores, predators alter the phytochemical landscape (Schmitz 2008b) (below). Moreover, because different herbivore species forage selectively on autotrophs of different chemistries, the diversity of herbivore foragers can have a profound influence on the structure and diversity of terrestrial (Greve et al. 2012) and aquatic (Rasher et al. 2013) autotroph communities.

In terrestrial systems, there are hundreds of published examples, dating back at least to the *Origin of Species* (Darwin 1859), that describe changes in the diversity and composition of plant communities from the arctic to the tropics caused by herbivores. As I noted in chapter 3, primary producers vary dramatically in their nutritional and defensive chemistry. Consequently, whenever herbivores cause changes in plant community structure, they inevitably cause changes in the standing chemistry of that plant community, with subsequent impacts on trophic interactions (chapter 4) and ecosystem processes (chapter 6). For example, rabbits can convert *Calluna* heath into grassland (Farrow 1925), thereby replacing tannin-rich heathers with silica-rich grasses. Similarly, elephants can convert tropical alpine vegetation into grassland (Mulkey et al. 1984). White-tailed deer dramatically reduce the diversity of plant species within the *Pinus resinosa* forests of Minnesota (Ross et al. 1970) while lesser snow geese change plant diversity on the west coast of Hudson Bay by foraging both above- and belowground (Kerbes et al. 1990). Three species of kangaroo rat maintain the transition zone between grassland and the Chihuahuan desert. When kangaroo rats are excluded from desert shrub habitat, it is transformed into grassland (Brown and Heske 1990). Even in hyperdiverse tropical forests, herbivory by vertebrate generalists can have a substantial impact on tree seedling recruitment and diversity (Clark et al. 2012).

Effects of herbivores on autotroph community composition are just as pervasive in aquatic ecosystems. Herbivory on seaweeds is a major factor affecting the distribution and abundance of marine algae (Lubchenco and Gaines 1981) and their ability to outcompete tropical corals (Rasher et al. 2013). Generalist fishes and sea urchins often play a dominant role in determining marine algal community structure (Carpenter 1986). For example, brown algae dominate the phytochemical landscapes of coastal North Carolina because omnivorous fish feed on red and green algae and on the amphipods that graze on brown algae (Duffy and Hay 2000). Likewise, the dramatic effects of sea urchin grazing on kelp forests off the coast of Alaska has become a community ecology paradigm (Estes and Duggins 1995); variation in kelp standing crop depends upon spatial variation in the strength of trophic cascades from killer whales through their sea otter prey to urchin herbivores and their effects on kelp (Estes et al. 2009). Perhaps less dramatically, there are reef fishes that manage "gardens" of palatable algae from which they selectively remove unpalatable species (Hixon and Brostoff 1983). In rocky intertidal communities, mollusk herbivores change the species richness (Williams et al. 2013) and relative abundance of algal species that grow on rocks, with the impacts of different mollusk species varying over the course of algal succession (Aguilera and Navarrete 2012). Finally, the persistence of eelgrass (*Zostera marina*) communities depends in part on mesograzers (e.g., isopods and amphipods) that remove epiphytic algae from eelgrass tissues (Whalen et al. 2013).

Similarly, herbivory influences the structure of algal communities and the phytochemical landscape of fresh water ecosystems. Grazing by midge larvae on algal epiphytes that colonize filamentous green algae in streams can influence algal community composition and ecosystem processes (Furey et al. 2012). In some mountain streams in the southern United States, palatable benthic diatoms form turfs that overgrow less palatable cyanobacterial colonies when grazing fish and invertebrates are excluded. If fish are reintroduced, the diatom turfs are stripped in a matter of minutes, and cyanobacterial colonies recover within two weeks (Power et al. 1988). Selective herbivory by grass carp and crayfish can mediate the community structure of freshwater vascular plants, ultimately favoring species that are physically and chemically defended, and low in nutritional quality (Parker et al. 2006, Dorenbosch and Bakker 2011).

Differential effects of herbivores on autotroph community composition, and therefore on the phytochemical landscape, often arise from cascading effects of predators on herbivore densities and behavior. It is often easiest to assess the effects of herbivores on autotroph community composition when spatial or temporal variation in predation pressure generates spatial and temporal variation in herbivory. An excellent illustration of this is on the island of Aldabra, where the lack of introduced predators has allowed a population of giant tortoises to persist. Giant tortoises have been extirpated by feral predators from most coral atolls in the Indian Ocean (Stoddart and Peake 1979) and greatly reduced throughout the Galápagos Islands. Loss of tortoises has caused substantial changes in plant community structure (Froyd et al. 2014). On Aldabra, where giant tortoise populations remain, tortoises cause the death of many trees and shrubs, and induce substantial soil erosion (Merton et al. 1976). Their herbivory generates a unique plant community known as tortoise turf, a close-cropped assemblage of dwarf grass, sedge, and herb species (Gibson and Hamilton 1983); giant tortoises generate a unique phytochemical landscape. The tortoise turf plant community is now absent from atolls on which predators have extirpated the tortoises. Moreover, within four years of experimentally excluding tortoises from sites within Aldabra, the 1–5 mm high tortoise turf is replaced by a sward 10–20 cm high with additional regeneration of shrub species (Gibson et al. 1983).

It is really a question of semantics whether we consider variation in the plant communities that are generated by the giant tortoises of Aldabra to be a product of predation pressure or a product of herbivory. Tortoises persist only in the absence of feral predators on islands, illustrating the importance of variation in predation pressure. At the same time, experimental tortoise exclusion demonstrates the radical effect that tortoises have on plant community composition. Rather than debating the primacy of predators or herbivores in generating variation in plant community structure, it is much more important to understand the range of trophic interactions among consumers at multiple levels that contribute to spatial and temporal variation on the phytochemical landscape. As a footnote to the matter of giant tortoises, recent conservation efforts include attempts to introduce functionally equivalent giant tortoises to islands in the Galápagos that have lost their native giant tortoise species. Studies show that even closely related species of giant tortoise may not be sound ecological equivalents, illustrating that the effects of consumers on plant community structure are species-specific (Hunter et al. 2013).

5.2.2 Predators

While extirpation of herbivores by feral predators is an extreme case, the ecology literature is replete with examples in which herbivore density and foraging behavior change with variation in predation pressure. Natural enemies, from pathogens (Peduzzi et al. 2014) to large carnivores (Estes et al. 2009), exert important effects on the structure and function of ecological communities by changing the density and behavior of their prey. Key to the thesis of this book is that any predator-mediated changes in autotroph communities caused by trophic cascades also represent changes in the nutritional and defensive phytochemical landscapes upon which trophic interactions (chapter 4) and nutrient dynamics (chapter 6) take place. Simply put, predators both respond to (section 4.2) and generate (this section) spatial and temporal variation on the phytochemical landscape.

Studies of the effects of predators on autotroph communities have focused primarily on trophic cascades. Trophic cascades were originally defined as the process by which predation pressure on herbivores could influence the standing stock or biomass of primary producers in ecosystems (Paine 1980, Carpenter et al. 1985). Over time, use of the term trophic cascade has broadened to include species cascades, which describe effects of predators on the performance of single autotroph species within communities (Dyer and Letourneau 1999), and diversity cascades that include effects of diverse predator communities on primary producer diversity (Dyer and Letourneau 2003). These broader definitions can be useful, because they move the focus away from autotroph biomass and toward predator-mediated changes in autotroph community composition and the phyto-chemical landscape. Unfortunately, because the nutritional and defensive chem-istry of primary producers are rarely measured in studies of trophic cascades, evidence that trophic cascades influence the phytochemical landscape is generally indirect. Here, I focus on the effects of predation pressure on the structure and (where possible) phytochemistry of autotroph communities.

Yellowstone National Park has served as a living laboratory for the study of trophic cascades in terrestrial systems. Wolves were extirpated from the park by the mid 1920s and remained absent for 70 years until their reintroduction in the mid 1990s. In the period during which wolves were absent, elk populations were culled extensively, yet culling could not prevent elk-mediated declines in the recruitment of aspen, cottonwood, and willow trees (Beschta and Ripple 2013). It therefore seems possible that wolf-mediated changes in elk behavior, rather than in elk density, prevented the declines in tree recruitment prior to wolf extirpation. However, elk populations have also declined since wolf reintroduction, suggest-ing that any recovery of tree recruitment since wolf reintroduction may result from a combination of behavioral and density changes in elk (Winnie 2012).

Whether mediated by density or behavior, can trophic cascades from wolves through elk to trees influence plant chemistry at the landscape scale? Although we have no phytochemistry samples from aspen or cottonwood that predate wolf extirpation, work in other systems demonstrates that the phenolic chemistry of these foundation tree species is central to both the structure of local food webs and to soil ecosystem processes (Wimp et al. 2004, Schweitzer et al. 2012a). It therefore seems almost certain that 70 years of wolf removal changed the phyto-chemical landscape in Yellowstone National Park, with consequences for com-munity structure and nutrient cycling. While the reintroduction of wolves into Yellowstone has not yet caused dramatic increases in tree recovery (Kauffman et al. 2010), there is evidence for reduced browsing by elk and increases in tree recruitment, particularly in riparian areas (Beschta and Ripple 2013). In addition, wolf reintroduction may have facilitated increases in berry-producing shrubs such as serviceberry, *Amelanchier alnifolia*. This is noteworthy because serviceberry can serve as an important source of nutrition for other mammal species, including grizzly bears (Ripple et al. 2014a).

Beyond the iconic Yellowstone National Park, the loss of large carnivores has had pervasive effects on plant community structure in other ecosystems too. For example, carnivore declines in and around Elk Island National Park in Alberta have caused substantial increases in ungulate density and grazing pressure. In turn, high ungulate density has changed the structure and composition of plant communi-ties, with deleterious effects on the densities of warblers and butterflies (Teichman et al. 2013). Similarly, a reduction in cougar densities in Zion National Park has increased the densities of mule deer, limiting the recruitment of cottonwood trees. As a consequence, the structure of riparian vegetation has changed, and densities of the organisms that it supports have declined (Ripple and Beschta 2006).

It is not always large mammalian carnivores and their mammalian prey that influence the structure and phytochemistry of terrestrial plant communities. Oswald Schmitz and colleagues have made substantial contributions to this research area by studying how grasshopper foraging behavior changes with predation pressure by spiders, and the consequences for old-field community structure (this section) and ecosystem processes (chapter 8). In this system, the presence of spiders causes grasshoppers to change their foraging strategies, even when spiders are prevented from catching their prey; it is predation risk rather than declines in grasshopper density that have the greatest impact on the relative biomass of grass and forbs in old fields (Schmitz 1998). For example, when exposed to risk from the nursery web spider, *Pisaurina mira*, the grasshopper *Melanoplus femurrubrum* reduces the average proportion of grass in its diet from 70 percent to 42 percent, preferring to forage instead on *Solidago rugosa* (Beckerman et al. 1997). As a consequence, spiders act to maintain the evenness of old-field plant communities because grasshoppers prevent *S. rugosa* from outcompeting other species of old-field plants (Schmitz 2003) (figure 5.3A). Importantly, effects of spiders on plant community structure are much stronger than the effects of spiders on plant biomass (Schmitz 2006). As I noted earlier, high species diversity in plant communities, and the high phytochemical diversity that accompanies it, have effects on both trophic interactions (chapter 4) and ecosystem processes (chapter 6). Effects of spiders on nutrient dynamics in old fields are discussed fully in chapter 8.

In Schmitz's old-field system, the life history traits of both herbivores and predators influence the form and strength of trophic cascades, with differential effects on the phytochemical landscape. Predation risk may have a greater impact on the behavior of generalist herbivores than specialist herbivores. By definition, specialists are restricted to a small subset of the plant community, and can only change their feeding rates, not their choice of plant species (Schmitz 1998) (figure 5.3B). In contrast, generalist herbivores may radically alter the species composition of their diet when under risk from predation (above). For example, stressed herbivores may need to consume food with higher concentrations of digestible carbohydrates to meet the increased energy demands of higher metabolic rates (Hawlena and Schmitz 2010a). As a consequence, predators can drive changes in the relative abundance of plants species, their standing variation in nutritional and defensive phytochemistry, and the subsequent rates at which plant litter decomposes (Hawlena and Schmitz 2010b) (chapter 8). The fear induced in grasshoppers by *P. mira* is estimated to increase the C:N ratio in remaining plants

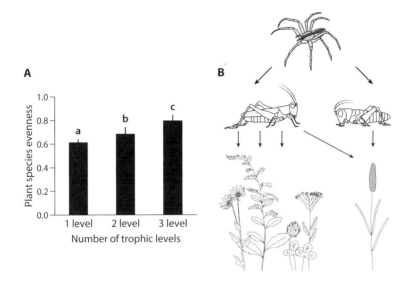

FIGURE 5.3. Spider predators influence the structure of plant communities. (A) Experimental treatments with spiders, herbivores, and plants (3 level) exhibit greater plant community evenness than do treatments with only herbivores and plants (2 level) or just plants alone (1 level). (B) The mechanism by which spiders influence plant community structure depends upon whether their herbivore prey are generalists (left-hand side) or specialists (right-hand side). Modified from Schmitz (1998, 2003).

by about 10 percent (Hawlena and Schmitz 2010a). In contrast to *P. mira*, which is a sit-and-wait predator, the hunting spider, *Phidippus rimator*, directly reduces the density of its grasshopper prey (Schmitz 2008a). Reductions in grasshopper density decrease the diversity of the old-field plant community, favor dominance by *Solidago rugosa*, and decrease the average C:N ratio of remaining plant tissue. In other words, Schmitz and colleagues have demonstrated that two predators with different foraging strategies both impose significant changes in the phytochemical landscape upon which they forage. While data on plant quality are limited to C:N ratio at present, changes in the relative representation of other indices of plant quality (lignin, cellulose, silica, and toxins) seem highly likely (chapter 4).

Cascading effects of predators on autotroph community structure are just as pervasive in aquatic ecosystems. I have mentioned previously the dramatic trophic cascades, from killer whales through sea otters and sea urchins, which influence the biomass and diversity of kelp forests (Estes and Duggins 1995). Predators also influence the community structure of primary producers in salt marshes, seagrass beds, rocky shores, and coral reefs. For example, in salt marsh

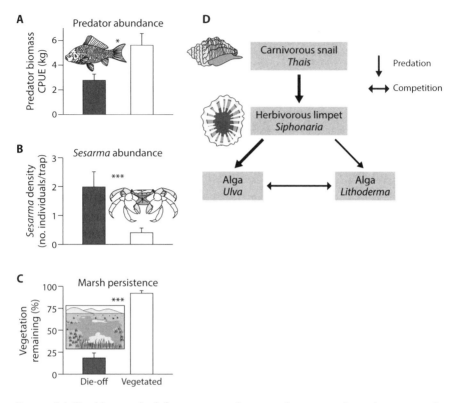

FIGURE 5.4. Trophic cascades influence autotroph community structure in marine systems. In A, B, and C, the dark bars represent sites of overfishing by recreational anglers in salt marshes (labeled "die-off") while white bars represent areas of low angling pressure (still vegetated). The decline in large fish predators (A) increases the density of herbivorous crabs (B) that severely reduce the densities of saltmarsh plant species (C). (D) On rocky shores, predatory snails keep herbivorous limpet densities low, thereby allowing palatable algae (*Ulva*) to outcompete less palatable cyanobacteria (*Lithoderma*). Modified from Altieri et al. (2012) and Wada et al. (2013).

ecosystems, recreational angling has reduced the densities of predatory fish, facilitating increases in the densities of herbivorous crabs (*Sesarma*), and subsequent declines in the relative abundance of salt marsh plant species (Altieri et al. 2012) (figure 5.4A–C). In some seagrass communities, tiger sharks influence the foraging behavior of large grazers (sea cows and sea turtles) that, in turn, modify the relative abundance of marine plant species and their foliar nutrient concentrations (Burkholder et al. 2013). Seagrasses are particularly susceptible to overgrowth by epiphytic algae when nutrient levels are high. Off the central coast of California, sea otters promote the recovery of seagrass communities from eutrophication by suppressing crab populations. When crab populations are low, mesograzer populations (isopods and sea slugs) can grow large enough to remove epiphytic algae from seagrass foliage (Hughes et al. 2013).

On rocky shores, preferential consumption by herbivores of palatable algae can release less palatable algae from competition, increasing their relative cover on rocks. Such interactions form the basis of a trophic cascade off the shores of Japan, where predatory snails (*Thais*) reduce the densities of limpet herbivores (*Siphonaria*), in turn favoring competitive dominance of palatable green algae (*Ulva*) over less palatable cyanobacteria (*Lithoderma*) (Wada et al. 2013) (figure 5.4D). In other words, zones of palatability on the phytochemical landscape depend upon the spatial distribution of predatory snails. Also on rocky shores, bird predators may keep populations of some sea urchins at low density, so favoring diverse algal communities. Bird exclusion can reduce algal species diversity by 6-fold and evenness by an order of magnitude (Wootton 1995). Importantly, in the absence of birds, urchins remove all but the most poor-quality and unpalatable algal species, demonstrating again that predators can change the phytochemical landscape of rocky shores. Predators also influence the algal communities on rocks in freshwater streams. For example, brook trout in boreal forest streams influence foraging of invertebrate herbivores. As a consequence, trout impose strong indirect effects on the species composition and physical structure of epilithic algal communities (Bechara et al. 2007).

Finally, predation pressure also influences phytoplankton community structure. Most often, the ratio of edible planktonic algae to inedible algae increases when fish predators keep zooplankton populations low (Leibold 1989). When zooplankton herbivores feed preferentially on palatable phytoplankton, they generate a low-diversity community of unpalatable primary producers. However, when fish predators reduce zooplankton populations, phytoplankton diversity increases to include a range of palatable and unpalatable forms (Tessier and Woodruff 2002). As a consequence, variation in the identity or life history traits of fish populations can have cascading effects on the palatability of phytoplankton communities (Weis and Post 2013). The phytochemical seascape is, in part, a product of predation pressure.

5.2.3 Don't Forget the Roots

In terrestrial systems, when we focus only on the chemistry of plant modules aboveground, we are missing half of the picture (chapter 4). There has been a recent surge of interest in the chemistry of root tissue, the ecology of root herbivores, and in soil community processes (Bardgett and Wardle 2010), which is helping to establish the fundamental links between above- and belowground processes in terrestrial ecosystems. There are two brief points that I would like to make here about the chemistry of plant roots (I discuss herbivore-induced changes in root chemistry in some detail later in this chapter).

First, any effects of herbivores and their natural enemies aboveground on plant community structure will influence the phytochemical soilscape generated by roots belowground. The nutritional and defensive chemistry of roots vary widely among plants species (Hunter 2008, Rasmann et al. 2011). Therefore, the pervasive herbivore- and predator-induced changes in plant community structure that I have just described above will also change the relative representation of root chemical phenotypes belowground. For example, we might predict that the reintroduction of wolves into Yellowstone National Park should increase the relative representation of aspen root chemistry in the soil. Likewise, selective foraging by moose on Isle Royale for deciduous trees over conifers (Pastor et al. 1993) should increase the relative representation of conifer roots in the soil; conifer roots often contain lower nutrient concentrations than do the roots of deciduous trees (Pregitzer et al. 2002). Unfortunately, while there have been numerous studies of the effects of aboveground herbivores on the root chemistry of individual plants (see section 5.4, below), community-wide studies of root chemistry remain remarkably rare. The spatial mapping of root phytochemical diversity in soil should be a high priority for future research.

Second, any effects of root herbivores and their enemies belowground on plant community structure will influence the phytochemical landscape aboveground (Van Der Putten 2003). Root feeders can reduce plant species richness (Brown and Gange 1989b), and accelerate succession by reducing rates of colonization or persistence of early successional species (Brown and Gange 1992, De Deyn et al. 2003). As a result, root feeders mediate patterns in the phytochemical landscape both above- and belowground.

For example, root-feeding ghost moth caterpillars can cause over 40 percent mortality of bush lupine in California and contribute to the long-term dynamics of lupine populations (Strong et al. 1995). In this system, nematode enemies of the caterpillars cause cascading effects on root herbivores that influence plant distribution and abundance (Strong et al. 1999). Because bush lupine associates with N-fixing bacteria, and N is a limiting resource in these soils, the cascade from nematodes to root-feeding moths can influence the expression of foliar N on the phytochemical landscape and the cycling of nutrients (Maron and Jeffries 1999). The potential for root-feeding herbivores to exploit particular plant species, and to change their relative abundance in plant communities, has led to the use of root herbivores as agents of biological control of weed pests (Hunt-Joshi et al. 2004). More broadly speaking, root-feeding herbivores and their natural enemies can change the phytochemical landscape above- and belowground by their important effects on plant community structure.

5.3 CONSUMER EFFECTS ON SUCCESSION—TEMPORAL CHANGE ON THE PHYTOCHEMICAL LANDSCAPE

In the examples that I described above, herbivores and their natural enemies influence both spatial and temporal variation in autotroph community structure and therefore phytochemistry. In some cases, trophic interactions influence patterns of succession and therefore the phytochemical succession that accompanies temporal changes in autotroph community structure. Of course, there are multiple biotic and abiotic factors that influence the successional process (Glenn-Lewin et al. 1992) (chapter 9), and the importance of trophic interactions in determining the rate and direction of succession likely varies substantially among systems. In my view, we are a long way from being able to predict where and when trophic interactions will have their greatest impact on succession. Here, I provide just a few examples to illustrate that trophic interactions influence the temporal trajectory of autotroph community structure, recognizing that accompanying datasets of phytochemical change remain sorely lacking.

Ecologists often study successional processes during recovery from disturbance. For example, storms generate bare patches on rocky shores, and their recolonization by algae can inform our understanding of autotroph succession. In a mid-high rocky intertidal zone in northern California, both herbivores and local algal diversity influence the rate of recovery of algal communities from disturbance. The presence of herbivores increases the rate of algal succession by preventing the fast-growing early colonists from preempting space from slower-growing late-successional algae (Aquilino and Stachowicz 2012) (figure 5.5). Key to this interaction, of course, is that the fast-growing early-successional algae are more palatable to herbivores than are the late-successional species (Lubchenco 1983). In other words, differential chemical quality of the algae contributes to temporal change in the phytochemical landscape during recovery from disturbance. As with any other effects of trophic interactions on autotroph community structure (above), effects of herbivores on phytochemical succession depend upon preexisting variation in chemical quality among potential community members and the differential consumption that variable phytochemistry engenders (Feng et al. 2009).

When trophic interactions have strong effects on succession, they may do more than simply alter the rate at which some predefined successional process takes place. Rather, they may alter the final outcome of succession by mediating alternative successional trajectories (ASTs) (Suding et al. 2004). ASTs are likely to occur when trophic interactions alter patterns of colonization during earlier stages of succession that subsequently determine species replacements during later stages of succession. In terrestrial systems, ungulates may be particularly

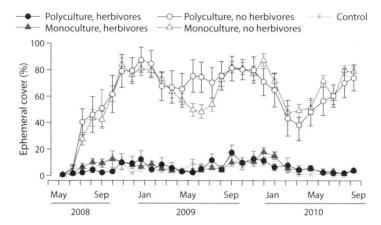

FIGURE 5.5. Herbivores (black symbols) prevent rapidly growing (ephemeral) algal species from preempting space during 28 months of succession following disturbance in a rocky intertidal community. Herbivores impose powerful effects on algal cover in either monocultures (triangles) or polycultures (circles) of algae. Modified from Aquilino and Stachowicz (2012).

important agents for generating ASTs (Tanentzap et al. 2012). For example, foraging by deer in boreal forests can substantially reduce colonization by palatable plant species, such as paper birch, during early stages of recovery from disturbance. The legacy of early deer browsing is apparent during later stages of succession, including large reductions in the densities of aspen, cherry, and balsam fir, and dominance by white and black spruce trees (figure 5.6) (Hidding et al. 2013). Given the substantial differences in phytochemistry between spruce trees and the broad-leaved species that they replace, this system provides a compelling example of herbivore-mediated variation in phytochemical succession (Feng et al. 2009). To the extent that high ungulate populations result from substantial declines in top predators (Estes et al. 2011), this system can also be viewed as one that has changed from predator-mediated to herbivore-mediated phytochemical succession (Feng et al. 2012).

There is now substantial evidence that trophic interactions belowground influence successional processes in terrestrial systems, and that variation in root chemistry is both a cause and a consequence of ASTs (Van Der Putten 2003). First, root defenses influence the relative palatability of plants to belowground herbivores, and therefore levels of herbivore attack. Subsequently, herbivore-mediated changes in plant succession alter the phytochemical landscape both above- and belowground. For example, primary succession on European coastal dunes of pioneer grasses and nitrogen-fixing shrubs is limited in part by root-feeding nematodes and the pathogenic fungi with which they interact. Nitrogen-fixing shrubs

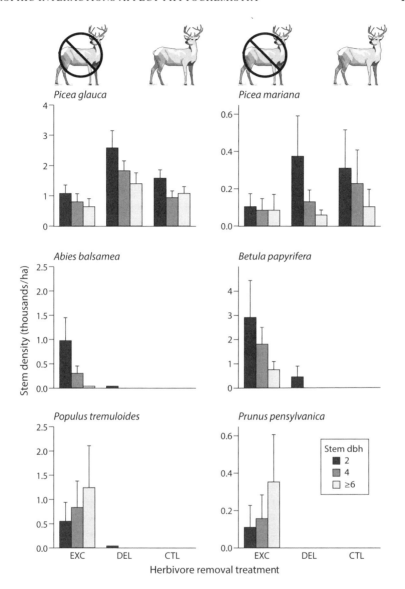

FIGURE 5.6. Levels of deer herbivory generate alternative successional trajectories in boreal forests of Quebec. Areas from which deer are excluded (EXC) contain a mixture of coniferous and deciduous tree species whereas areas with deer (CTL) are dominated by less palatable spruce trees. Note that forest community structure in delayed exclosures (DEL) approximates that in control areas, suggesting that the effects of deer on successional change are most important during early seedling stages. Modified from Hidding et al. (2013).

such as *Hippophae rhamnoides* (sea buckthorn) mediate important fluxes of N from the atmosphere into terrestrial ecosystems, and root-feeding herbivores limit the spatial and temporal distribution of *H. rhamnoides* (Van Der Putten 2003). Similarly, during secondary succession in old-field ecosystems, soil-dwelling insect herbivores can accelerate rates of succession by feeding preferentially on early successional forbs (Gange and Brown 2002). Experimental addition of soil invertebrates to grasslands also favors later successional plant species at the expense of early successional species, so increasing the rate of secondary succession and overall plant diversity (De Deyn et al. 2003). The key point is that standing chemical variation on the phytochemical landscape belowground establishes patterns of differential herbivore attack. In turn, differential herbivory influences plant succession and the overall phytochemical trajectory of the plant community.

5.4 PHYTOCHEMICAL INDUCTION—A MULTIPLIER OF VARIATION IN AUTOTROPH CHEMISTRY ON THE PHYTOCHEMICAL LANDSCAPE

Autotrophs are not passive recipients of damage by herbivores. Rather, primary producers are masters of adaptive phenotypic plasticity, expressing a wide variety of defensive responses when consumers attack them or their neighbors. Pervasive among these responses are induced changes in the chemistry of autotroph tissues (Green and Ryan 1972, Baldwin and Schultz 1983), many of which act to reduce the subsequent performance of herbivores (Karban and Baldwin 1997). Critically for the purposes of this book, defense induction is a multiplier of standing spatial and temporal variation (Cipollini et al. 2014) on the phytochemical landscape. Moreover, phytochemical induction provides an important mechanism of feedback among diverse consumers in communities (Faeth 1986, Li et al. 2014) and a mechanism linking trophic interactions with nutrient dynamics in ecosystems (Findlay et al. 1996) (figure 5.7).

In chapter 4, I described the ubiquitous and powerful effects of constitutive variation in autotroph chemistry on trophic interactions. Chemical induction magnifies this variation in autotroph chemistry, and it does so at spatial scales ranging from millimeters (Edwards and Wratten 1983) to hundreds of hectares (Baltensweiler and Fischlin 1988), and on time scales from seconds (Zagrobelny et al. 2004) to years (Haukioja and Niemela 1977). The consequences include changes in the abundance, behavior, and distribution of herbivores and concomitant changes in trophic interactions at multiple trophic levels, including manifold indirect effects on predators and parasites (Poelman and Dicke 2014). Induced defenses also influence subsequent rates of decomposition and nutrient flux

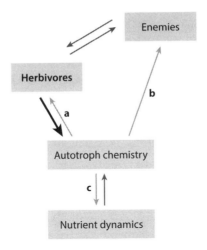

FIGURE 5.7. Herbivores induce changes in phytochemistry (black arrow) that (a) feed back to influence herbivores and their trophic interactions, (b) attract natural enemies of herbivores, and (c) influence nutrient dynamics. Therefore, phytochemical induction can provide a mechanistic link between trophic interactions and nutrient dynamics.

from senesced plant materials (Chapman et al. 2006) (chapter 8). Simply put, phytochemical induction generates important variation on the phytochemical landscape.

What types of changes in autotroph chemistry do herbivores induce? Terrestrial plants may express localized or systemic changes in the concentration of PSMs that make them more toxic (Karban and Baldwin 1997) or more difficult to digest (Reynolds et al. 2012). Alternatively, the concentrations of limiting nutrients may decline in plant tissues following herbivore damage. For example, damage by aphids to *Asclepias viridis* reduces foliar N concentrations (Zehnder and Hunter 2007b). Plants may hide critical resources from herbivores by reallocating nutrients such as nitrogen or photosynthate away from sites of herbivore attack (Orians et al. 2011), reducing the subsequent quality of food for herbivores. Finally, herbivory (and herbivore oviposition) can induce increases in the indirect defenses of plants by signaling to the natural enemies of herbivores (Turlings et al. 1990) or by increasing rewards for predators such as extrafloral nectar (Ness 2003). These kinds of changes in the nutritional and defensive chemistry of primary producers on the phytochemical landscape are important because they (a) influence subsequent performance by herbivores, (b) mediate interactions between herbivores and their enemies, and (c) affect subsequent rates of tissue decomposition and nutrient cycling of senesced plant material (chapter 6).

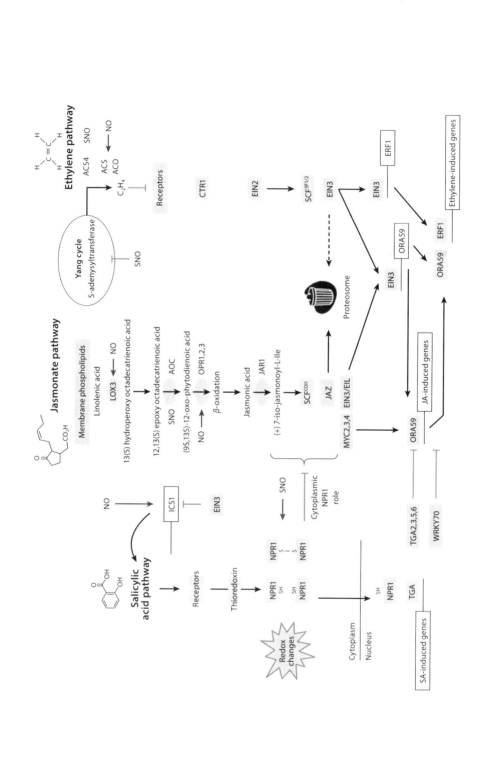

The molecular mechanisms underlying terrestrial plant responses to attack are now quite well known (Hu et al. 2013). Herbivore attack generally induces signaling pathways based on the phytohormones jasmonic acid (JA) and ethylene, which then induce the transcription of defense-related genes in plants (Kessler et al. 2004). Pathogen attack is more often associated with induction of the salicylic acid (SA) pathway, which can sometimes act as an antagonist of the JA pathway (Hunter 2000, Mur et al. 2013) (figure 5.8). Induction of JA (by the octadecanoid pathway) often causes significant reprograming of plant defenses. Defense induction may include increases in phenylpropanoid biosynthesis (the pathway leading to synthesis of flavonoids, coumarins, tannins, and lignin), alkaloid biosynthesis, and induction of the mevalonic acid pathway (leading to terpene synthesis). In other words, the induction of JA is associated with the synthesis of diverse PSMs that reduce the chemical quality of plants for herbivores (chapter 4). The JA pathway can also lead to the release of methyl jasmonate (MJ). MJ is a volatile signal that can induce emissions of additional volatile organic compounds (VOCs) that serve to attract natural enemies to sites of herbivore attack (Thaler 1999, Kessler and Baldwin 2001), with notable specificity in both herbivore induction and enemy response (McCormick et al. 2012). As we saw earlier in this chapter, natural enemies can have cascading effects that influence plant community structure, the phytochemical landscape, and (chapter 8) effects on nutrient dynamics.

In addition to inducing the production of toxic or recalcitrant defensive compounds, herbivores can induce changes in the concentrations and distributions of elements such as N and C among plant tissues (Babst et al. 2005), so changing the elemental stoichiometry of plant tissues. Reallocation of nutrients away from sites of damage following herbivore attack (resource sequestration) presumably helps to confer tolerance to herbivory (Rosenthal and Kotanen 1994), and may also protect precious resources from further consumption; plants and herbivores may play a form of hide-and-seek with elemental nutrients (Tao and Hunter 2011). For example, JA can induce rapid export of photosynthate from leaves to storage

FIGURE 5.8. The salicylic acid (SA), jasmonate (JA), and ethylene (ET) pathways of chemical defense induction. Note the potential for antagonism between the pathways (flat-tipped lines) whereby induction of the SA pathway can inhibit induction of the JA/ET pathway, and vice versa. Missing lines represent steps that are not yet well characterized. Nitric oxide (NO) can mediate signal induction in all pathways. Biosynthetic enzymes are represented as gray ovals and signaling components are gray rectangles. Abbreviations in the jasmonate biosynthetic pathway are as follows: LOX, lipoxygenase; AOC, allene oxide cyclase; OPR, oxo-phyto dienoate reductase; for the ethylene biosynthetic pathway: ACS, 1-aminocyclopropane-1-carbox ylic acids synthase; ACO, 1-aminocyclopropane-1-carboxylic acid oxidase. Genes and their regulatory promoters are represented as open boxes. Further details are in Mur et al. (2013).

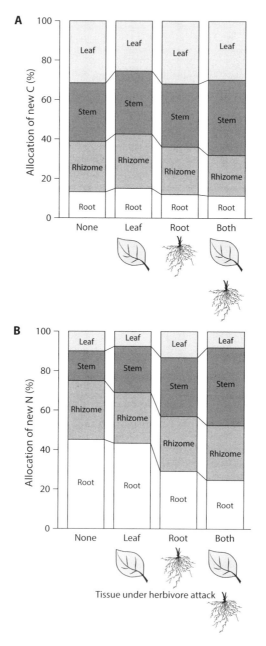

FIGURE 5.9. Damage by leaf and root herbivores affects the allocation of new carbon (C) and nitrogen (N) within *A. syriaca* plants. Damage by both leaf and root herbivores strongly favors allocation of new N away from sites of damage to the stems of milkweed plants. Data from Tao and Hunter (2012a).

organs in *Populus* (Babst et al. 2005). Similarly, oaks reallocate newly acquired N to storage tissues following attack by herbivores (Frost and Hunter 2008a). When attacked simultaneously by above- and belowground herbivores, *Asclepias syriaca* allocates newly acquired C and N away from roots and leaves, and preferentially allocates them to stems (Tao and Hunter 2012a) (figure 5.9). Changes in allocation are stronger for N than for C, perhaps reflecting differences in the ability of plants to replace those two resources; N may simply be a more limiting resource for milkweed than is C.

However, the resources that terrestrial plants sequester when under herbivore attack cannot be hidden forever; they are still required for fundamental metabolic processes. Accordingly, plants must balance protecting their resources with the requirements of replacing lost photosynthetic tissue, maintaining resource uptake, and the physiological and ecological costs of resource sequestration (Orians et al. 2011, Schultz et al. 2013). As a consequence, plants with a limited potential to sit and wait for safer times may actually increase allocation of resources to damaged tissues (Green and Detling 2000), presumably to repair lost tissue and maintain resource uptake. Whether plants increase or decrease the allocation of limiting resources to sites of damage, the key point is that herbivore activity changes the nutritional quality of plant tissue on the phytochemical landscape, with consequences for herbivore performance.

Evidence is now overwhelming that herbivore-induced changes in the chemistry of primary producers influence the subsequent performance of herbivores and their natural enemies, and I won't repeat that well-reviewed material here. However, in the sections that follow, I emphasize a few themes that have either received proportionately less attention in the literature or are particularly relevant to defining the feedback processes between phytochemistry and trophic interactions.

5.4.1 Phytochemical Induction in Marine Ecosystems

Despite over two decades of research in marine ecosystems, induced responses to herbivores are still much better characterized in terrestrial than in aquatic environments. Because seaweeds lack the typical vascular system of most terrestrial plants, we might expect systemic induction of PSMs in seaweeds to be rare or absent. Moreover, because unicellular bacteria and algae conduct most of the primary production in aquatic ecosystems, the typical mechanisms underlying induction in multicellular terrestrial plants appear unlikely to operate. Perhaps as a consequence, research on chemical induction in aquatic primary producers has provided variable results. For example, meta-analyses suggest that brown and green algae induce resistance to grazing by small crustacea and gastropods (mesograzers), but not to large gastropods or urchins. Moreover, red algae appear not to induce defenses to any grazer (Toth and Pavia 2007).

Because they often live on seaweeds over extended periods of time, meso-grazers such as amphipods represent good potential targets for induced PSMs in seaweeds. In support of this idea, feeding by the generalist amphipod *Ampithoe longimana* induces increases in the diterpene alcohol concentrations expressed by the brown alga *Dictyota menstrualis*. Secondary metabolite concentrations increase by up to 35 percent over those in ungrazed plants, and reduce subsequent grazing by amphipods by 50 percent (Cronin and Hay 1996b). Since the foundational work by Mark Hay and colleagues, induction of PSMs has been demonstrated in other seaweeds. For example, feeding by snails on *Ascophyllum nodosum* induces increases in the concentrations algal phlorotannins, which then act to deter further snail grazing (Pavia and Toth 2000). As shown first in maize (Alborn et al. 1997), the induced resistance in *A. nodosum* appears to be induced by snail saliva (Coleman et al. 2007b). Induced defenses then increase rates of snail movement (Borell et al. 2004) and reduce the subsequent fitness of consumers (Toth et al. 2005). Interestingly, grazing on the same alga by isopods does not induce phlorotannin increases, and does not deter subsequent isopod grazing (Pavia and Toth 2000). As in terrestrial systems (McCormick et al. 2012), induction responses in aquatic systems appear to be specific to the species of herbivore inflicting the damage.

Grazing-induced chemical defense might appear to be an unlikely strategy for single-celled phytoplankton that are consumed by filter-feeding herbivores. Nonetheless, chemical induction has been reported in unicellular primary producers. For example, in response to mechanical damage to their cells, some diatoms use lipoxygenase-mediated enzyme systems to convert fatty acids into potent oxylipin toxins. Grazing-induced release of oxylipins can reduce the subsequent performance of copepod herbivores (Ianora et al. 2003). Critically, the induced diatom toxins inhibit copepod recruitment, thereby reducing herbivore pressure on subsequent diatom generations. In other words, for single-celled phytoplankton, chemical induction may serve as a transgenerational defense (Ianora et al. 2004).

Moreover, as in terrestrial systems, damage by marine herbivores may also elicit signals that attract natural enemies to sites of damage (arrow b in figure 5.7). This is important because natural enemies have cascading effects on the phytochemical landscape (section 5.2) and therefore ecosystem processes (chapter 8). While it remains comparatively understudied in aquatic ecosystems, there is mounting evidence that aquatic primary producers can release chemical cues that attract natural enemies to sites of herbivore attack. For example, after snails have attacked the brown alga *Ascophyllum nodosum*, the algae are more attractive to enemies of the snails than are unattacked or artificially damaged algae (Coleman et al. 2007a). Additionally, feeding by periwinkles on *A. nodosum* releases a signal that causes unattacked neighboring seaweeds to increase their production of phlorotannins; apparently there are "talking algae" as well as "talking trees" (Baldwin and Schultz 1983, Karban et al. 2004). In turn, phlorotannins reduce the palatability of the seaweed to subsequent periwinkle grazing (Toth and Pavia 2000).

5.4.2 Phytochemical Induction and Plant Roots

Back on dry land, it would be a significant oversight to ignore the chemical changes in plants that are caused by root-feeding herbivores, and the subsequent effects on trophic interactions and ecosystem processes. Root feeders often cause a reallocation of nutrient resources within plants (Tao and Hunter 2012a) that may increase the performance of foliar-feeding herbivores (Gange and Brown 1989). Significant root damage can disrupt plant protein metabolism and amino acid synthesis, increasing the hydrolysis of existing proteins to free amino acids, and leading to higher availability of mobile N in shoots (Johnson et al. 2008). This response may explain why damage by root herbivores will sometimes favor the performance of phloem-feeding insects aboveground (Masters et al. 1993).

However, beyond simple nutrients, root-feeding herbivores may also induce either increases or decreases in shoot PSMs, influencing the phytochemical landscape, and subsequent performance of herbivores, aboveground (Bezemer et al. 2003). Root herbivores can induce increases in shoot concentrations of diverse secondary metabolites, including terpenoids, phenolics, alkaloids, glucosinolates, phytoectosteroids, and proteinase inhibitors (Erb et al. 2008). Root herbivores can also decrease concentrations of primary nutrients such as P in shoot tissue, with deleterious consequences for aboveground herbivores; overall, it is the balance between defensive and nutritional responses to root damage, and the life histories of the associated herbivore species, that determines the overall effects of root damage on community structure and trophic interactions aboveground (Johnson et al. 2013) (figure 5.10). Despite this variability in chemical change and herbivore response, the overall message is clear; we can no longer ignore the simple fact that the phytochemical landscape aboveground is generated in part by trophic interactions belowground (Wondafrash et al. 2013).

Reciprocally, the phytochemical landscape belowground depends in part on induced responses to herbivore damage aboveground. Early work suggested that shoot-feeding herbivores generally act to decrease the quality of root tissue and depress densities of root-feeding herbivores (Masters and Brown 1992) in part because damage aboveground can induce increases in the expression of chemical

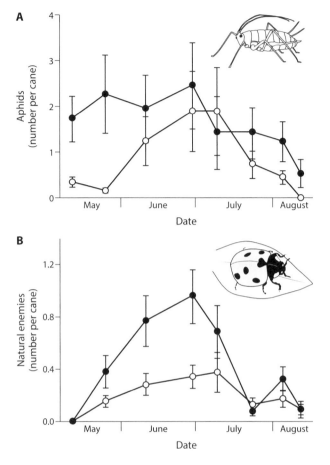

FIGURE 5.10. Chemical changes in plants imposed by root-feeding herbivores influence trophic interactions above ground. The presence (closed circles) or absence (open circles) of root-feeding weevils influences (A) the population density of aphids on blackcurrant shoots and (B) the abundance of their natural enemies. Modified from Johnson et al. (2013).

defense belowground (Baldwin et al. 1994). However, PSM expression belowground can increase, decrease, or remain unchanged in response to defoliation aboveground (Erb et al. 2008), and we are far from being able to generalize.

As we might expect, root-feeding herbivores also induce root chemical defenses. For example, damage to the roots of wild parsnip plants induces increases in root concentrations of furanocoumarins (Zangerl and Rutledge 1996). Similarly, herbivore damage to the roots of spinach plants induces increases in defensive phytoecdysteroids in root tissue (Schmelz et al. 1998). In other words, the phytochemical landscape belowground is, in part, a product of belowground herbivores. And just as herbivore damage aboveground can induce the release of VOCs that attract the natural enemies of herbivores, so does damage below-ground. When larvae of the beetle *Diabrotica virgifer* feed on the roots of maize, they induce release of the volatile sesquiterpene (E)-β-caryophyllene from roots. In turn, VOC release attracts entomopathogenic nematodes (EPNs) that attack the beetle larvae (Rasmann et al. 2005). Similarly, feeding by root weevils on the roots of citrus trees induces the release of pregeijerene, a VOC that attracts EPNs, which attack and kill the weevil larvae (Ali et al. 2012). Consensus is emerging that induced VOC release by plant roots, especially in the presence of CO_2 respired from roots, is an important cue by which foraging EPNs locate their herbivore prey (Turlings et al. 2012).

5.4.3 Manipulation of Autotroph Chemistry by Herbivores

Given that herbivores induce complex changes in the phytochemical landscape, we can ask whether herbivores ever manipulate the chemistry of primary produc-ers for their own benefit. Manipulation of host phenotype by other kinds of para-sites is well established, so we might expect that some herbivores induce changes in autotroph chemistry that increase their own fitness, or prevent the induction of defensive molecules that would reduce their fitness. For example, the caterpillar *Helicoverpa zea* can suppress the induction of nicotine in damaged tobacco plants by introducing glucose oxidase with saliva (Musser et al. 2002). Glucose oxidase may represent a common mechanism by which caterpillars attempt to mitigate defense induction in their host plants (Eichenseer et al. 2010).

The work by Musser and colleagues (2002) prompted a fundamental shift in our understanding of autotroph responses to herbivore damage. Rather than sim-ple victims of induced defenses, we began to realize the extent to which herbi-vores could suppress the defensive responses of primary producers, manipulating phytochemistry for their own fitness. Of course, gall-forming herbivores on ter-restrial plants (often nematodes, mites, flies, and wasps) are well known champi-ons of plant manipulation, inducing complex structures upon, and changing the chemistry of, their plant hosts (Stone and Cook 1998, Nabity et al. 2013). But a variety of other herbivores can suppress PSMs and so manipulate the chemistry of their food. For example, larvae of the Colorado potato beetle exploit bacterial symbionts in their saliva to induce the SA pathway in tomato plants. As a result of cross talk between the SA and JA pathways (figure 5.8), induction of the SA pathway prevents up-regulation of the JA pathway. JA-based defenses against the Colorado potato beetle cannot therefore be induced, thereby preventing deleteri-ous effects on larval performance (Chung et al. 2013).

Even insect eggs may provide signals that suppress PSMs and so favor development of hatching larvae. For example, eggs of *Pieris brassicae* induce the localized accumulation of salicylic acid (SA) in *Arabidopsis thaliana* (Bruessow et al. 2010). As described above, SA accumulation suppresses the jasmonic acid (JA) defense pathway through cross talk between the pathways. As a result, caterpillars perform better on plants upon which eggs are laid (Bruessow et al. 2010). Egg laying can even suppress the subsequent induction of volatile signals used by plants to attract the enemies of herbivores (Penaflor et al. 2011). In short, some herbivores appear able to manipulate for their own benefit the phytochemical landscape upon which they forage.

5.4.4 Effects of Chemical Induction on Herbivore Population Dynamics

In chapter 4, I described how the nutritional and defensive chemistry of primary producers influences the population dynamics of herbivores (section 4.1.3). In addition to the effects of constitutive phytochemistry on herbivore population dynamics, induced chemical defenses might impose negative feedback (density-dependence) on herbivore populations, thereby limiting herbivore population growth (Baltensweiler et al. 1977, Haukioja 1980). However, widespread evidence for population regulation by chemical induction remains elusive, perhaps because of the many other ecological forces that act simultaneously on herbivore dynamics. Effects of induction on dynamics are important to consider here because coupled cycles between herbivore populations and the chemistry of primary producers could generate predictable temporal patterns on the phytochemical landscape that propagate through ecological systems to influence other trophic levels (chapter 4) and nutrient dynamics in ecosystems (chapter 8).

I'm unaware of any research that has studied potential feedback between induced defenses and herbivore population dynamics in aquatic ecosystems; while short-term effects of induction on aquatic herbivore performance have been reported (above), there is a notable lack of work on potential long-term dynamical consequences of phytochemical induction for herbivores and autotrophs in aquatic habitats. In terrestrial systems, experimental and modeling work by Nora Underwood and colleagues has improved significantly our understanding of the dynamical consequences for herbivores of induced resistance in plants. In one of her focal systems, Underwood has explored defense induction in soybean by Mexican bean beetles, and the consequences of induction for beetle population dynamics. When beetles feed on soybean plants, there is a rapid increase in resistance to herbivores that peaks around three days after damage. This induced resistance decays over time and is followed by a period of induced susceptibility to further beetle damage (Underwood 1998). The strength of induction varies with the damage inflicted by herbivores (i.e., is density dependent) and also varies among soybean genotypes (Underwood 2000, Underwood et al. 2000) (figure 5.11A). Both of these results are important. First, density dependence is a prerequisite for induced resistance to bring herbivore populations toward an equilibrium density (Hunter 2001b). Second, genetic variation among plants provides the raw material for induction to evolve overtime. Notably, field experiments on beetle recruitment indicate that the long-term population dynamics of bean beetles should vary among these soybean genotypes (Underwood and Rausher 2000) (figure 5.11B), illustrating that (a) variation in phytochemistry can influence the population dynamics of herbivores, and (b) induced defenses may be responsible in part for those dynamical differences. Simulation models confirm that induced resistance can bring herbivore populations to equilibrium in the absence of other density-dependent forces, while time lags in the induction response can drive fluctuations in herbivore populations that vary from damped oscillations, through stable cycles, to irregular fluctuations (Underwood 1999).

In subsequent work, Underwood and colleagues have explored the relative importance of constitutive and induced resistance traits in soybean for the dynamics of Mexican bean beetle populations (Underwood and Rausher 2002). As expected from her previous work, induced resistance lowered beetle population densities, and generated density-dependent population growth in beetles. However, constitutive resistance had stronger effects on beetle dynamics than did induced resistance. Because induced and constitutive defense traits are uncorrelated among genotypes in this system (Underwood et al. 2000), it appears that both types of defense may operate simultaneously and independently to influence herbivore dynamics on the phytochemical landscape. More recent work has underscored that nonlinear effects of herbivores on the chemistry and biomass of primary producers can influence the form of density-dependent population growth and the population dynamics that result (Abbott et al. 2008). For example, thresholds in the amount of damage required to induce resistance, and saturation in the degree of resistance that can be induced, both influence the population dynamics of herbivores (Underwood 2010).

While Underwood's work provides the most complete picture to date of induction-mediated dynamics of herbivore populations, phytochemical induction influences herbivore dynamics in other systems too. For example, the grass *Deschampsia caespitosa* responds to small mammal grazing by increasing the uptake of silica from soil, and depositing that silica in granules in its foliage (figure 5.11C). In turn, foliar silica reduces the performance of small mammals that feed on the grass. The magnitude of induced silica deposition is a function of small

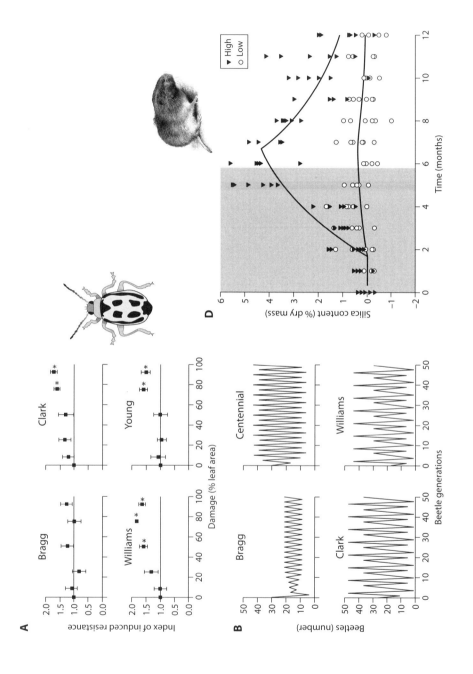

A

Index of induced resistance

Bragg

Clark

Williams

Young

Damage (% leaf area)

B

Beetles (number)

Bragg

Centennial

Clark

Williams

Beetle generations

D

Silica content (% dry mass)

Time (months)

▼ High
○ Low

mammal density, and foliar silica concentrations take around a year to decay after herbivore feeding stops. Modeling work illustrates that the delayed induction and decay of foliar silica is capable of generating cycles in the populations of voles that feed on the grass (Reynolds et al. 2012).

In other systems, induced defenses may interact with additional ecological forces to influence herbivore population dynamics. Mountain birch trees in Finland respond to defoliation by the autumnal moth with changes in their phenolic and nutritional chemistry (Haukioja et al. 1985). While these herbivore-induced changes in phytochemistry are associated with population cycles of the autumnal moth (Haukioja 1980, 1991), time-lagged changes in phytochemistry may not be sufficiently strong on their own to drive cyclic dynamics in this system. Rather, interactions with parasitoids appear to be likely candidates for generating population cycles (Klemola et al. 2010). Similarly, gypsy moth caterpillars induce increases in the polyphenolic concentrations of their host plants (Rossiter et al. 1988) that help to protect the caterpillars from a polyhedrosis virus (Hunter and Schultz 1993). Rather than induction effects alone driving dynamics, it is the tri-trophic interaction between phytochemical induction, herbivore, and pathogen that influences temporal dynamics of the gypsy moth.

In summary, two key results emerge from studies of phytochemical induction. First, herbivore-induced changes in the chemistry of primary producers affect the foraging of natural enemies and the density-dependent processes operating between herbivores and autotrophs on the phytochemical landscape (figure 5.7, arrows a and b). In other words, phytochemical induction is a pathway of feedback between autotroph chemistry and trophic interactions (loop A in figure 1.1). Second, some of the PSMs induced by herbivore damage are recalcitrant to enzymatic degradation. Consequently, they can reduce subsequent rates of litter decomposition once plant tissues senesce (Chapman et al. 2006) (arrow c in figure 5.7). In chapter 6, I consider in more detail the role of autotroph chemistry in ecosystem nutrient dynamics and, in chapter 8 I focus on the effects of herbivore-induced changes in the chemistry of autotroph residues on decomposition and nutrient dynamics.

FIGURE 5.11. Herbivore-induced changes in phytochemistry influence population dynamics of herbivores. (A) Most soybean genotypes exhibit density-dependent induction responses that (B) can contribute to the long-term population dynamics of bean beetles (Bragg, Clark, Williams, Young, and Centennial are soybean genotypes). Similarly, the rise and fall of induced silica deposition in grasses caused by vole grazing (C) can feed back to generate cycles in vole populations. Black rectangles are high vole densities while open circles are low vole densities. Modified from Underwood (2000), Underwood and Rausher (2000), and Reynolds et al. (2012).

CHAPTER 6

Effects of Autotroph Chemistry on Nutrient Dynamics

6.1 THE ELEMENTS OF LIFE

We are stardust, and primary producers form the base of the food chains through which we eat the stars. The big bang produced the lighter chemical elements (H, He, Li) around 13.7 billion years ago, with the heavier elements generated from fusion reactions in the centers of collapsing stars at least a billion years later (Schlesinger and Bernhardt 2013). Around 30 of the chemical elements created by collapsing stars occur today in living organisms (Sterner and Elser 2002) and are considered essential for life. What separates geochemistry from biogeochemistry is that these essential elements pass through living organisms as they cycle through ecosystems (Schlesinger and Bernhardt 2013). Life, then, is the organized assembly and disassembly of stardust. This chapter focuses primarily on rates of assembly and disassembly, and how the chemistry of autotrophs on the phytochemical landscape influences those rates.

So far, I have described how variation on the phytochemical landscape influences trophic interactions among plants, herbivores, and diverse natural enemies (chapter 4). Trophic interactions then feed back to influence the phytochemical landscape at diverse spatial and temporal scales (chapter 5). Feedback loops between primary producer chemistry and trophic interactions are ubiquitous and fundamental forces in terrestrial and aquatic ecosystems (figure 1.1, loop A) and generate pathways that link trophic interactions with nutrient dynamics at the ecosystem level. These links emerge because autotroph chemistry is also a fundamental driver of the cycling of matter in ecosystems (figure 6.1).

Critically, organisms exist both as a cause and a consequence of elemental fluxes in ecosystems, and the rates of biogeochemical cycles are linked to the rates of life processes. At the global level, aquatic and terrestrial primary producers combine to fix approximately 190 gigatons of C each year from the atmosphere through photosynthesis, reducing inorganic CO_2 into organic molecules. As a first approximation, about an equal amount of C is returned to the atmosphere each year

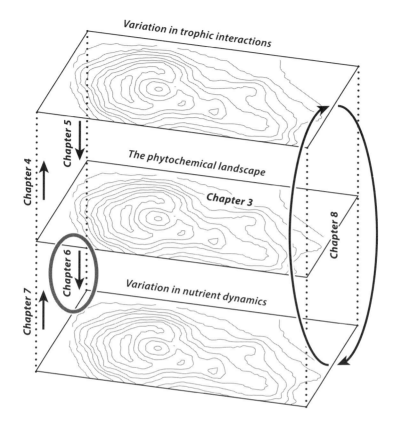

FIGURE 6.1. Variation on the phytochemical landscape is the nexus linking trophic interactions and nutrient dynamics. In this chapter (heavy circle), I focus on the role of variable phytochemistry in generating spatial and temporal variation in nutrient dynamics.

through the oxidation of organic molecules, which powers the growth, reproduction, and activity of most living organisms. Most autotrophs and heterotrophs return CO_2 to the air or water as they respire, and the decomposition of their remains completes the flux of C back to the atmosphere.

The cycling of the other elements critical to life is linked tightly to the cycling of C. Carbon makes up the backbone of the organic molecules (e.g., proteins, lipids, carbohydrates) from which life is built and from which chemoorganoheterotrophs (e.g., fungi, animals, many bacteria) get their energy. In other words, fluxes of elements such as N and P from the living to the nonliving world and back again are tied to the fluxes of C (Sterner and Elser 2002). While the abiotic environment and the structure of decomposer communities both influence profoundly rates of decomposition and nutrient flux (Strickland et al. 2009a, García-Palacios et al. 2013), the chemical structure of organic molecules synthesized by primary producers also influences nutrient cycling in a number of important ways. Figure 6.2 provides a simplified summary of an arbitrary elemental cycle, in which the arrows represent fluxes of the element and boxes represent the pools. After primary producers take up inorganic nutrients from the environment, and convert them into organic forms, the given element may be recycled from dead plant or algal material (generally called litter on land and phytodetritus in aquatic systems) by passing through the microbial pool, wherein it is mineralized into inorganic forms, and released back into the environment (heavy arrows a–d in figure 6.2A, B). As we shall see in this chapter, the rate at which this cycle turns depends fundamentally on the chemistry of autotroph detritus on the phytochemical landscape, particularly the difficulty with which the bonds in organic molecules are broken by microbial enzyme activity.

Rather than entering the detrital pool, elemental nutrients may instead pass from primary producers through the green food web composed of herbivores, their predators, and parasites (figure 6.2A). The waste products and cadavers of these higher trophic levels provide additional resources for mineralization by the microbial pool or consumption by detritivores. Additionally, decomposers and detritivores are consumed by microbivores and predators, respectively, so generating the microbial loop and brown food web of aquatic and terrestrial ecosystems (Azam et al. 1983, Myer 1990, Bardgett and Wardle 2010) (figure 6.2B). As in green food webs, microbivores and detritivores pass nutrients to higher trophic levels for assimilation, egestion, and eventual decomposition. The waste products and cadavers of consumers are generally far more labile during decomposition than is autotroph detritus, and so the waste products of trophic transfer provide one fundamental pathway by which trophic interactions influence nutrient dynamics (Crossley et al. 1988, Lovett and Ruesink 1995).

In a critical contribution to the above pathways, the foraging behaviors of both predators and herbivores influence which autotrophs are consumed, and which are not (chapter 5); the latter are left to rot on the phytochemical landscape. Consequently, predators and herbivores influence the chemistry of autotroph detritus that enters the decomposer pool. All herbivores are choosy in what they eat, and predation risk modifies herbivore choices yet further, leaving only a subset of available primary producer cells and tissues to enter the detrital pool. In a very real sense, predators influence ecosystem nutrient dynamics by altering the ratios of recalcitrant organic molecules to the availability of microbial enzymes capable of degrading them. I discuss the cascading effects of consumers on nutrient dynamics in detail in chapter 8. In this chapter, I focus on arrows a, b, and c in figure 6.2. In chapter 7, I consider arrow d, and its role in generating variation on the phytochemical landscape.

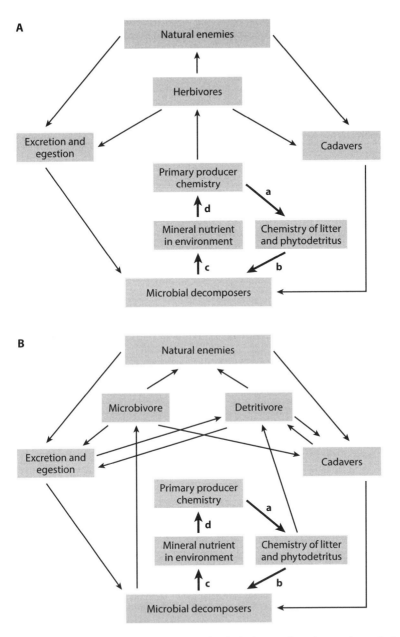

FIGURE 6.2. Simplified cycling of an arbitrary chemical element, focusing on the cycle from primary producers, through their residues, to microbial decomposers, the subsequent release of inorganic nutrients, and their uptake by autotrophs (heavy arrows a–d). For clarity, cycling of the element through the green food web (A) and brown food web/microbial loop (B) are shown separately, although both occur simultaneously in most ecosystems.

6.2 RECALCITRANT ORGANIC CHEMISTRY

Globally, about 56 percent of primary production enters the detrital pool without being consumed by herbivores (Cebrian and Duarte 1995) and this value can reach 90 percent in some terrestrial ecosystems. The rate of decay of that detritus is a measure of the rate at which C cycles back to inorganic forms; in both aquatic and terrestrial ecosystems, fast-growing autotrophs generally produce labile detritus that decomposes rapidly whereas slow-growing autotrophs produce recalcitrant detritus that decomposes slowly. Importantly, recalcitrant detritus is characterized by recalcitrant organic molecules and low concentrations of mineral nutrients, both of which impede microbial activity and favor the accumulation of auto-troph residues in terrestrial soils and aquatic sediments. The topic of recalcitrant chemistry can be controversial. Certainly, the catabolism of organic molecules is context-dependent (below), and no molecules have static, unchanging levels of recalcitrance; they will decompose much faster in some environments than in others (below). However, there is still no doubt that some autotroph chemicals on the phytochemical landscape are much more easily disassembled than are others, and influence disproportionately the rate at which nutrients are recycled.

It is no coincidence that the most abundant organic polymers on Earth's phy-tochemical landscapes are also among the most difficult to break apart. Cellulose, the most abundant natural polymer of all, is the principle component of terres-trial plant cell walls, and is also synthesized by many algae, some bacteria, and oomycetes. Cellulose is a linear polysaccharide, composed of glucose monomers that are joined together by 1–4 beta glycosidic bonds (figure 6.3A). Compared to the 1–4 alpha glycosidic bonds of starch, the beta glycosidic bonds of cellu-lose are notably difficult to break. Some bacteria and fungi produce the cellulases (e.g., β-glucosidase) necessary to hydrolyze cellulose polymers into monomers, whereas most other organisms are either unable to catabolize cellulose, or else rely on microbial symbionts to do so.

In the cell walls of terrestrial plants, microfibrils of cellulose generally occur in a matrix that includes lignin and hemicellulose, two other organic polymers that combine with cellulose to reduce the digestibility of plant tissues (Ding et al. 2012). After cellulose, lignin is the second most abundant organic polymer on Earth, occurring in plant cell walls, vascular tissue, wood, and in red algae.

FIGURE 6.3. Cross-links among organic molecules such as cellulose (A), lignin (B), and hemi-cellulose (C) reduce the decomposition rate of organic residues from primary producers. Stable organic matter may include humic acids (D) that likely include residues of bacterial, fungal, and autotroph origin.

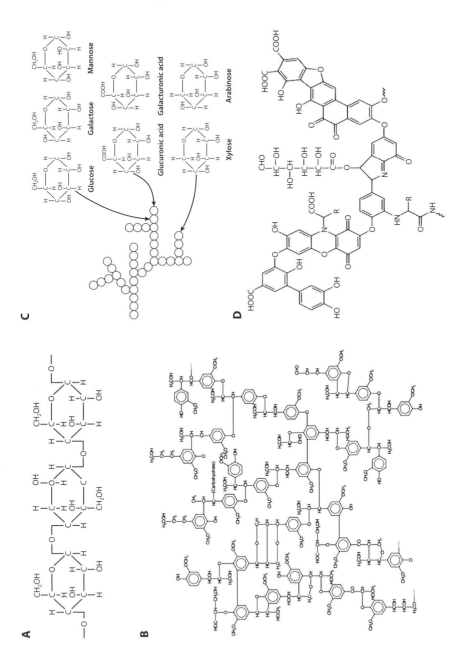

A

B

C

Glucose
Glucuronic acid
Galactose
Galacturonic acid
Mannose
Arabinose
Xylose

D

Lignin is derived from three phenolic-like monolignols from the phenylpropanoid pathway and, unlike cellulose, lignin is highly branched and cross-linked, with no predefined structure (figure 6.3B). Lignin is notoriously resistant to microbial decomposition (Meentemeyer 1978), requiring the activity of extracellular enzymes such as phenol oxidase and peroxidase for degradation (Sinsabaugh 2010). Hemicellulose polymers are composed of a number of different sugars, including glucose, xylose, mannose, galactose, and arabinose (figure 6.3C). Although not particularly resistant to enzymatic degradation on their own, hemicellulose polymers are branched, and serve with lignin polymers to cross-link cellulose microfibrils in plant cell walls. The cross-linked matrix of cellulose, hemicellulose, and lignin in plant litter is responsible in large part for the slow rate at which plant residues decay relative to those of the animals that eat them (Hunter 2001a, Cornwell et al. 2008). Specifically, the ability of different lignin types to cross-link with other cell wall constituents is a strong predictor of decomposition rate (Talbot et al. 2012). As we shall see, spatial and temporal variation in the cross-linked matrix of cellulose, hemicellulose, and lignin contribute in fundamental ways to functional variation on the phytochemical landscape.

As with cellulose, bacterial and fungal decomposers are the principal sources of enzymes capable of lignin degradation. In forest soils, where saprophytic fungi mediate the majority of plant litter decomposition (Osono 2007), the abundance and activity of fungal lignocellulolytic enzymes can place a fundamental constraint on the rate at which plant litter decomposes (Zak et al. 2008). Suppression of fungal enzymes, including lignin oxidase, cellobiohydrolase, and cellobiose dehydrogenase, results in the accumulation of undecomposed organic matter in forest soils (Edwards et al. 2011). Suberin, an irregular waxy polymer common in plant roots and bark, has a polyaromatic domain similar to that of lignin, which is attached to a polyaliphatic domain (Bernards 2002). While the ecological properties of suberin are less well known than are those of lignin, its structure likely makes it recalcitrant to decomposition. The concentration of aromatic domains in soil organic matter, typical of those found in lignin and suberin, are associated negatively with C cycling and N mineralization in some soils (Paré and Bedard-Haughn 2013).

While links between the initial decomposition rate of detritus (first few years) and its chemistry are well established, other factors contribute to the long-term (multidecadal) stabilization and storage of organic molecules in soils and sediments (so-called stable soil organic matter, or humus) (Dungait et al. 2012, Cotrufo et al. 2013). Additional factors that contribute to humus stability include the formation of soil aggregates, strong chemical bonds with minerals, the condensation of low molecular weight molecules of microbial origin, and the availability of oxygen. The incomplete mineralization of recalcitrant organic molecules, such as lignocellulose complexes, alkanes, and suberin, contributes to the formation of humic substances, including humic and fulvic acids, that participate in long-term C storage (Kramer et al. 2012). But these acids represent complex mixtures of degradation products of both plant and microbial origin, often linked together to form complex branching molecules of highly variable molecular weight and chemical structure (figure 6.3D). The relative contribution of recalcitrant molecules such as lignin and its degradation products to long-term C storage remains open to debate (Dungait et al. 2012, Kramer et al. 2012), although it appears that microbial and detritivore activity generate novel recalcitrant molecules that strongly resist decay (Allison 2006, Prescott 2010). Roots, and their associated mycorrhizal fungi, also contribute substantially to humus formation, at least in boreal forests (Clemmensen et al. 2013). In short, humus is more than just undecomposed phytochemistry.

Traditionally, the recalcitrance of soil organic matter was thought to arise from biochemical inaccessibility to most enzymes and the acidic nature of humic substances, which tends to retard enzyme activity (Sinsabaugh et al. 2008). In other words, the same characteristics that retard the early stages of litter decomposition were assumed to facilitate the long-term storage of organic matter. However, recent work emphasizes the importance of physical inaccessibility to enzyme attack (in aggregates and organomineral complexes) for long-term humus stability (Dungait et al. 2012, Kramer et al. 2012).

The phytochemical landscapes of aquatic ecosystems include organic residues from both terrestrial and aquatic primary producers. In many streams, wetlands, and lakes, terrestrial inputs dominate, and most such ecosystems are net heterotrophic, respiring more C than they fix through their own photosynthesis; they eat the energy captured by terrestrial plants. Even the largest lakes, such as Lake Superior, may be net heterotrophic, with substantial C inputs from terrestrial sources (Urban et al. 2005). Allochthonous inputs occur in both particulate form (plant litter) and in dissolved organic form, leached from terrestrial soils. Both the quality and quantity of detrital inputs influence aquatic ecosystem processes (Marcarelli et al. 2011), and the chemical traits of terrestrial detritus and dissolved organic carbon (DOC) are crucial for understanding nutrient flux in heterotrophic aquatic ecosystems (the role of detritus from unicellular algae in nutrient flux of large lakes and oceans is considered later in this chapter). As with litter decomposition in terrestrial systems, terrestrially derived organic matter in aquatic ecosystems is composed of both labile and recalcitrant fractions. In the latter case, humic and fulvic acids (humic substances) in lake DOC can persist for many months (figure 6.4) (da Cunha-Santino and Bianchini 2002).

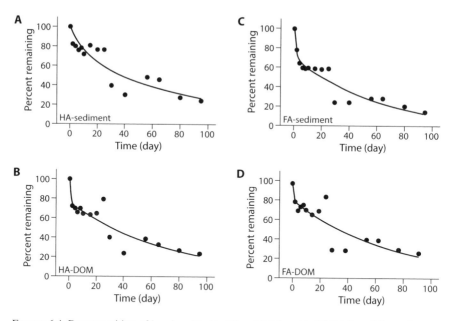

FIGURE 6.4. Decomposition of humic acids (A, B) and fulvic acids (C, D) from lake sediments (A, C) and lake dissolved organic matter (DOM) (B, D) in a tropical oxbow lake. Modified from da Cunha-Santino and Bianchini (2002).

In both terrestrial and aquatic ecosystems, the microbial community can be primed to increase the decomposition rate of stable organic matter by the addition of labile C (Kuzyakov 2010). This is important, because the form and availability of labile C varies on the phytochemical landscape (chapter 3), and likely contributes to observed variation in the strength of priming among ecosystems (Paterson and Sim 2013). In terrestrial systems, root exudation provides an important source of labile C that can prime the microbial community to increase the degradation of recalcitrant organic matter (Hamilton and Frank 2001). In headwater streams, in which energy inputs are dominated by allochthonous C from terrestrial leaf litter, products from benthic unicellular diatoms can serve to prime the decomposition of recalcitrant terrestrial litter (Danger et al. 2013). In addition to priming with labile C, the availability of nutrients such as N in new labile inputs of plant detritus can influence the rate at which older, stable organic matter decomposes (Chen et al. 2014). Again, the N concentration of autotroph cells and tissues varies markedly on the phytochemical landscape (chapters 3 and 4). In other words, across diverse ecosystems, variation in the chemistry of new plant detritus influences the rate at which older organic matter is recycled.

The carboxyl and phenolate groups in humic acids also form complexes with biologically important cations, including ions of calcium, iron, and magnesium (Schlesinger and Bernhardt 2013). By reducing the leaching of cations, humic acids may help to maintain the availability of nutrients in the environment for primary producers (e.g., maintain the cation exchange capacity of soil). The critical point here is that autotroph chemistry and the ecological factors that contribute to its variation determine in part the storage and fluxes of elemental nutrients in the environment. When herbivores and their natural enemies influence the ecology and evolution of PSMs on the phytochemical landscape, they indirectly mediate the cycling of nutrients in ecosystems (chapter 8).

6.2.1 Plant Polyphenols—Important Modifiers of Decomposition Dynamics

Tannins and other polyphenols are organic molecules that can occur in substantial concentrations in terrestrial plant detritus and in waters draining vegetated watersheds (Myer and Edwards 1990). Tannins are generally defined as polyphenolic molecules with the capacity to bind (or tan) proteins (Haslam 1989). Although most tannins are not nearly as recalcitrant to decomposition as are cellulose and lignin, some tannins clearly reduce the rate of nutrient mineralization from plant residues. In the same way that tannins bind to proteins in the digestive tracts of vertebrate herbivores (chapter 4), their ability to decrease rates of decomposition appears to arise largely from a combination of binding with nutrient sources in the detritus and with microbial enzymes engaged in litter degradation (Goldstein and Swain 1965, Hart and Hillis 1974). Of course, tannin concentrations vary many-fold over time and space on the phytochemical landscape (chapter 4) (Hunter 1997).

The effects of tannins on the activity of extracellular enzymes released by microbes may be particularly important to nutrient dynamics. Polyphenols can strongly inhibit the hydrolase enzymes that catalyze the hydrolysis of covalent bonds in organic molecules, including cellulose, hemicellulose, and protein. In contrast, oxidative enzymes (phenol oxidases, peroxidases) degrade aromatic nuclei, including those present in lignin and tannin (Sinsabaugh 2010). Because phenol oxidase enzymes can degrade tannin molecules, tannin inhibition of decomposition may be particularly important in low-oxygen environments in which phenol oxidase activity is low (Freeman et al. 2001). In short, polyphenols in terrestrial plant detritus inhibit some microbial enzymes while serving as a substrate for others. Because the decomposition of plant detritus requires intimate physical and temporal associations among diverse microbes and their extracellular enzymes (Sinsabaugh et al. 2009), inhibition of one stage of the process by plant tannins can inhibit overall decomposition rates and favor C storage (Freeman et al. 2001).

By inhibiting the activities of enzymes such as β-1,4-glucosidase, N-acetylglucosaminidase, and acid phosphatase in soils (Joanisse et al. 2007, Triebwasser et al. 2012), plant polyphenols influence rates of nutrient cycling. For example, *Kalmia angustifolia* is an ericaceous shrub that can spread rapidly following disturbance to boreal forests and reduce subsequent rates of nutrient cycling and tree growth. Experimental addition of condensed tannin extracts from *K. angustifolia* litter to humus from black spruce soils reduces the leaching of mineral N from those soils (Bradley et al. 2000), apparently by inhibiting microbial enzyme activity (Joanisse et al. 2007). Moreover, field surveys confirm a negative relationship between the cover of *K. angustifolia* among forested sites and the activity of β-glucosidase in their soils. Likewise, high polyphenol concentrations in the DOC of peatland streams and lakes significantly inhibit microbial enzyme activity and contribute substantially to the storage of C in peatland waters (Fenner and Freeman 2013). Experimentally reducing the polyphenol concentration of dissolved organic matter (DOM) in streams causes a dramatic increase in the activity of extracellular enzymes, increasing phosphatase activity by more than 28-fold (Mann et al. 2014).

In summary, plant tannins are critical components of the phytochemical landscape for understanding patterns of decomposition and nutrient dynamics. Through their effects on microbial enzyme activity, litter tannin concentrations often associate negatively with rates of detritus decomposition in aquatic and terrestrial ecosystems (Driebe and Whitham 2000, Hättenschwiler and Vitousek 2000), with consequences for nutrient dynamics and primary productivity at ecosystem scales (Northup et al. 1995, Bradley et al. 2000). Tannins may also reduce the diversity of fungi in leaf litter (Winder et al. 2013). Finally, tannins and other polyphenols may be toxic or deterrent to some members of decomposer and detritivore communities. For example, phenolics in the litter of salt marsh cordgrass both inhibit feeding by detritivorous snails and reduce their rates of growth (Rietsma et al. 1988).

6.3 NUTRIENTS, STOICHIOMETRY, AND THE DECOMPOSITION OF AUTOTROPH RESIDUES

While lignocellulose complexes resist decomposition in large part because they are difficult for enzymes to access, they are also very low in nutrients such as N and P that are vital for the metabolism of bacterial and fungal decomposers (Hobbie

TABLE 6.1. Ratios of C, N, and P in leaf detritus and fine particulate organic matter (FPOM) in a forested stream, compared to ratios from living phytoplankton and terrestrial plants.

Ratio	Leaf Detritus	FPOM	Phytoplankton	Terrestrial Plants
C:P	4858	1015	307	968
C:N	73	34	10	36
N:P	67	28	30	28

Source: Data are from Cross et al. (2003)

1992, Berg and McClaugherty 2003). The C:N and C:P ratios of terrestrial leaf litter, root litter, and woody material are very high (Cross et al. 2003) (table 6.1) and as we saw in chapter 4, low-nutrient food is generally of poor quality for consumers (Sterner and Elser 2002), including microbes (Woodward et al. 2012). For example, the C:N ratio of wood can reach 800:1, and that of abscised leaves 80:1, both substantially higher than the 8:1 ratio of a typical protein. As a consequence, the decomposition rate of terrestrial plant litter is generally inversely correlated with its initial C:N ratio (Cornwell et al. 2008). Likewise, the initial C:P ratio of leaf litter can exceed 5,300 (Hättenschwiler et al. 2008), such that P availability may limit litter decomposition rate (Kaspari et al. 2008a).

The C:N:P ratio of plant detritus changes during the decomposition process, as microbes exploit labile C, leaving increasingly recalcitrant forms behind. Although individual species vary widely, the average C:N:P ratio of terrestrial plant litter is around 3000:46:1. As labile C is used up, and nutrients are immobilized by microbial activity, the C:N:P ratio of soil organic matter reaches an average of 186:13:1. The microbial biomass itself exhibits a C:N:P ratio of 60:7:1 (Sinsabaugh et al. 2009). The differences between detritus C:N:P ratios and microbial C:N:P ratios therefore influence the flux of all 3 elements during the decomposition process (Mooshammer et al. 2012).

Both N and P concentrations are higher in aquatic phytodetritus than in terrestrial plant litter, which may explain the greater consumption rates of aquatic than terrestrial detritus per unit detrital mass (Cebrian 2004). Nonetheless, low concentrations of N and P relative to C appear to limit the rate at which aquatic microbes decompose aquatic detritus, and cause turnover in the microbial community; successive microbial species or strains appear better matched to the changing stoichiometry of their detrital food source (Danger et al. 2008). Likewise, in terrestrial environments, turnover in the microbial community during litter decomposition appears to enhance the stoichiometric match between microbial consumer and detrital resource. Specifically, the stoichiometry of the microbial community changes over time to best match that of the soluble fraction of plant detritus (Fanin et al. 2013). Interactions among microbes, and among the enzymes that they produce to degrade plant residues, are now considered central to the process of residue decomposition (Kjøller and Struwe 2002, Allison 2012), with substantial evidence for adaptation by microbes to varying chemistries of autotroph detritus (Strickland et al. 2009b, Pearse et al. 2013). Moreover, when nutrients are low in detrital tissue itself, microbes may immobilize nutrients from the surrounding environment to meet their metabolic requirements, illustrating how the chemistry of autotroph detritus influences nutrient fluxes among environmental pools.

Nutrients in addition to N and P, including concentrations of manganese, magnesium, calcium, and other micronutrients, can limit rates of residue decomposition, (Hobbie et al. 2006, Berg et al. 2010, Makkonen et al. 2012). For example, in a study of litter decomposition in a tropical forest, potassium addition increased the rate of detrital cellulose breakdown while the addition of a mixture of micronutrients increased the overall rate of leaf litter decay (Kaspari et al. 2008a) (figure 6.5). These results support the hypothesis that the collective synthesis of many different metal-bearing enzymes by diverse microbial communities is necessary for the decomposition of recalcitrant autotroph detritus. Moreover, densities of important detritivores such as termites can be limited by the availability of sodium, with the consequence that sodium addition increases the rate of litter decay (Kaspari et al. 2009).

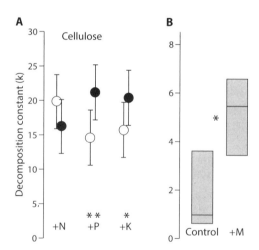

FIGURE 6.5. Nutrient availability constrains plant residue decomposition in a tropical forest. (A) Cellulose degradation is limited by the availability of both P and K whereas (B) litter degradation increases following addition of a mixture (+M) of micronutrients (B, Ca, Cu, Fe, Mg, Mn, Mo, S, Zn). Modified from Kaspari et al. (2008a).

However, the apparent requirement of diverse nutrients to facilitate the degradation of autotroph detritus does not minimize the fundamental constraint imposed on residue decomposition by recalcitrant chemical bonds. For example, in laboratory experiments with purified compounds, decomposition rates of glucose are about 10 times faster than those of cellulose and about 100 times faster than those of lignin (Barnes et al. 1998). All three compounds in purified form lack N, P, K, and micronutrients, and therefore the differences in their decomposition rates reflect inherent differences in the recalcitrance of their chemical bonds to enzyme activity, not nutrient stoichiometry. While the availability of mineral nutrients in autotroph tissues or in the environment can certainly increase decomposition rates, they cannot overcome completely the fundamental constraint of breaking strong chemical bonds. This is why microbial decomposers often appear to be limited by carbon availability (Bardgett and Wardle 2010), despite the high C:nutrient ratios in autotroph residues; labile sources of C are rapidly exhausted during decomposition, leaving only recalcitrant forms like lignin in later stages (figure 6.6A). As a consequence, lignin concentration is a powerful predictor of decomposition rate across diverse plant species, plant tissues, and diverse biomes (figure 6.6B) (Freschet et al. 2012). Although we know less about the decomposition of algal residues than we do about terrestrial litter decomposition, there is some indication that labile forms of C are mineralized first, leaving less-labile forms for slower decomposition over time (Cole et al. 1984).

In summary, the chemical composition of autotroph detritus on the phytochemical landscape has important effects on subsequent residue decomposition and nutrient dynamics. Concentrations of lignin, cellulose, and tannins, as well as the ratios between these recalcitrant molecules and residue N (Melillo et al. 1982), residue P (Kurokawa et al. 2010), and other nutrients (Hobbie et al. 2006, Kaspari et al. 2008a) mediate the short-term storage and flux of nutrients in terrestrial and aquatic environments (Ostrofsky 1997, Treseder and Vitousek 2001). Interactions among these detrital traits also influence decomposition and nutrient flux. For example, N availability increases the rate of lignin degradation, whereas a high lignin concentration decreases the rate at which cellulose and hemicellulose are catabolized in detritus (Talbot and Treseder 2012). While long-term (centennial and longer) C storage in soil organic matter may depend more on microbial byproducts and organomineral complexes than it does on initial autotroph chemistry, the chemistry of new labile residue inputs influences the decomposition of old organic matter (Chen et al. 2014). Overall, consensus has emerged that the chemical quality of autotroph detritus is a fundamental determinant of nutrient flux in biomes worldwide (Hobbie 1992, Hättenschwiler and Vitousek 2000, Cornwell et al. 2008).

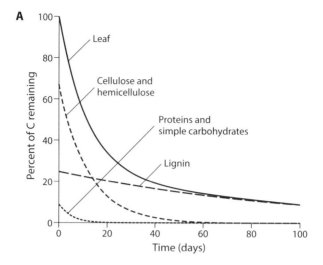

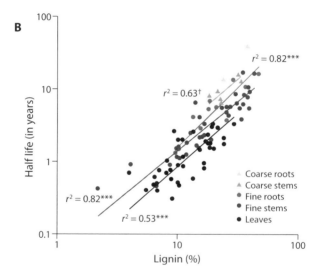

FIGURE 6.6. (A) The rapid loss of labile C during the decomposition of terrestrial plant residues leaves behind increasingly recalcitrant organic molecules like lignin. Consequently, (B) lignin is a generalizable predictor of residue decomposition across diverse plant species and tissues. Modified based on Barnes et al. (1998) and Freschet et al. (2012).

6.4 EFFECTS OF AUTOTROPH IDENTITY AND DIVERSITY ON NUTRIENT DYNAMICS

In chapter 4, I described how the identity and diversity of primary producers generates diversity on the phytochemical landscape. In turn, phytochemical diversity influences interactions among higher trophic levels. Likewise, given the relationships I have just described between the chemistry of autotroph detritus and the rate at which it decomposes, we should expect to see relationships in terrestrial and aquatic ecosystems between plant identity, plant community structure, the chemistry of detritus, and nutrient dynamics.

Plant identity is a well-known driver of nutrient dynamics in both aquatic and terrestrial systems (Binkley and Giardina 1998). For example, the presence of *Rhododendron maximum* in the understory of southern Appalachian forests contributes to the accumulation of recalcitrant humic material in forest soil (Horton et al. 2009). Likewise, invasive species of *Typha* in the wetlands of the Great Lakes produce recalcitrant litter that decomposes more slowly than does the litter from most native wetland plants. The accumulation of recalcitrant *Typha* litter likely contributes to declines in the density and diversity of native wetland species (Vaccaro et al. 2009). In some ways, we might consider species such as *Rhododendron* and *Typha* to be phytochemical engineers whose presence dominates the quality and diversity of the phytochemical landscapes in which they occur. Beyond *Typha*, invasive autotrophs in general may influence ecosystem structure and function at least in part because of the differential effects of their residue phytochemistry on decomposition processes (Castro-Díez et al. 2013). For example, in the western United States, recalcitrant litter from invasive *Elaeagnus angustifolia* has increased C storage in stream sediments by 4-fold over preinvasion conditions (Mineau et al. 2012).

6.4.1. Terrestrial Litter Identity and Diversity

On land, the phytochemical traits of individual plant species that have the greatest impact on their rates of decomposition are often those that I mentioned above: lignin, cellulose, hemicellulose, tannin, and elemental nutrient concentrations. The importance of these chemical traits is sufficiently strong that it can maintain a consistent rank among plant species in the rates at which their litters decompose, even if those species' litters are moved among biomes with very different climates (Makkonen et al. 2012). As others have noted, plant species that grow quickly and that are poorly defended from herbivores generally produce litter that decomposes more rapidly than does the litter produced by slow-growing, well-defended plant

species (Hobbie 1992, Cornelissen et al. 2004). In other words, there are fundamental links between plant physiology, growth, chemistry, and decomposition.

Of course, it is not the names of plants that matter to nutrient dynamics, but
rather the phytochemical traits associated with those names. Just as with the effects
of identity and diversity on trophic interactions (chapter 4), we use the identity
and diversity of autotroph species as proxies for variation in their important phenotypic traits. The "leaf economics spectrum" represents one way of categorizing
foliar traits of terrestrial plant species along a continuum (Wright et al. 2004). The
spectrum categorizes plants on their levels of investment in, and rates of return
from, the nutrients and dry mass that they use to construct foliage. Traits of interest include chemical, structural, and physiological properties that are associated
with strategies that vary from a fast to a slow rate of return on investment. Most
important for the current discussion, foliage with low N and P concentrations also
tends to be thicker (greater mass per unit area) and to live longer than does foliage
with high N and P concentrations (Wright et al. 2004). In short, one end of the leaf
economic spectrum is characterized by thick, long-lived, nutrient-poor foliage
(supporting low rates of photosynthesis and respiration). After abscission, such
foliage tends to decompose slowly and engender low rates of nutrient cycling,
which feeds back to select for slow-growing, nutrient-poor plant species (Hobbie
1992) (chapter 7). Although complex organic molecules such as lignin, cellulose,
hemicellulose, and tannin are not included explicitly in the leaf economic spectrum, they contribute to the structural toughness and longevity of low-nutrient
tissue, and play a causal role in their low rates of decomposition. Many of these
same traits contribute to the recalcitrance of root, wood, and stem tissue to microbial attack (Freschet et al. 2012) (figure 6.6B).

While it is important to understand the chemical traits of individual plant species that influence the decomposition of their detritus, even numerically dominant plant species do not decompose in isolation, and the decay of their litter is
context-dependent. The diversity of chemical forms on the phytochemical landscape contributes to the decomposition of residues from any single autotroph species. Recall that priming of microbial communities with labile C can increase
the rate at which recalcitrant C decomposes (Kuzyakov 2010) (Section 6.2). This
explains why there are no absolute levels of molecular recalcitrance, and all rates
of catabolism are context-dependent; even the addition of recalcitrant cellulose
can speed up the degradation of even more recalcitrant substrates such as lignin (Talbot and Treseder 2012). Moreover, diverse litter mixes contain diverse
nutrient stoichiometries that may provide complementary nutrient resources
for decomposers (Barantal et al. 2014). This emphasizes the key point that the
decomposition rate of any one species of plant detritus is contingent upon the
diversity of detrital chemical resources available on the phytochemical landscape.

As we have already seen in chapter 5, the relative abundance of plant species and their residues depends to a significant degree upon the actions of herbivores and their natural enemies. This provides a pathway by which trophic interactions influence nutrient dynamics in ecosystems (figure 1.1, chapter 8).

Recent research has stressed how the diversity and composition of litter species entering the detrital pathway influences nutrient dynamics during decomposition (Gessner et al. 2010) and the structure of decomposer communities (Rodrigues et al. 2012, Eisenhauer et al. 2013, Eissfeller et al. 2013). When individual species of plant litter decompose in monoculture, rates of nutrient release do not generally predict those that result when the same litter species decompose together. In other words, rates of nutrient release from decomposing litter species are often nonadditive (Hättenschwiler et al. 2005), perhaps reflecting complementary use of detrital resources by microbes during microbial succession (Chapman et al. 2013). Using litter from four common tree species in the southern Appalachians, we have found that estimates of nutrient release based on monoculture experiments grossly overestimate rates of nutrient release from the same species in mixture. Diverse litter mixtures retain much more N and especially P in the soil than is predicted from monoculture experiments (Ball et al. 2009b), perhaps because microbes can transfer nutrients from one litter type to another to offset nutrient limitation (Schimel and Hättenschwiler 2007). This work has three important implications. First, mixtures of litter of different chemical qualities tend to retain nutrients within forested ecosystems, presumably by providing complimentary resources to microbial decomposers that immobilize the nutrients. Second, trophic interactions among predators, herbivores, and plants that influence plant species diversity (chapter 5) will influence ecosystem nutrient flux (chapter 8). Third, loss of plant species from ecosystems as a result of human activity, and the resultant simplification of the phytochemical landscape, could result in substantial increases in nutrient export from forested watersheds. In other words, the maintenance of forest fertility depends in part upon the presence of chemically diverse litter types entering the detrital food web on a diverse phytochemical landscape.

Despite the nonadditive effects of litter mixing on N and P flux in our experiments, effects on C dynamics (Ball et al. 2008) and on the community structure of decomposers and detritivores (Ball et al. 2009a) are largely additive. The only nonadditive effects that we have observed are on the abundance of some nematode taxa. This suggests that the degree of chemical heterogeneity in litter can have impacts on N and P dynamics that are larger than might be expected by observed changes in C dynamics or in decomposer and detritivore community structure. That being said, our level of taxonomic resolution was not strong. We pooled together all bacteria (from counts under epifluorescent microscopy) and all fungi (from measures of the fungal steroid ergosterol). Nematodes were categorized into feeding groups, and microarthropods were identified to order. Consequently, we very likely missed changes in the relative abundance of decomposer or detritivore species within our broad taxonomic categories. Modern molecular techniques can provide much more taxonomic resolution within the brown food web than we were able to achieve.

It is important to note that even when effects of litter species diversity on decomposition are additive, variation in litter phytochemistry matters enormously to rates of C cycling and the structure of decomposer and detritivore communities. In our litter experiments, high-quality litter (low C:N ratio, low polyphenols) such as *Liriodendron tulipifera* supported a higher biomass of both fungi and bacteria, and more diverse microarthropod communities, than did low-quality litter (high C:N ratio, high polyphenols) such as *Rhododendron maximum* (Ball et al. 2009a). This is important because the presence of high- or low-quality litter species can have a dramatic effect on the abundance, diversity, and biomass of soil flora and fauna, whether effects on decomposers and detritivores are additive or nonadditive. In other words, there are strong bottom-up effects of litter chemistry that influence microbial populations and those organisms that graze upon them. It also emphasizes the point that, in a world of declining biodiversity, it matters which particular species are lost from ecosystems and not just that species richness declines. Because species loss from ecosystems will be nonrandom (some species are more likely to go extinct than are others), an effect of species identity on ecosystem processes matters just as much as an effect of species richness on ecosystem processes. For example, the loss of *Rhododendron* from the southern Appalachians would dramatically increase rates of nutrient cycling whereas the loss of *Liriodendron* would dramatically decrease rates of nutrient cycling, even though the change in species richness is the same (Ball et al. 2008).

We replicated the entire experiment that we had conducted on the forest floor in a headwater stream in the same forest in the southern Appalachians. We were therefore able to compare effects of litter chemical heterogeneity on nutrient dynamics in the phytochemical landscapes of soils and streams. In the stream, we found nonadditive effects of plant species on litter breakdown rates, with both species identity and species richness contributing to those effects. In other words, litter decomposition in streams is a function of which litter species are present, and how many species occur (Kominoski et al. 2007); phytochemical identity and phytochemical richness both matter. As in the terrestrial environment, accelerated rates of litter breakdown with increasing species richness likely arise from complimentary effects of litter quality on detritivores and decomposers (Fernandes et al. 2013). Yet there were strong effects of litter identity on both litter breakdown and microbial biomass. As in the terrestrial environment, the presence of *Rhododendron* (high C:N ratio, high polyphenols) reduced the biomass of both fungi and bacteria whereas the presence of *Liriodendron* (low C:N ratio, low polyphenols) increased their biomass.

Similar studies in other systems confirm that species-specific effects of litter chemistry on decomposition are as pervasive in aquatic systems as they are in their terrestrial counterparts (Schadler et al. 2005, Swan and Palmer 2006, Gessner et al. 2010). For example, in estuarine systems, both the identity and species richness of macrophyte detritus influence a suite of ecosystem processes, including N and P flux, and subsequent primary production (Kelaher et al. 2013).

Recent reviews and meta-analyses suggest that identity and species composition of detritus may have more powerful effects on decomposition and nutrient flux than does species richness per se. Again, identity is just a proxy for a particular chemical phenotype, and it is the composition of chemical phenotypes on the landscape that influence nutrient dynamics (figure 6.7). While high species richness of litter appears to accelerate decomposition of plant detritus in both streams and on land, and from subarctic to tropical biomes (Cardinale et al. 2011, Handa et al. 2014), the effects of species identity (or functional types) and species composition often dominate over simple effects of species richness on decomposition and nutrient flux. Diverse communities may be more likely to contain taxa with unusually powerful (perhaps keystone) chemical traits that influence residue degradation (figure 6.7). For example, decomposition and nutrient flux are particularly enhanced by the combination of litter from N-fixing species with litter from rapidly decomposing deciduous plants (Handa et al. 2014). Again, it is the mixture of chemical types on the phytochemical landscape, not the simple diversity of taxa, which drives subsequent effects on ecosystem function.

As noted in chapters 3 and 4, it is not just plant species that differ in their chemical phenotypes. There exists substantial genetic variation in the chemistry of plants within and among populations of the same species that may influence ecosystem processes (Driebe and Whitham 2000, Treseder and Vitousek 2001) and the structure of decomposer and detritivore communities (Peiffer et al. 2013, Wang et al. 2013). We have explored the effects of genotypic and phenotypic variation in litter chemistry on nutrient dynamics in soils under oak trees. We compared the effects of litter from nine different turkey oak, *Quercus laevis*, genotypes on C and N flux in soils. Litters from the nine oak genotypes varied in their concentrations of lignin, condensed tannin, hydrolysable tannin, total phenolics, and C:N ratios. In other words, genotypes differed in the chemical traits that are known to influence nutrient dynamics during decomposition. Consequently, genotypes differed in the rates at which their chemistries changed during the decomposition process, and the end result was genetic variation in rates of mass loss and fluxes of C and N from litter (Madritch and Hunter 2002).

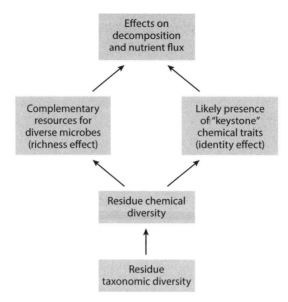

FIGURE 6.7. Pathways by which the taxonomic richness, identity, and community composition of residues produced by primary producers can influence nutrient dynamics. Diverse chemical resources may simply provide complementary resources for microbial communities that exploit autotroph residues. Alternatively, a diverse autotroph community may be more likely to include taxa with dominant (keystone) chemical traits that influence nutrient dynamics. Such traits might include unusually high polyphenol concentrations, or unusually low C:N ratios typical of N-fixing symbioses.

However, effects of genetic variation in litter chemistry on nutrient dynamics were not restricted to the litter itself. We also measured processes in the soil underlying the litter in our experiments. We found that fluxes of soil ammonium were influenced by the genotype of litter above the soil. Moreover, changes in soil C, N, respiration, and pH were all correlated with the concentrations of lignin and condensed tannin in the litter treatments above the soil (Madritch and Hunter 2002). For example, leaching of dissolved organic N from soil increased with litter lignin concentration (Madritch and Hunter 2003), presumably because rates of microbial immobilization of N declined. Indeed, rates of soil respiration also declined as concentrations of foliar lignin and hydrolysable tannin increased. In turkey oak forest, the effects of litter chemical phenotype on litter and soil nutrient flux can persist for at least three full years after the litter falls (Madritch and Hunter 2005).

When we explicitly manipulated the intraspecific chemical diversity of oak litters above the soil, we found that rates of respiration increased with litter chemical diversity (Madritch and Hunter 2003). This result parallels those that I described earlier, from interspecific variation in litter chemistry in forests (Ball et al. 2008) and streams (Kominoski et al. 2007). Again, high diversity on the phytochemical landscape may provide complimentary resources for decomposers and detritivores, increasing rates of nutrient cycling, or include chemistries that are particularly important during decomposition (figure 6.7). A potential criticism of studies that link ecosystem processes with intraspecific diversity in litter chemistry is that they ignore potential effects of other litter species that might overwhelm more subtle effects of intraspecific chemical variation. We explored this possibility by expanding our studies of turkey oak litter to include litter from longleaf pine, a co-occurring species with very different litter chemistry (Madritch and Hunter 2004). We observed that some effects of intraspecific chemical diversity in oak litter on nutrient fluxes were maintained in the presence of longleaf pine litter. However, our results also confirmed that litter chemistry, irrespective of taxonomic source, is the most important variable driving soil nutrient fluxes. Whether emerging from intraspecific or interspecific mixes, condensed tannin, hydrolysable tannin, and litter C:N ratios were all excellent predictors of soil N and C fluxes (Madritch and Hunter 2004). This is why mapping chemistry on a phytochemical landscape is much more useful mechanistically than is mapping the identities of autotrophs (chapter 2).

Taken together, our results illustrate that genetic variation at multiple taxonomic levels in the chemistry of litter influences rates of decomposition and nutrient flux both within the litter and in the mineral soil beneath. Such effects are now well established in a number of terrestrial ecosystems (Schweitzer et al. 2012b). For example, intraspecific genetic variation in condensed tannin concentrations in the litter of cottonwood trees influences decomposition and nutrient cycling in riparian ecosystems (Schweitzer et al. 2004). This genetic variation predicts both the rate of litter decomposition and the rate of N mineralization in soil. However, across common garden experiments, significant genotype by environment interactions illustrate how the local environment mediates the influence of cottonwood genotypes on nutrient dynamics (Pregitzer et al. 2013). This provides a further justification for creating maps of the phytochemical landscape rather than simply of genetic identity; genotype-by-environment interactions (chapter 3) essentially guarantee that autotroph chemistry does not map exactly onto genetic identity. Because it is the chemical phenotype that matters most to nutrient dynamics (sections 6.2 and 6.3), it has more explanatory and predictive power than does genetic identity.

Nonetheless, genetic identity remains a useful proxy for phytochemistry. For example, genetic variation among clones of aspen influences soil nutrient dynamics from landscape (Madritch et al. 2009) to continental (Madritch et al. 2014) scales. Moreover, links between genetic variation and ecosystem processes are particularly interesting given current rates of environmental change. Heritable genetic variation is a prerequisite for evolution by natural selection, and we should therefore expect to see links between climate change, evolutionary change in plant populations, and effects on nutrient dynamics in ecosystems (Fischer et al. 2014).

Genetic variation in litter chemistry also influences decomposition and nutrient flux in streams. For example, genetically based variation in condensed tannin concentration of *Populus* species and their hybrids influences rates of decomposition and nutrient release from litter that falls from the terrestrial environment into streams (Driebe and Whitham 2000). Moreover, *Populus* genotype interacts with the availability of nutrients in soil to influence litter chemistry and subsequent nutrient dynamics in associated stream habitats (LeRoy et al. 2012). The importance of genetic variation in litter chemistry for stream nutrient dynamics is beautifully demonstrated by the local adaptation of stream detritivore communities to the chemistry of the trees in their adjacent riparian zones (Jackrel and Wootton 2013). In streams of the Olympic Peninsula in Washington State, leaf litter from red alder trees growing in the local riparian zone decomposes on average 24 percent faster than does red alder litter from nonlocal sources. Fine scale adaptation by stream decomposers to the chemistry of local inputs has increased the efficiency with which nutrients are recycled from plant detritus (Jackrel and Wootton 2013).

6.5 EFFECTS OF PHYTOPLANKTON RESIDUE CHEMISTRY ON NUTRIENT DYNAMICS IN AQUATIC ECOSYSTEMS

As we have just seen, allochthonous plant litter and its degradation products have important effects on rates of nutrient cycling in stream, wetland, and lake ecosystems. However, unicellular algae and bacteria are responsible for the vast majority (> 90%) of autochthonous primary production in large lakes and marine ecosystems, with angiosperms and macroalgae contributing primarily to production in shoreline ecosystems (Smith and Hollibaugh 1993, Duarte and Cebrian 1996). As noted in chapters 3 through 5, phytoplankton vary greatly in their phytochemistry, and we might therefore expect that algal secondary metabolites should influence nutrient dynamics during decomposition in lakes and oceans. Unfortunately, the effects of phycotoxins on nutrient dynamics in these ecosystems remain rather poorly understood (Hay and Kubanek 2002, Hay 2009). Although it is abundantly clear that DOC from algae is a fundamental C source for aquatic bacteria and helps to power the microbial loop (Thomas 1997), we know remarkably little about how variation in DOC chemistry on the phytochemical landscape influences nutrient flux (Nelson and Wear 2014).

Fluxes from living and dying phytoplankton cells into the pool of DOC are substantial, and marine DOC represents one of the largest active C pools on Earth. Dominant marine phytoplankton such as *Prochlorococcus* are constantly releasing vesicles containing complex organic molecules including lipids, proteins, DNA, and RNA (Biller et al. 2014). These vesicles can support the growth of heterotrophic bacteria and may contribute substantially to marine nutrient flux. At present, however, there is no reliable measure of the importance of phytoplankton PSMs to nutrient flux during the release of vesicles, exudates, or during cell lysis.

Most components of DOC in lakes and oceans are described by their size class, reactivity (how labile they are), or by their broad molecular type such as protein, carbohydrate, or lipid (Hansell and Carlson 2014). So-called labile DOC is metabolized within hours or days of release, semilabile DOC is removed seasonally, and recalcitrant pools of DOC persist from years to millennia (Nelson and Wear 2014) (figure 6.8). As in terrestrial ecosystems, the recalcitrance of DOC is related both to its inherent chemical structure and to local ecological context, including the availability and enzymatic capacity of the heterotrophic microbes that serve as decomposers (Carlson et al. 2004). While single isolates of opportunistic bacteria may be sufficient to metabolize labile DOC, diverse bacterial communities appear necessary to degrade more recalcitrant fractions (Pedler et al. 2014), presumably because they express diverse enzymatic products. Unfortunately, the molecular

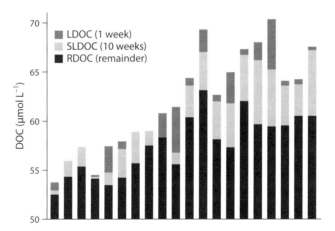

FIGURE 6.8. Variation in pools of dissolved organic carbon (DOC) in 20 samples from the phytochemical landscape within the Santa Barbara Channel, California. DOC pools are defined by their rate of catabolism by bacteria over 1 week (labile = LDOC), 10 weeks (semilabile = SLDOC), and the recalcitrant remainder (RDOC). Note the significant levels of variation in DOC recalcitrance on the phytochemical landscape. Modified from Nelson and Wear (2014).

basis of these DOC reactivity classes, and the enzymes necessary to degrade them, remains unclear.

Nonetheless, there are some data to suggest that the complex and variable chemistry of algal secondary metabolism influences aquatic nutrient flux. For example, when cell lysis releases microcystins from cyanobacterial cells into lake water, the toxins can persist for weeks. Their cyclic structure and unusual amino acid composition makes them resistant to many bacterial proteases, and hence to bacterial metabolism (Hay and Kubanek 2002). Moreover, microcystins readily adsorb to clay and humic substances, increasing substantially their persistence in aquatic sediments (Miller et al. 2001, Chen et al. 2008), where anoxia may further decrease rates of microcystin degradation (Grutzmacher et al. 2010). As a consequence, phytoplankton microcystins may decrease the rate of N cycling in aquatic ecosystems by sequestering N. However, the different chemical forms of microcystin are differentially labile to bacteria. In a shallow lake in Ontario, microcystin (MC)-RR accumulates in surface sediments whereas MC-LA is degraded more rapidly, with a half-life of 1.5 to 8.5 days. MC-LR is degraded most rapidly of all (Zastepa et al. 2014). However, even MC-LR can persist in lake sediments for considerable periods of time, presumably bound to clay or humic substances; it has been found in cores from lakes in the Nebraska Sand Hills that date back to 1832 (Efting et al. 2011).

Similarly, domoic acid is an N-containing neurotoxin produced by some red algae and diatoms (chapter 3). Studies off the coast of Southern California have explored the dynamics of diatom blooms and the fate of the domoic acid that they produce (Sekula-Wood et al. 2009). Results illustrate that domoic acid is rapidly transported to deeper waters on sinking particles, where it can persist in sediments long after the algal bloom is over. Clays in sediments also adsorb other potent N-containing neurotoxins such as anatoxin-a and saxitoxin (Rapala et al. 1994, Burns et al. 2009). The key point is that phytoplankton neurotoxins may sequester the limiting nutrient N over extended periods in lake and marine sediments, linking the secondary metabolites of primary producers with nutrient dynamics at the ecosystem scale.

Intact phytoplankton cells may also support diverse communities of heterotrophic bacteria (Cole 1982) attracted to chemical exudes on the phytochemical landscape (Seymour et al. 2009). For example, over 200 strains of bacteria have been found in association with the dinoflagellate *Alexandrium minutum* (Prol et al. 2009), which synthesizes potent neurotoxins. At least some of the heterotrophic bacteria that associate with phytoplankton make use of the vast pool of organic compounds released by phytoplankton (Sieg et al. 2011). Such bacteria contribute to the recycling of nutrients in the phycosphere (Bell et al. 1974) in ways analogous to the rhizosphere bacteria associated with the roots of terrestrial plants (Hamilton and Frank 2001). This appears to be the case with the marine bacterium *Marinobacter* that facilitates iron uptake by the photosynthetic dinoflagellates with which it associates (Amin et al. 2009).

More important for the current discussion is to consider what happens to the exudates and dead cells of aquatic primary producers. About 45 percent of marine phytoplankton production enters the detrital pathway without being consumed by herbivores (Duarte and Cebrian 1996), although a significant fraction of production may be attacked by marine viruses (Avrani et al. 2011). Algal exudates and dead/lysed cells represent phytodetritus that may be recycled in surface waters (the pelagic pathway) or be transported to deeper waters and sink to the benthos (benthic pathway) (Lindegren et al. 2012). We might consider these as pelagic and benthic phytochemical landscapes, respectively. On coral reefs, where nutrient dynamics are very tightly conserved, sponges appear to assimilate much of DOC released by algae and photosynthetic corals. In turn, sponges continuously shed their choanocytes, rereleasing the C as particulate organic carbon (POC) that is then consumed by detritivores. This sponge loop provides a mechanism by which algal exudates on the phytochemical landscape enter the coral reef food web and support its notably high secondary production (de Goeij et al. 2013).

In the pelagic zone, phytodetritus contributes to powering the microbial loop, by which C and nutrients are recycled from DOM, through heterotrophic bacteria, to their unicellular consumers, and on up the food chain (Azam et al. 1983). The microbial loop recycles about 50 percent of the primary production of the oceans, and so any factor (including DOC chemistry) that influences the rate of bacterial activity will have a profound affect on marine nutrient dynamics. As DOM is released by phytoplankton, water turbulence transforms it into networks of thin filaments, and motile bacteria use chemotaxis to exploit these filament networks (Taylor and Stocker 2012) that run throughout pelagic phytochemical landscapes. Phytodetritus also contributes to the larger aggregates of "marine snow" (Suzuki and Kato 1953, Alldredge and Silver 1988) and "lake snow" (Grossart and Simon 1993) that transfer C (and energy) from the atmosphere to deep-water ecosystems as part of the biological pump (Etter and Mullineaux 2001, Burd et al. 2010). Nutrients such as N, P, and silica are transported with the C to deeper waters where they may be recycled, stored, or returned to the surface by upwelling or seasonal mixing of water. If primary production is particularly high, for example during algal blooms, the C and nutrients that sink to deep waters and sediments can significantly reduce nutrient availability in the photic zone, and may sequester C and nutrients within the benthic phytochemical landscape for decades or centuries (Smetacek et al. 2012).

However, the organic molecules in aggregates host an active community of heterotrophic bacteria that utilize C and nutrients as the aggregates descend in the water column. The densities of heterotrophic bacteria may be 100 times greater on detrital aggregates than in typical water samples, generating mineral nutrient concentrations 1,000 times greater than background levels. These numbers illustrate that marine and lake snow serve as critical hot spots of decomposition and nutrient cycling as they descend through the water column (Grossart and Simon 1993). As a result, there are substantial changes in the organic chemistry of marine snow as it sinks, representing the effects of microbial activity on algal detritus, such that the remaining material becomes increasingly recalcitrant with time and depth (Wakeham et al. 1997), analogous to patterns in terrestrial leaf litter (figure 6.6A). A key point is that inputs into the benthic phytochemical landscape have been modified from their pelagic origins by the action of decomposers during their descent.

If algal densities are very high, for example during algal blooms, their subsequent decomposition by heterotrophic bacteria can cause substantial zones of hypoxia in surface and in deeper waters (Michalak et al. 2013). In turn, hypoxia influences nutrient dynamics by favoring the use of terminal electron acceptors other than oxygen (e.g., nitrate) by microbial decomposers, leading to increased rates of denitrification and the loss of nitrate from aquatic ecosystems (Chen et al. 2012). Anoxic conditions are also prevalent in deeper layers of marine sediments, where archaea may play a dominant role in nutrient recycling (Lloyd et al. 2013).

In terrestrial ecosystems, a substantial fraction of plant detritus can enter long-term storage in soils, particularly in cold or waterlogged environments that inhibit aerobic decomposition processes. Is there an equivalent storage process in lake and marine environments, and is it related to phytodetrital chemistry? Estimates suggest that 10 to 17 percent of net primary production (NPP) from higher plants in marsh and mangrove ecosystems may be stored in sediments (Duarte and Cebrian 1996). Similarly, most seagrass tissues enter the detrital pathway without being consumed by herbivores, and contribute to local C storage in sediments (primarily dead roots and rhizomes) or C export beyond the seagrass meadows (primarily dead leaves) (Cebrian and Duarte 2001). As with terrestrial angiosperms, the decomposition of seagrass detritus is inhibited by its high lignin, cellulose, and phenolic content (Harrison 1989). Remarkably, seagrass litter has been found floating over 1,000 km from any known source and, after it sinks, may support significant densities of detritivores (e.g., urchins and sea cucumbers) and decomposers at depths approaching 4,000 m (Suchanek et al. 1985) (figure 6.9). The recalcitrant chemistry of seagrass detritus facilitates its export and subsequent use on the benthic phytochemical landscape by deep-sea communities, in which C and nutrients are recycled into living biomass.

In contrast, while settling can account for anywhere from 5 percent to 95 percent of population losses of phytoplankton from surface waters (Heiskanen and

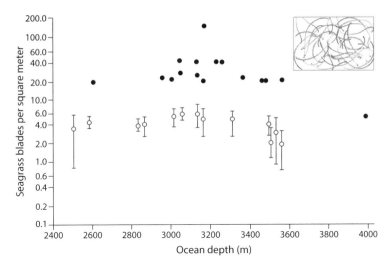

FIGURE 6.9. Senesced blades of seagrass, *Syringodium*, can be observed to depths approaching 4,000 m in the ocean. Black circles represent visual observations whereas open circles represent estimated averages. Data from Suchanek et al. (1985).

Kononen 1994, Hansson 1996), the vast majority of phytoplankton production is either consumed by herbivores, lysed by viruses, or decomposed as DOC by microbes in the water column or sediment surface; less than 1 percent of phytoplankton production enters sediment storage in open marine systems (Duarte and Cebrian 1996) (figure 6.10). If algal detritus does become buried in sediment, decomposition processes continue, albeit slowly, just as they do in terrestrial soils. For example, the predominant archaea in anoxic marine sediments appear to play an important role in protein degradation (Lloyd et al. 2013). By the time it is buried in sediments, "algal" organic matter likely represents a complex mixture of humic substances derived from algae, bacteria, archaea, marine macrophytes, and even terrestrial plants (Arndt et al. 2013). In that regard, it parallels the multiple origins of stable organic matter in terrestrial soils (section 6.2). While multiple physical, chemical, and biological factors influence the rate at which organic matter decomposes in marine sediments, increasing chemical recalcitrance with age contributes to C storage in sediments (Arndt et al. 2013), just as it does in soils. Compounds such as algaenan, an aliphatic polymer produced by some algal taxa (de Leeuw et al. 2006), may even persist in sedimentary rock (Kodner et al. 2009). However, chemical characterization of the humic substances in marine sediments remains problematic, with most studies reporting variation in gross chemical types rather than providing more precise data on molecular structure. In other words, we remain greatly limited in our ability to

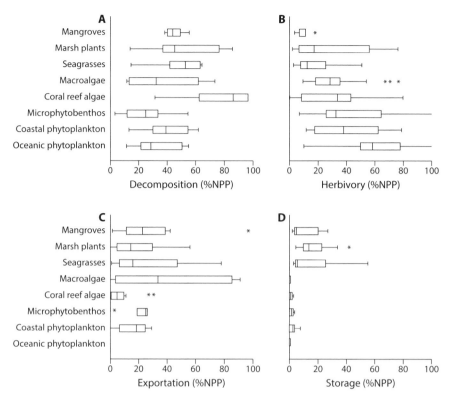

FIGURE 6.10. The fraction of NPP that is (A) decomposed, (B) consumed by herbivores, (C) exported, or (D) stored in sediments varies markedly among marine autotrophs. Note that only a small proportion of oceanic phytoplankton resists consumption or decomposition to enter sediment storage. Modified from Duarte and Cebrian (1996).

assess the effects of detrital chemistry on nutrient flux during the later stages of algal decomposition (de Leeuw et al. 2006).

However, a majority of dead phytoplankton C may be utilized by bacteria within a few hours or days of cell lysis (Hansen et al. 1986). Even potent phycotoxins can be metabolized rapidly; concentrations of neurotoxins produced by some dinoflagellates often rise and fall in concert with the density of cells that produce them (Hakanen et al. 2012), suggesting that they do not persist long in the water column. Compared to their angiosperm counterparts, phytoplankton detritus is relatively rich in lipids and nitrogen-containing compounds (Burdige 2007), which may be comparatively labile for microbes. In contrast, because mangrove and salt marsh plant species require tough structural tissues to support their vertical architecture, the chemistry necessary to produce those tissues (lignin, cellulose, hemicellulose)

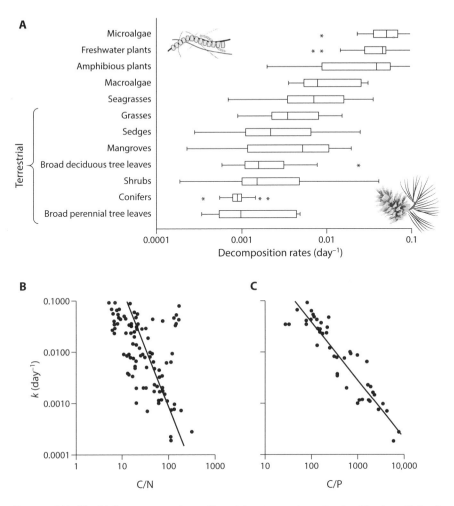

FIGURE 6.11. The higher concentrations of recalcitrant organic molecules (lignin, cellulose) required for structural support by terrestrial autotrophs tends to decrease their rates of decomposition (A) relative to aquatic producers. Across terrestrial and aquatic ecosystems, C:N (B) and C:P (C) ratios are significant predictors of decomposition rate (k). Modified from Enriquez et al. (1993).

likely reduces rates of herbivory (Shurin et al. 2006) and decreases rates of detrital decomposition (Cornelissen et al. 1999) in comparison to phytoplankton. Therefore, despite their much lower biomass, marine angiosperms have a disproportionately large effect on C storage in marine systems. Additionally, the relatively low C:N and C:P ratios in phytoplankton, compared with those of most terrestrial plants, may contribute to their faster rates of decomposition (Enriquez et al. 1993) (figure 6.11).

To date, then, evidence suggests that C:N:P stoichiometry is a principal driver of phytoplankton decomposition rate (Enriquez et al. 1993), rather than the more complex organic chemistry associated with structural and defensive compounds that increase litter recalcitrance in terrestrial ecosystems (Melillo et al. 1982, Cornwell et al. 2008). However, by interfering with the ecology of pelagic consumers and detritivores, phytoplankton toxins may result in relatively more materials being decomposed in benthic communities and sediments (Christoffersen 1996). And even species in the benthos can be affected deleteriously by phycotoxins in materials that sink to the floor of aquatic ecosystems. Blooms of the toxic flagellate *Chrysochromulina polylepis* may reduce the density and diversity of benthic invertebrates for nearly two years (Olsgard 1993), with likely consequences for the decomposition of animal and algal detritus (Hay and Kubanek 2002).

Moreover, although phytoplankton cells lack structural tissues (stems, twigs, petioles) in the fashion of terrestrial plants, they are not without structural components. The silica frustule of diatoms and the cellulose theca of armored dinoflagellates represent structural elements that can reduce consumption by herbivores and influence the cycling of nutrients. For example, diatoms often dominate aquatic algal blooms. At death, settling and sedimentation of diatoms sequesters the element silicon, at least temporarily, from aquatic ecosystems, generating periods of silica limitation (Dugdale and Wilkerson 1998, Ardiles et al. 2012). Silica sequestration by large-shelled diatoms in the Southern Ocean may even limit silica in northern oceans (Assmy et al. 2013) because of the importance of global water currents to nutrient distributions. In turn, diatom-induced silica limitation (whether in oceans or in lakes) may favor succession of algal species to groups that are limited by alternative nutrients, such as N and P (Horne and Goldman 1994) (figure 6.12).

6.6 EFFECTS OF PHYTOPLANKTON STOICHIOMETRY ON NUTRIENT DYNAMICS IN AQUATIC ECOSYSTEMS

While silica plays a structural and defensive role in diatoms, this example also helps to emphasize a more fundamental link between the cellular stoichiometry of phytoplankton, their nutrient uptake rates, and the subsequent availability of nutrients in aquatic ecosystems (Redfield 1958, Reynolds 1984). Because of the rapid recycling of most nutrients in aquatic ecosystems, phytoplankton community dynamics are both a cause and a consequence of nutrient dynamics in the environment. Interactions among phytoplankton with diverse life history traits

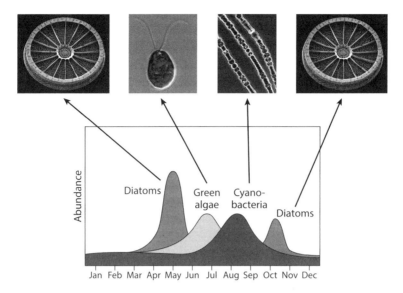

FIGURE 6.12. The sequestration of silica by diatoms influences algal community succession in aquatic ecosystems. In lakes, spring and fall mixing of water returns silica from sediments to the photic zone, encouraging diatom blooms. Sequestered silica sinks back to sediments as diatoms die, favoring phytoplankton species that are limited by alternative nutrients.

combine with seasonal resource limitation and water mixing to influence patterns of algal succession and biogeography. In turn, the succession and biogeography of phytoplankton generates important spatial and temporal heterogeneity on the phytochemical landscape. Key functional traits of phytoplankton that influence their ecology, succession, and impact on nutrient dynamics include cell size, ability to fix N, nutrient resource use, degree of mixotrophy, and resistance to grazing (Zalewski and Wagner-Lotkowska 2004, Barton et al. 2013a). Unfortunately, the types of chemical defenses used by phytoplankton still remain notably absent from most trait-based approaches to understanding algal distribution and abundance.

Critically for the theme of this chapter (figure 6.1), successional and biogeographic patterns in phytoplankton also drive changes in nutrient availability in the environment. The famous Redfield Ratio (Redfield 1934) of 106:16:1 of C:N:P in marine environments reflects the tight links between nutrients in water and in living and dead algal tissues. The N:P (16:1) component of the Redfield Ratio likely emerges from a fundamental ratio between protein and rRNA in living organisms (Loladze and Elser 2011), aptly confirming the tight linkage between tissue quality of primary producers and ecosystem nutrient dynamics.

In turn, the C:N ratio of marine organic matter appears to determine the relative importance of denitrification and anaerobic ammonium oxidation to the loss of N from oxygen depleted zones of the ocean. In other words, the nutritional chemistry of algal residues mediates the cycling of N, controlling the fertility of ocean waters (Ward 2013).

The Redfield Ratio is actually quite variable among phytoplankton from various regions of the globe and among taxonomic groups that vary in their metabolic requirements for N and P (and other mineral nutrients) (Klausmeier et al. 2004, Martiny et al. 2013). Recent work illustrates that diversity in algal stoichiometry is actually necessary to reconcile the marine N budget (Weber and Deutsch 2012). Biological diversity in cell chemistry, and therefore in uptake requirements, drives major patterns of N dynamics and, by extension, other limiting nutrients. Overall, consensus is emerging that the ecology, evolution, and biogeography of phytoplankton with species-specific resource needs mediates spatial and temporal variation in global oceanic nutrient dynamics (Assmy et al. 2013, Daines et al. 2014). In short, oceanic nutrient chemistry is a product of the life that it supports, with phytoplankton biology and chemistry driving global nutrient flux.

The requirement for tough structural tissues, and their subsequent recalcitrance to decomposers, serves to weaken the linkage between the nutrient content of terrestrial plants and the availability of mineral nutrients in the environment. Without those same structural constraints, nutrients in water and in phytoplankton are free to covary more closely, with interactions among algal species that vary in nutrient demands averaging to approximate the nutrient content of seawater. Notably, diatom species that have evolved the thickest silica shells to reduce grazing by copepods may decouple typical nutrient cycles in the ocean (Assmy et al. 2013), akin to the effects of the structural and defensive chemistry of tree leaves on soil nutrient dynamics (above). In other words, the contributions of diatom silica to the phytochemical landscape, and to subsequent nutrient cycling, are a function of grazer activity, thereby linking trophic interactions with nutrient dynamics (chapter 8).

Interestingly, the decomposition of unicellular algae may still require a succession of bacterial decomposers in much the same way that microbes with different stoichiometric requirements and enzymatic capabilities sequentially attack terrestrial detritus. For example, in the North Sea, diatom blooms collapse through a combination of nutrient limitation, zooplankton grazing, and viral attack. A succession of heterotrophic bacterial decomposers degrade the algal remains, with coexistence of the bacterial species mediated by their differential use of degradation enzymes (Teeling et al. 2012). This study clearly illustrates that the chemical quality of phytodetritus influences the identity and diversity of microbial decomposers during the degradation of algal remains, so mediating the recycling of mineral nutrients.

Dimethylsulfoniopropionate (DMSP) is a fascinating contributor to the pelagic phytochemical landscape of marine ecosystems. DMSP is synthesized by some marine primary producers and is broken down by marine microbes to produce the volatile compound dimethylsulfide (DMS). DMS facilitates aerosol production in the atmosphere, and may influence cloud formation, albedo, and even the temperature of the earth (Seymour et al. 2010). Moreover, by attracting consumers from considerable distances to areas of high production in the ocean (Nevitt 2008, Seymour et al. 2010), DMS may facilitate the transport of nutrients and energy among geographic regions, serving as an important mediator of ecosystem subsidies (Hay 2009). In other words, DMSP in the phytochemical landscape influences nutrient dynamics and ecosystem processes from local to global scales.

The effects of DMSP on terrestrial and aquatic ecosystems highlight the complex indirect pathways by which phytoplankton chemistry can influence nutrient dynamics. For example, some green algae secrete compounds that interfere with cyanobacterial fitness (Kearns and Hunter 2000), including heterocyst production. Heterocysts are specialized cells produced by some species of filamentous cyanobacteria during periods of N limitation. Heterocysts engage in N fixation, synthesizing ammonia from dinitrogen, and therefore contributing to the N cycle in aquatic environments. A chemical signal produced by the green alga *Chlamydomonas reinhardtii* suppresses heterocyst formation in the cyanobacterium *Anabaena flos-aquae* reducing N-fixation by about one third (Kearns and Hunter 2002). Heterocyst frequency is over 3-fold higher in the absence of algal secretions than in their presence (figure 6.13). Consequently, chemical secretions by green algae into the phytochemical landscape have the potential to reduce substantially rates of N fixation in aquatic ecosystems.

In this section, I have described how the chemistry of phytoplankton detritus, including its elemental stoichiometry and secondary metabolite concentration, influences nutrient dynamics in lake and marine systems. We know much more about the effects of algal nutrient stoichiometry on nutrient flux than we know about effects of detrital secondary metabolites, and this imposes a profound limitation on our current understanding of aquatic ecosystems (Hansell and Carlson 2014, Nelson and Wear 2014). However, we do know that phycotoxins can have substantial impacts on pelagic food webs, influencing the distribution and abundance of animals at higher trophic levels (arthropods, gastropods, fish, mammals) (Hay and Kubanek 2002) (see chapter 4). I will explore potential cascading effects of phycotoxins, from predators to nutrient dynamics, in chapter 8.

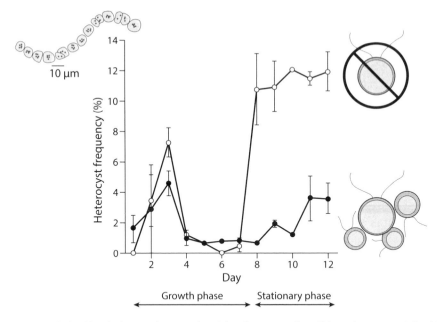

FIGURE 6.13. Chemical secretions produced by the green alga *Chlamydomonas reinhardtii* reduce nitrogen fixation by the cyanobacteria *Anabaena flos-aquae*. Algal secretions (closed circles) inhibit the production of N-fixing cells (heterocysts) in comparison to secretion-free controls (open circles). Based on Kearns and Hunter (2002).

Overall, this chapter has explored how variation on the phytochemical land-scape influences the dynamics of nutrients that are available in the environment. In chapter 7, I complete the feedback loop between the chemistry of primary pro-ducers and environmental nutrient availability by considering how the chemistry of autotrophs is, in part, a product of the nutrients available in soil and water.

Effects of Nutrient Availability
on the Chemistry of Primary Producers

7.1 INTRODUCTION

In chapter 6, I described how the secondary metabolites and nutrients in autotroph detritus influence the availability and dynamics of nutrients in the soils and waters of Earth. By influencing the abundance, diversity, and enzymatic activity of heterotrophic microbes, the chemistry of autotroph detritus influences the rate at which organic molecules are mineralized into inorganic forms. To complete the feedback loop between the phytochemical landscape and nutrient dynamics (figure 1.1, loop B; figure 2.6), I focus in this chapter on the relationship between environmental nutrient availability and subsequent autotroph chemistry (figure 7.1).

There are three fundamental processes by which environmental nutrient availability influences variation on the phytochemical landscape. First, as I mentioned in chapter 3, the phenotype of individual autotrophs arises from the combined and interactive effects of genotype and the environment in which those individual autotrophs grow (Fritz and Simms 1992, LeRoy et al. 2012). Nutrient availability is a critical component of the environment of all primary producers, and direct nutrient uptake (Bryant et al. 1983, Herms and Mattson 1992), or indirect uptake from associated nutritional symbionts (Vannette and Hunter 2011b), determines in part the nutritional and defensive phenotype of autotrophs. Second, the availability of N, P, and other nutrients in the environment influences the species richness and relative abundance of primary producers in ecological communities (Isbell et al. 2013b, Jochimsen et al. 2013). The identity and diversity of primary producers contributes fundamentally to the phytochemical landscape upon which trophic interactions take place (chapter 4). Finally, environmental nutrient availability influences autotroph chemistry through evolution by natural selection (Coley et al. 1985), generating matches between the availability of nutrients in the environment and the growth and defensive strategies of primary producers that grow in those environments (Fine et al. 2013). I will consider each of these three processes briefly in this chapter.

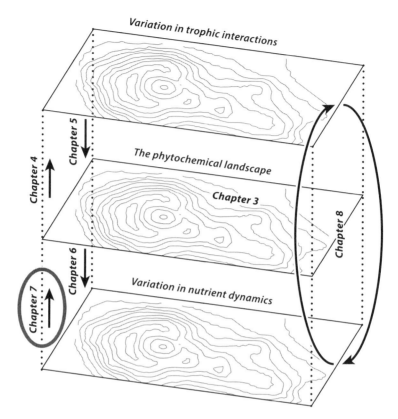

FIGURE 7.1. Variation on the phytochemical landscape is the nexus linking trophic interactions and nutrient dynamics. In this chapter (heavy circle), I focus on the role of nutrient availability in generating spatial and temporal variation on the phytochemical landscape.

But first it is worth noting that, in terrestrial ecosystems, the feedback loop from plant chemistry to soil nutrient availability (chapter 6) and from soil nutrient availability back to plant chemistry (this chapter) represents one component of what has come to be known as plant-soil feedback (Van Der Putten 2003, Bardgett and Wardle 2010). The mechanisms underlying plant-soil feedback include a much wider variety of processes than those described here, including changes in the abundance of nutritional symbionts, plant pathogens, and specialist herbivores (Bever 1994, 2002, Mangan et al. 2010, Coley and Kursar 2014). The diverse causes and effects of positive and negative plant-soil feedback fall beyond the scope of this book, and readers are encouraged to explore some excellent texts on the subject (Wardle 2002, Bardgett and Wardle 2010). Here, I limit my attention to feedback mechanisms operating through phytochemistry and nutrient dynamics.

7.2 EFFECTS OF NUTRIENT DYNAMICS ON THE CHEMICAL PHENOTYPE OF INDIVIDUAL AUTOTROPHS

Many primary producers exhibit substantial phenotypic plasticity, in which their traits vary depending upon the environment in which the autotrophs are growing (Denno and McClure 1983). Plastic traits include the nutritional and defensive chemistry of autotroph cells and tissues, which vary markedly with resource availability (Hunter et al. 1992). Primary production in both aquatic and terrestrial ecosystems often appears to be limited by both N and P availability (Elser et al. 2007, Harpole et al. 2011) (figure 7.2), and we might therefore expect that these nutrients would have particularly important effects on the chemical phenotype of primary producers.

Environmental nutrient availability varies at diverse scales, from small patches of high or low nutrient availability, through variation at landscape scales, to systematic change along gradients such as elevation (Reynolds and Hunter 2001, Fine et al. 2004, Taylor and Stocker 2012). Of course, there are multiple causes of spatial and temporal variation in environmental nutrient

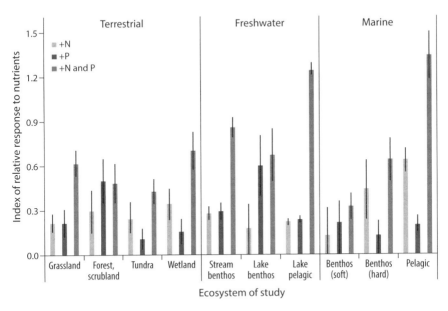

FIGURE 7.2. The relative responses of primary production in diverse terrestrial and aquatic ecosystems to the experimental addition of N, P, and both N and P. Note the evidence that primary production is strongly colimited by both N and P in diverse ecosystems. Modified from Elser et al. (2007).

availability, and the chemistry of autotroph detritus (chapter 6) is only one of them; nutrients recycled from detritus combine with new inputs from biological fixation, weathering, and human activity to provide the nutrients that support global primary production (Cleveland et al. 2013). This spatial and temporal variation in nutrient availability is an important force generating variation on the phytochemical landscape.

7.2.1 Nutrient Availability and the Stoichiometry of Autotroph Cells and Tissues

Perhaps the most obvious expectation of increasing nutrient availability in the environment is an increase in the nutrient concentration of autotroph cells and tissues. Such increases would be important because the nutrient content of primary producers has powerful effects on trophic interactions among producers, herbivores, and predators (chapter 4). The theory of ecological stoichiometry leads us to expect some degree of stoichiometric homeostasis among organisms, albeit with a broad match to environmental availability (Sterner and Elser 2002). At least among terrestrial plants, tissue nutrient concentrations vary widely with the availability of nutrients in the environment. In turn, although some nutrients are resorbed before plant tissues senesce (Killingbeck 1996), the effects of nutrient availability on terrestrial plant leaves persist as afterlife effects during litter decomposition (Tully et al. 2013) (chapter 6).

Our own work confirms the observations of many others that the foliar nutrient concentrations of terrestrial plants vary with soil nutrient availability. For example, in natural populations of oak (*Quercus*) trees, foliar protein concentrations can increase by 50 percent (Hunter 1997) and foliar N concentrations by 23 percent (Zehnder et al. 2010) in patches of high soil N availability. Moreover, experimental fertilization of oak trees causes comparable increases in foliar N concentrations to those observed upon natural gradients in soil N (Forkner and Hunter 2000) (figure 7.3A, B). Similarly, fertilization of milkweed (*Asclepias*) species with mineral N increases foliar N concentrations by 30 to 50 percent (Zehnder and Hunter 2009, Tao and Hunter 2012b). Interestingly, the ability of milkweed to take up additional N from soil depends in part on the availability of P in soil (figure 7.3C), confirming the synergistic effects of nutrient availability on plant performance (Elser et al. 2007) and phytochemistry. While increasing the availability of mineral P in soil also increases foliar P concentration in milkweed, P uptake does not appear to vary with soil N availability (figure 7.3D). Global increases in mineral N availability resulting from anthropogenic N deposition (Vitousek et al. 1997) provide further evidence of the positive relationship

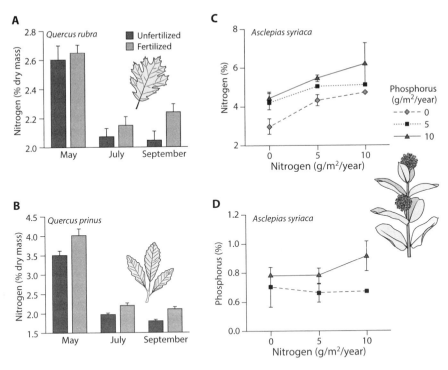

FIGURE 7.3. Foliar nutrient concentrations increase with experimental addition of nutrients to (A, B) oaks and (C, D) common milkweed. Note that the ability of milkweeds to take up N increases with both N and P availability, while the ability to take up P depends only on P availability. Data from Forkner and Hunter (2000) and Zehnder and Hunter (2009).

between soil N availability and the N concentration of terrestrial plant tissues. For example, in China, increases in anthropogenic N deposition during the last 30 years have caused increases in the foliar N concentrations of plants in natural, seminatural, and agricultural ecosystems (Liu et al. 2013a) (figure 7.4).

Human influences on environmental nutrient availability are perhaps most notable in the rivers, lakes, and coastal ecosystems of Earth, where N and P inputs from agricultural fertilizer, sewage, and aquaculture have dramatic and well known effects on aquatic ecosystems (Schindler 1974, Michalak et al. 2013). The average increase of 3- to 10-fold in N and P inputs into coastal ecosystems has provided compelling evidence of the role of nutrient availability in the ecology of algae (Anderson et al. 2012). Among the phytoplankton, most of the research linking nutrient availability in water to algal chemistry has focused upon harmful algal blooms (HABs). This body of work has tended to focus on changes in phytoplankton production and species composition under various

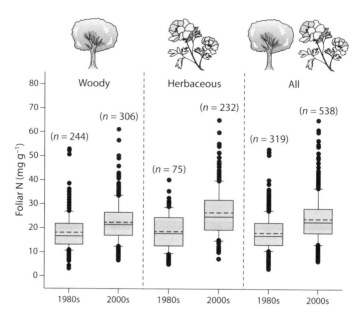

FIGURE 7.4. Increases in anthropogenic N deposition in China between the 1980s and the 2000s are associated with increases in the foliar N concentrations of woody and herbaceous plant species across terrestrial ecosystems. Modified from Liu et al. (2013a).

levels of nutrient loading, rather than any phenotypic plasticity in cell nutrient concentration or per capita toxin production by a given strain or species (Anderson et al. 2002). I'll consider the effects of nutrient availability on plankton community dynamics, and the subsequent phytochemical landscape, in the section on autotroph diversity (below). Here, I focus on what we know about phenotypic plasticity in algal nutrient content and secondary metabolism under various levels of nutrient availability.

In the same way that the leaves of terrestrial plants vary in their N and P concentrations with soil nutrient availability, we can ask whether the cellular nutrient concentrations of phytoplankton covary with the availability of nutrients in water. It has proven challenging to measure the physiological and chemical responses of individual phytoplankton to changes in resource availability, in part because of their small size. Additionally, nutrient uptake by phytoplankton varies with other abiotic variables including light availability, turbulence, and temperature (Anderson et al. 2002). Most models of nutrient uptake by phytoplankton follow a Monod model, and focus on subsequent effects on cell division and the rate of population growth (Smayda 1997). In other words, in response to nutrient enrichment, phytoplankton may often make more cells rather than increase significantly the nutrient concentration of individual cells.

Nonetheless, the Redfield Ratio (Redfield 1934) suggests that phytoplankton stoichiometry is both a cause and a consequence of environmental nutrient availability and there is clear evidence of intraspecific variation in algal nutrient content based on resource availability (Boersma and Kreutzer 2002, Malzahn et al. 2007). Recent work confirms that the nutrient concentrations of phytoplankton correlate with a global latitudinal gradient in nutrient availability, exhibiting C:N:P ratios of 195:28:1 in warm nutrient-depleted waters, 137:18:1 in warm nutrient-rich waters, and 78:13:1 in cold nutrient-rich waters (Martiny et al. 2013). The ability of some individual phytoplankton cells to store nutrients in excess of current demand suggests that at least some of the covariation between nutrient availability and phytoplankton stoichiometry is based on plasticity of individual cells rather than species turnover. For example, cyanobacterial cells can sequester N that exceeds immediate demand in the storage molecule cyanophycin (Allen 1984). Similarly, P limitation in the abundant cyanobacterium *Synechococcus* favors low overall P concentrations, replacement of nonmembrane phospholipids with sulfolipids and changes in the dominant mechanism of P storage toward polyphosphate (polyP) (Martin et al. 2014). PolyP is an inorganic polymer, and, according to Martin et al. (2014), its relative dominance as a storage molecule under low environmental P availability may favor rapid and highly conserved cycling of P in the P-depleted Sargasso Sea. This research emphasizes the tight links among environmental nutrient availability, the chemical quality of phytoplankton cells, and subsequent recycling of nutrients (figure 1.1, loop B).

However, differential effects of nutrient availability on population growth rates (Elser et al. 2003), resource competition (Klausmeier et al. 2004), and the turnover of phytoplankton species (Martiny et al. 2013) may have a greater impact on overall phytoplankton C:N:P ratios than does phenotypic plasticity (section 7.3, below). In turn, the relative abundance of phytoplankton species will influence nutrient dynamics through variation in the chemistry of their phytodetritus (chapter 6) and the differential ways in which they use and fix resources such as C and N (Mills and Arrigo 2010).

Recent work has emphasized the importance of phytoplankton cell size as a predictor of nutrient uptake rates, metabolic rates, and population growth rates (Marañón et al. 2013). Nitrogen uptake rates scale isometrically with cell volume, whereas minimum nitrogen requirements scale allometrically with cell volume to the power 0.84. In other words, increasing cell volume provides larger phytoplankton with the potential to store N beyond their minimum demands, and become somewhat buffered against short-term N fluctuations. However, large-celled species are less efficient at turning assimilated resources into new phytoplankton cells than are phytoplankton species of intermediate size. At the other end of the size spectrum, small phytoplankton species are unable to take

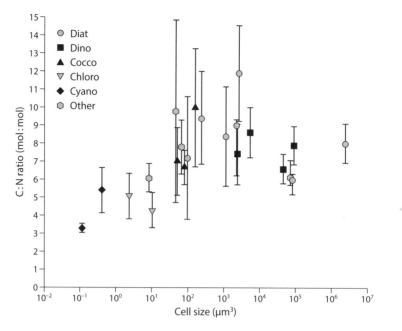

FIGURE 7.5. The C:N ratio of phytoplankton cells peaks at intermediate cell sizes, associated with size-dependent variation in N uptake and C fixation. Modified from Marañón et al. (2013).

up N much beyond their immediate metabolic needs, and also have lower population growth rates than do species of intermediate size. Carbon fixation rates peak at intermediate size classes in phytoplankton, and the overall combination of size-dependent N uptake, C fixation, and resource use results in cellular C:N ratios that peak at intermediate cell volume (Marañón et al. 2013) (figure 7.5). In short, size structure influences the phytochemical landscape within plankton communities.

Because larger phytoplankton species are relatively more resistant to grazing than are smaller phytoplankton species (Kiorboe 1993), the ultimate size structure of phytoplankton communities depends on the combined effects of nutrient availability and predation pressure. Low nutrient availability and high predation pressure should favor larger phytoplankton species with higher C:N ratios, a perfect example of how ecological forces from both the bottom up and the top down combine with metabolic constraints to determine the chemistry of primary producers and variation on the phytochemical landscape. It is interesting to note that terrestrial plants that grow in areas of high grazing pressure and low nutrient availability also tend to have tissues with high C:N ratios (Endara and Coley 2011).

Overall, experiments and observations in diverse ecosystems confirm the strong relationships between environmental nutrient availability and autotroph nutrient content. As described in chapter 4, low nutrient concentrations in primary producers can limit the performance of herbivores, predators, and parasites, establishing fundamental links between nutrient availability in the environment and trophic interactions (chapter 8).

7.2.2 Nutrient Availability and Autotroph Secondary Metabolism

In addition to their effects on the elemental stoichiometry of cells and tissues, nutrients also influence the expression of plant secondary metabolites (PSMs) by autotrophs. Plasticity in defense expression with nutrient availability has been discussed for decades in the ecological literature (Bryant et al. 1983, Herms and Mattson 1992, Koricheva et al. 1998), and yet consensus on any mechanistic relationships between the two has not emerged (Hamilton et al. 2001, Stamp 2003). We have already considered the relative merits of the hypotheses relating environmental nutrient availability to plant defense in detail elsewhere (Speight et al. 2008, Vannette and Hunter 2011b), and I won't repeat those arguments here. Given the diversity of PSMs synthesized by primary producers, and the diversity of functions that they serve, it is unrealistic to expect a single mechanistic pathway to link defense production with nutrient availability across all plant species (Berenbaum 1995). Phylogeny and local evolutionary pressures likely fine-tune the chemical responses of plant species to nutrient availability.

Critically, however, there remains overwhelming evidence that nutrient availability in the environment influences the expression of PSMs, even if our mechanistic theory to explain why remains incomplete. Across diverse growth forms in diverse terrestrial biomes, experiments illustrate again and again that increasing or decreasing the nutrients available for plant growth causes changes in allocation to secondary metabolism. Consistent with early predictions (Bryant et al. 1983), the C-based defenses of woody plants often decline with increasing N fertilization, particularly those such as condensed tannin and lignin that are derived from the phenylpropanoid pathway (Koricheva et al. 1998). For example, the addition of N to soil generally decreases both constitutive and induced condensed tannin concentrations in the foliage of oak trees (Hunter and Schultz 1995, Forkner and Hunter 2000) (figure 7.6A). This is important because C-based defenses such as condensed tannin and lignin inhibit the microbial degradation of plant litter (chapter 6). The reduction in recalcitrant molecules in woody plants that results from high nutrient availability establishes a clear route of feedback between soil fertility, plant and litter chemistry, and the recycling of nutrients back to soil (Hobbie

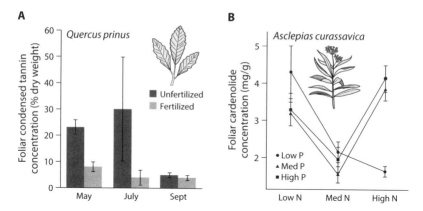

FIGURE 7.6. High nutrient availability generally results in declines in the foliar polyphenolic concentrations of woody plants like chestnut oak trees (A). However, other classes of C-rich chemical defenses, such as cardenolides in tropical milkweed (B), respond in nonlinear or idiosyncratic fashion to nutrient enrichment. Data from Forkner and Hunter (2000) and Tao and Hunter (2015).

1992, Treseder and Vitousek 2001) (figure 1.1, loop B); the phytochemical landscape is both a cause and a consequence of nutrient dynamics.

While the synthesis of tannin and lignin molecules may respond predictably to resource availability, the same cannot be said for other C-based defenses (Koricheva et al. 1998). For example, various C-rich defenses in milkweed respond in completely opposite directions to changes in resource availability. Under carbon fertilization (with elevated CO_2) milkweed leaves become tougher and exude more latex (Vannette and Hunter 2011a), as might be expected; both the lignocellulose complexes that determine leaf toughness and the polyisoprene in latex are very rich in C. However, the same milkweed plants express substantial declines in foliar cardenolides, also rich in C, under elevated CO_2. That various C-rich defenses respond in opposite directions to C fertilization confirms that there is no simple relationship between relative nutrient availability and defense expression (Hamilton et al. 2001).

Whether or not their responses are predictable from current theory, all of the chemical traits in milkweed foliage exhibit substantial plasticity in response to changing resource availability. Nitrogen fertilization causes nonlinear responses in the foliar concentrations of cardenolides in milkweed; cardenolides decline from low to intermediate N availability, and then rise again with high N availability (figure 7.6B). By its effects on foliar cardenolide concentration, soil N availability influences the growth of monarch (*Danaus plexippus*) caterpillars and their ability to sequester cardenolides for their own defense against vertebrate predators and parasites (Tao and Hunter 2015). In other words, by influencing the phytochemical landscape, nutrient availability has effects that cascade up to influence rates of predation and disease (chapter 8).

Of course, terrestrial plants also express PSMs in their roots (chapter 3), and root chemical defenses vary with nutrient availability. For example, the C-based PSMs, iridoid glycosides, are potent chemical defenses in the tissues of Dalmatian toadflax, *Linaria dalmatica*. In contrast to the patterns described previously for C-based foliar tannins in oak, the concentrations of iridoid glycosides in the roots of Dalmatian toadflax increase by more than 4-fold under N fertilization (Jamieson et al. 2012). Interestingly, concentrations decline in flowers and remain unchanged in shoots, demonstrating that responses of a single chemical defense to nutrient availability can vary markedly among tissues of the same individual plants.

Environmental N availability also influences the expression of N-containing PSMs on the phytochemical landscape, although not always in the direction predicted by simple theory. Fertilization of yaupon holly, *Ilex vomitoria*, with mineral N can increase foliar caffeine and theobromine concentrations by 5- to 10-fold, suggesting that N availability may limit alkaloid production (Palumbo et al. 2007) (figure 7.7A). Caffeine is well known for its biological activity, again serving to illustrate the link between soil N availability and trophic interactions in ecosystems. However, contrary to this pattern, we have observed declines in the tissue alkaloid concentrations of bloodroot, *Sanguinaria canadensis*, following experimental application of nutrients (Salmore and Hunter 2001b) (figure 7.7B). In wild tobacco, high soil N availability is associated with low constitutive concentrations of nicotine in foliage, but high induced nicotine concentrations following herbivore damage (Lou and Baldwin 2004). In other words, N-containing secondary metabolites can respond strongly to environmental nutrient availability, but not necessarily in a consistent direction among plant species.

As in terrestrial ecosystems, variation in nutrient availability also influences the per capita toxin production of phytoplankton. In fact, the magnitude of change in phycotoxin synthesis that occurs when nutrients vary in availability often dwarfs those reported for terrestrial angiosperms. For example, saxitoxin production by the dinoflagellate *Alexandrium tamarense* increases by 5- to 10-fold in P-limited environments (Anderson et al. 1990). Similarly, either silica or P limitation induces substantial increases in domoic acid production by the diatom *Pseudo-nitzschia multiseries*; significant toxin accumulation does not occur until the onset of nutrient limitation (Pan et al. 1996) (figure 7.8). Cyanobacterial toxin production also depends on nutrient availability; allelopathic effects of *Cylindrospermopsis raciborskii* PSMs increase markedly under P limitation (Antunes et al. 2012).

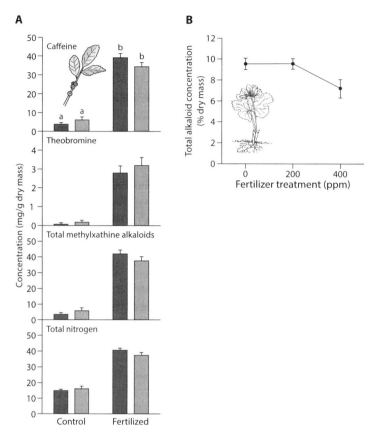

FIGURE 7.7. Fertilization does not have consistent effects on the expression of N-containing PSMs. Concentrations of alkaloids and total nitrogen in the foliage of female (dark bars) and male (light bars) yaupon holly plants, *Ilex vomitoria*, increase in response to N fertilization (A), whereas concentrations of benzophenanthridine alkaloids in fertilized bloodroot, *Sanguinaria canadensis*, decline (B). Modified from Palumbo et al. (2007) and Salmore and Hunter (2001b).

A recent synthesis (Van de Waal et al. 2014) illuminates strong links between the nutrient stoichiometry of phytoplankton cells and the production of PSMs within those cells. While such simple relationships between nutrient availability and secondary metabolite production have not held up to close inspection in terrestrial plants (Hamilton et al. 2001), the evidence provided by Van de Waal and colleagues in their study of 31 phytoplankton species across 19 genera appears compelling (table 7.1). Specifically, cellular N limitation reduces the per-cell production of N-rich algal toxins whereas P limitation favors the synthesis of some N-rich toxins, including paralytic shellfish poisons (PSPs) such as

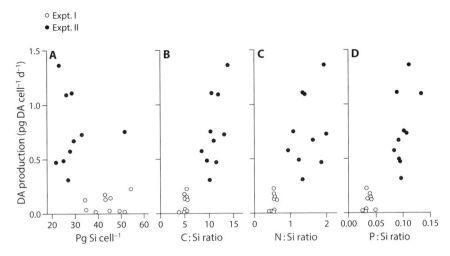

FIGURE 7.8. (A) Domoic acid (DA) production by the diatom *Pseudo-nitzschia multiseries* is highest at low cellular concentrations of silica (Si). Note the positive correlations between DA production and the ratios of carbon (B), nitrogen (C), and phosphorus (D) to silica in phytoplankton cells. Modified from Pan et al. (1996).

saxitoxin. High cellular N:P ratios also favor the synthesis of N-rich secondary metabolites whereas both N and P limitation promote higher cellular concentrations of C-rich algal toxins. These patterns are consistent with ecological stoichiometry theory (Sterner and Elser 2002), and confirm the strong relationships in aquatic systems among environmental nutrient availability, cellular nutrient concentrations, and algal secondary metabolite production on the phytochemical landscape. As we saw previously, algal nutrients and secondary metabolites both influence the rate at which phytodetritus decomposes or enters storage in aquatic sediments (chapter 6).

7.3 EFFECTS OF NUTRIENT AVAILABILITY ON PRIMARY PRODUCER DIVERSITY

As described above, environmental nutrient availability influences the elemental stoichiometry and per capita toxin production of primary producers. However, in addition to effects on the chemistry of individual autotrophs, nutrient availability also influences the richness and relative abundance of primary producers in communities. Associations between nutrient availability and autotroph diversity have been recognized for decades, and research in this area has accelerated with growing concerns over anthropogenic increases in nutrient availability (Penuelas et al.

TABLE 7.1. Phytoplankton toxins grouped by their molecular C:N ratios.

Group	Toxins		Molecular formula	C:N ratio*
N-rich				
	PSP	Paralytic Shellfish Poisoning Toxins (e.g., Saxitoxin)	$C_{10}H_{19}N_7O_4$	1.5
	CYN	Cylindrospermopsin	$C_{15}H_{21}N_5O_7S$	3.0
	MC	Microcystin (e.g., MC-LR)	$C_{49}H_{74}N_{10}O_{12}$	4.3
	NOD	Nodularin	$C_{41}H_{60}N_8O_{10}$	5.1
C-rich				
	ANTX	Anatoxin A	$C_{10}H_{15}NO$	10
	ASP	Amnesic Shellfish Poisoning Toxin (e.g., Domoic Acid)	$C_{15}H_{21}NO_6$	15
	GYM	Gymnodimine	$C_{32}H_{45}NO_4$	32
	SPX	Spirolide	$C_{42}H_{64}NO_7$	42
	PLTX	Palytoxin	$C_{129}H_{223}N_3O_{54}$	43
	OVTX	Ovatoxin	$C_{129}H_{223}N_3O_{52}$	43
	DSP	Diarrhoeic Shellfish Poisoning Toxins (e.g., Okadaic Acid)	$C_{44}H_{68}O_{13}$	—
	NSP	Neurotoxic Shellfish Poisoning Toxin (e.g., Brevetoxin)	$C_{49}H_{70}O_{13}$	—
	KMTX	Karlotoxins (e.g., KMTX-1-1)	$C_{67}H_{120}O_{34}$	—
	CTX	Ciguatoxin	$C_{64}H_{88}O_{19}$	—
	MTX	Maitotoxin	$C_{164}H_{256}O_{68}S_2Na_2$	—
	HA	Haemolytic Activity Based on Bioassays		

Source: Data from Van de Waal et al. (2014).
Note: Cellular N limitation appears to limit the expression of N-rich toxins.
*Values refer to the described toxins, or to the average of the shown analogues.

2012). In the context of this book, relationships between environmental nutrient availability and the structure of primary producer communities are important because plant community composition contributes in part to the phytochemical landscape upon which trophic interactions take place (chapter 4). Indeed plant community composition is among the strongest predictors of phytochemistry at the landscape scale (Dahlin et al. 2013).

7.3.1 Nutrient Availability and Autotroph Communities in Terrestrial Ecosystems

Autotrophs partition niche space based in part on variation in resource availability, with particular species favored over others under certain ratios of nutrients, light, and water availability (Tilman 1987). Consequently, variation in the availability and chemical form of soil nutrients influences the stability and structure of terrestrial plant communities (Song et al. 2012). Certainly, plant species exhibit evolutionary adaptations to conditions in the soil that influence their distribution and relative abundance (Casper and Castelli 2007) and may be competitively excluded if conditions change (Gough et al. 2000). For example, the relative availability of N and P in Eurasian herbaceous communities associates strongly with life-history traits. Plants in P-limited communities invest relatively less in sexual reproduction and exhibit more conservative leaf-economic traits than do plants in N-limited communities (Fujita et al. 2014). As a consequence, plant community composition is determined in part by nutrient dynamics in soil. Because of strong feedback processes (chapter 6), soil nutrient availability itself depends upon the chemistry of the local plant community (Hobbie 1992, Treseder and Vitousek 2001). Associating particular plant communities with particular soil characteristics is a central focus of plant community ecology, and there are many excellent texts that describe these associations in detail (Gurevitch et al. 2006, van der Maarel and Franklin 2013).

That the composition of terrestrial plant communities varies with nutrient availability has been forcefully emphasized by declines in plant species richness caused by anthropogenic nutrient inputs. Perhaps because a relatively small number of species are strong competitors in high-nutrient environments, or under the changes in soil pH that nutrient addition can cause, most studies report declines in plant species richness with terrestrial nutrient additions (Suding et al. 2005). For example, in Minnesota, long-term chronic N addition to prairie grasslands results in substantial declines in plant species richness relative to plots receiving only ambient levels of N deposition (Clark and Tilman 2008) (figure 7.9A). Similarly, the diversity of plants in 68 grasslands across the United Kingdom correlates negatively with rates of N deposition, with systematic reductions in species adapted to low N environments (Stevens et al. 2004) (figure 7.9B). This is important because plant species richness correlates directly with phytochemical diversity on the landscape (chapter 4).

Moreover, the negative effects of N addition on grassland biodiversity persist for at least two decades after fertilization ends, suggesting that levels of nutrient availability can entrain plant communities into characteristic stable states (Isbell et al. 2013b). N deposition causes particularly serious declines in lichens and bryophytes, which compete poorly under high N availability (Phoenix et al. 2012, Wardle et al. 2013). Lichens and bryophytes are often critical members of boreal and arctic communities, with important effects on ecosystem structure and function. However, a recent study in alpine tundra reported that fertilization increased diversity in infertile habitats dominated by plant species with high foliar C:N ratios and high condensed tannin concentrations (Eskelinen et al. 2012). In this

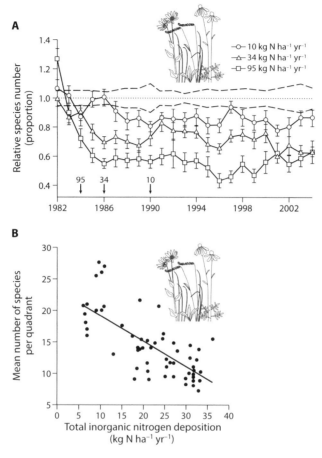

FIGURE 7.9. Plant species richness declines in US prairie (A) following experimental additions of mineral N to soil. The pattern is similar across 68 European grasslands (B) under varying levels of anthropogenic N deposition. Modified from Clark and Tilman (2008) and Stevens et al. (2004).

case, fertilization may have allowed plants with higher nutrient demands (herbs, grasses) to compete more favorably with the dominant slower growing (shrub) species, a common response to N addition in tundra vegetation (Wardle et al. 2013). In contrast, in more productive tundra habitats, fertilization decreased plant diversity in line with typical results from grassland habitats.

In addition to the effects of N availability on terrestrial plant diversity, there is increasing evidence that P availability influences plant diversity. As with N pollution, anthropogenic additions of P into ecosystems threaten ecosystem structure and function (Elser and Bennett 2011). For example, across 501 grassland

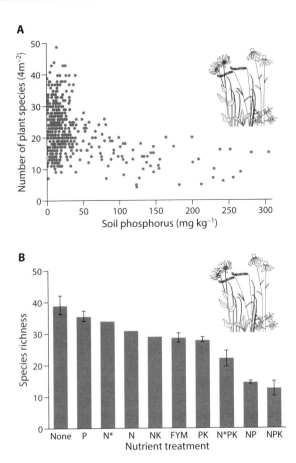

FIGURE 7.10. (A) Plant species diversity declines with P loading across 501 European grasslands, reflecting a general negative response (B) of plant diversity to excess nutrient inputs. In (B), FYM = farmyard manure, N* = sodium nitrate, N = ammonium sulfate, P = phosphorus, K = potassium. Modified from Ceulemans et al. (2014) and Crawley et al. (2005).

plots throughout Europe, increasing P availability in soil is associated with significant declines in plant species richness (Ceulemans et al. 2014) (figure 7.10A). Moreover, many more endangered plant species across Eurasia persist in habitats that remain under P limitation, suggesting that anthropogenic P additions may be responsible for local extinctions of plant species and general losses of biological (and therefore chemical) diversity (Wassen et al. 2005). Declines in plant species richness appear to be a common response to excess nutrient enrichment. In the Park Grass Experiment in the south of England, experimental inputs of N, P, and K have occurred since 1856; all have reduced plant species richness, with the greatest declines under combined nutrient enrichment (Crawley et al. 2005) (figure 7.10B). Plots without any nutrient addition retain diverse herbaceous communities that simultaneously compete well for N, P, and K. In short, nutrient enrichment causes declines in plant species richness, an associated decline in diversity on the phytochemical landscape (chapter 4), with significant consequences for trophic interactions (chapter 4) and nutrient dynamics (chapter 6).

7.3.2 Nutrient Availability and Autotroph Communities in Aquatic Ecosystems

Differential responses to the availability of key nutrients (particularly N, P, Si, Fe), in both their inorganic and organic forms, play a major role in determining the structure of phytoplankton communities (Anderson et al. 2012). In lake ecosystems, P is traditionally considered to limit the biomass of some phytoplankton groups (Schindler 1974), although N and P colimitation may be more common than once thought (Harpole et al. 2011). However, the link between P availability and cyanobacterial production is sufficiently strong in some lakes that spring concentrations of total P are good predictors of late-season cyanobacterial abundance (Wetzel 1983) and the concentrations of toxins that they produce (Chorus and Bartram 1999). This illustrates a key point for the current discussion; some phytoplankton species are favored disproportionately over others by nutrient inputs, changing their relative abundances in the community (Jochimsen et al. 2013). In turn, because of their differential production of PSMs, changes in phytoplankton community structure engender changes in the phytochemical landscape.

For example, the relative abundance of the diatom *Pseudo-nitzschia* increases markedly off the coast of Louisiana during periods of high nutrient loading from the Mississippi River, particularly when inorganic N availability is high (Parsons et al. 2013). *Pseudo-nitzschia* produces the potent neurotoxin domoic acid which influences trophic interactions (chapter 4) and the sequestration of N in marine sediments (chapter 6). Similarly, dinoflagellate blooms in Tolo Harbour, Hong Kong, occur whenever the N:P ratio of the water drops below about 10:1. The dinoflagellates are more P than N limited, with physiological optima well below the Redfield N:P Ratio (Hodgkiss and Ho 1997) (figure 7.11).

In a recent and dramatic example, a 2011 algal bloom in Lake Erie was stimulated by elevated P inputs from agricultural soils, transported into the lake by very heavy spring rains (Michalak et al. 2013). By its peak in October, the algal bloom had grown to 2,000 square miles in size, more than three times larger than any bloom observed previously in Lake Erie. The bloom was composed almost entirely of the cyanobacterium *Microcystis* and concentrations of the

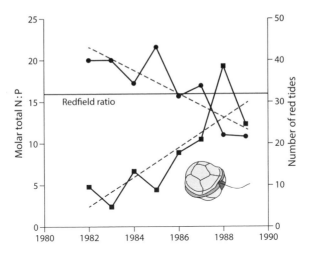

FIGURE 7.11. Anthropogenic P additions to Tolo Harbour, Hong Kong, have systematically lowered the molar N:P ratio of seawater (circles) thereby favoring dinoflagellate blooms (squares) responsible for toxic red tides. Based on Hodgkiss and Ho (1997) and Anderson et al. (2002).

toxin microcystin may have reached levels over 220 times higher than World Health Organization guidelines. In short, by changing phytoplankton community structure, variation in nutrient availability drives variation in the chemistry of algal toxins in water (Anderson et al. 2002) and the ecological consequences that follow.

In addition to N and P, the availability of other nutrients such as silica and iron influence the structure and chemistry of phytoplankton communities (Assmy et al. 2013). For example, experimental fertilization of the Southern Ocean with iron results in blooms of diatoms that are sufficiently large to cause export of C from the atmosphere to deep waters (Smetacek et al. 2012). Natural diatom blooms in these regions of the ocean are stimulated by iron-rich dust from terrestrial soils and volcanic activity. Likewise, iron can limit the productivity of N-fixing phytoplankton, including some cyanobacteria, particularly in iron-depleted basins of the Pacific Ocean. Iron is critical for diazotrophs because it serves a component of the nitrogenase enzyme system. In addition to iron limitation at local scales, N depletion, caused by competing algae and denitrifying bacteria, favors N-fixing phytoplankton at larger spatial scales (Weber and Deutsch 2014). As a group, these studies all illustrate the tight link between environmental nutrient availability, differential resource use, algal community structure, and the phytochemical landscape.

Moreover, theories based on the supply ratios of nutrients (Tilman 1977) predict differential performance of algal species under different nutrient regimes, linking algal physiology with community structure and the phytochemical land-scape (Smayda 1997). Differential performance under varying nutrient availabil-ity contributes in part to the seasonal turnover of algal classes that I described in chapter 6 (figure 6.12), although with significant contributions from grazers and pathogens as well (chapter 8). In some lakes, diatoms compete well for resources until silica concentrations are depleted, after which green algae may emerge as dominant competitors. Cyanobacteria may flourish as N levels fall, until P limitation (and grazing) limit cyanobacterial population growth. Water mixing may then replenish silica supplies, allowing a further increase in diatom populations (Wetzel 1983). While this sequence represents a simplification, it illustrates the fundamental idea that variation in nutrient supply, itself a partial product of algal nutrient uptake, leads to variation in algal community struc-ture and the expression of algal secondary metabolism at the community level (Anderson et al. 2002).

7.4 EVOLUTIONARY EFFECTS OF NUTRIENT AVAILABILITY ON AUTOTROPH CHEMISTRY

So far, I have described effects of nutrient availability on cellular and tissue stoi-chiometry, or the production of PSMs, that result from phenotypic plasticity or changes in autotroph community composition. Additionally, over evolutionary time scales, variation in nutrient availability selects for growth and defense strat-egies in primary producers that contribute to their chemical phenotype. Many authors have reviewed the relationships among environmental nutrient availabil-ity, growth, reproduction, and defense in plants (Koricheva 2002, Stamp 2003, Leimu and Koricheva 2006), and there is no need to repeat those discussions here. Rather, for completeness, I want to provide a brief reminder that the nutritional and defensive chemistry of autotrophs evolve in part based on nutrient availability in the environment.

   Assumed trade-offs between growth and defense would suggest that the strat-egies of primary producers to combat herbivores should fall somewhere along a continuum. At one end of the continuum, autotrophs may evolve substantial investment in resistance traits, but little capacity to regrow and compensate for herbivore damage. At the other end of the continuum, autotrophs may exhibit low investment in resistance traits but a substantial capacity for regrowth and compensation following damage (Van der Meijden et al. 1988). Key for the topic at hand is that regrowth and compensation strategies should evolve more read-ily in high-resource environments, where nutrients, light, and water are in good supply. High-resource environments should select for rapidly growing autotroph species in which the costs of diverting resources from growth to defense should outweigh the benefits. In contrast, low-resource environments will favor slow-growing autotroph species and should select for physical and chemical resistance traits that protect tissues that are hard to replace. These alternative evolutionary strategies form the basis of the resource availability hypothesis (RAH) of plant defense (Coley et al. 1985). The RAH is important for the topic of this book because it posits that autotroph chemical traits should evolve in response to environmental nutrient availability. As we have already seen, these same chemical traits have powerful effects on trophic interactions (chapter 4) and the recycling of nutrients in ecosystems (chapter 6). In short, the RAH provides an evolutionary mechanism linking nutrient dynamics with trophic interactions through the nexus of the phytochemical landscape.

A recent meta-analysis confirms that the basic predictions of the RAH are often met, particularly in systems dominated by woody plants (Endara and Coley 2011). Specifically, high-resource environments select for fast-growing plant species (figure 7.12A) with rapid rates of leaf turnover. These fast-growing plant species invest less in resistance traits (toughness, chemical toxins) and suffer higher rates of herbivory than do slow-growing plants with long-lived leaves in resource-poor environments (figure 7.12B). In other words, the effects of nutrient availability cascade up to influence trophic interactions (figures 1.1 and 2.6), with this phytochemically mediated cascade occurring over evolutionary time scales. The importance of such evolutionary responses in plant defense to resource availability can be illustrated by reciprocal transplant experiments. When fast-growing tree species in the Amazon are transplanted from their native nutrient-rich clay soils to nutrient-poor white sands, they are unable to maintain positive leaf growth rates in the presence of herbivores, and suffer higher rates of mortality. As a result, they are outperformed by the well-defended, slow-growing species that are native to white sand soils (Fine et al. 2004).

In short, high-nutrient environments select for chemical traits in fast-growing plants that increase rates of consumption by herbivores, support diverse predator communities (chapter 4), and increase rates of nutrient recycling in soils (chapter 6). Low-nutrient environments select for chemical traits in slow-growing plants that decrease consumption by herbivores, support lower predator diversity (chapter 4), and decrease rates of nutrient recycling (chapter 6). The phytochemical landscape links trophic interactions with nutrient dynamics on evolutionary, as well as ecological, time scales. I consider some further consequences of evolutionary processes on the phytochemical landscape in chapter 9.

While theory linking resource availability with the evolution of defense strategies has proven generally predictive for terrestrial angiosperms, there is no such consensus linking environmental nutrient availability to the evolution of toxin

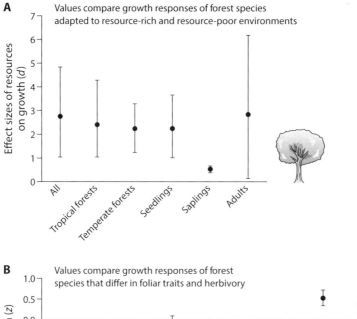

FIGURE 7.12. (A) In forested ecosystems, plant species adapted to resource-rich environments grow faster than do species adapted to resource-poor environments. (B) In turn, rapidly growing species are less well defended, have higher foliar nutrient concentrations, and suffer higher levels of herbivore damage than do slow-growing species. Modified from Endara and Coley (2011).

production in phytoplankton. And yet there are profound differences in toxin production among species and strains of unicellular algae (Anderson et al. 2012),
including genetically based variation in toxin production within co-occurring
populations (Alpermann et al. 2009). Strains within a given algal species can
be nontoxic whereas others produce high toxin concentrations. Among toxin-
producing strains of the same species, the synthesis of secondary metabolites
can vary many-fold. However, within phytoplankton clades, genetically based
physiological traits, and evolved strategies for nutrient uptake, do not correlate
clearly with differential toxin production (Smayda 1997). For example, the so-
called affinity, growth, and storage strategies of phytoplankton (Sommer 1989) do
not appear to reflect differential investment in chemical defense in the same way
that trade-offs among growth, storage, and defense occur among terrestrial plants
(Coley et al. 1985, Herms and Mattson 1992).

In other words, while there has been considerable progress in identifying new
strains of algae, and their phylogenetic relationships with one another (Casteleyn
et al. 2008), the ecological forces that maintain genetic variation in toxin production remain elusive (Alpermann et al. 2010). As I noted in chapter 5, variation in
selection pressure imposed by grazers and parasites likely plays a significant role
in maintaining genetic variation in toxin production (Yang et al. 2011), although
some secondary metabolic pathways appear to predate the evolution of metazoan
grazers (chapter 3). Likewise, allelopathic interactions among competing species may select for differential toxin production among algal species (Smayda
1997, Kearns and Hunter 2000, 2001). However, transport by advection (currents) (Krock et al. 2013) and recolonization by algae from cysts in sediments
(McGillicuddy et al. 2011) may weaken the association between local ecological variables and selection on toxin production. At present, we are left with the
unfortunate generalization that variation in diverse ecological factors, including
resource availability, competition among species, abiotic variables, and trophic
interactions, likely interact to maintain genetic variation for toxin production
among the phytoplankton (Anderson et al. 2012).

7.5 CONCLUSIONS

Two major goals of this chapter were to (a) describe the effects of nutrient availability on the nutritional and defensive chemistry of primary producers, and therefore (b) complete the feedback loop (figure 1.1, loop B) between nutrients in the
environment, variation on the phytochemical landscape, and the recycling of nutrients during decomposition of autotroph residues. This second goal is illustrated
nicely in some recent work on the foliar and litter chemistry of aspen, *Populus*

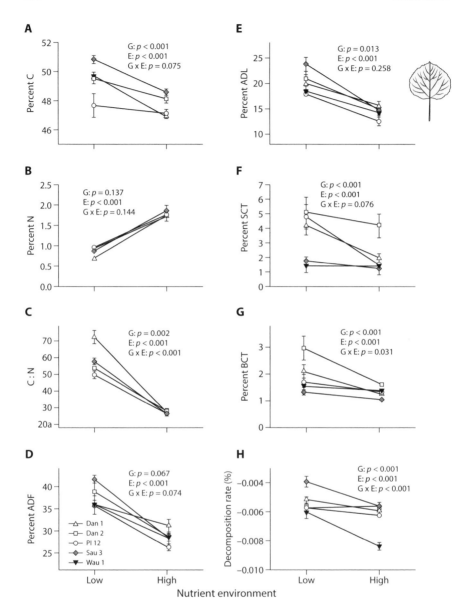

FIGURE 7.13. Reductions in the chemical recalcitrance of five genotypes of aspen litter (A–G), generated by high nutrient availability, feed back to increase subsequent rates of litter decomposition (H). Note that increasingly negative values in (H) represent increasing rates of decomposition. ADF = acid detergent fiber, ADL = acid detergent lignin, SCT = soluble condensed tannins, BCT = bound condensed tannins. Modified from LeRoy et al. (2012).

tremuloides (LeRoy et al. 2012). Across most aspen genotypes, litter C:N ratio, fiber content, lignin content, and condensed tannin content decline as environmental nutrient availability increases (figure 7.13A–G). Critically, the subsequent decomposition rates of litter from plants grown under high nutrient availability are faster than are litter decomposition rates from plants grown under low nutrient availability (figure 7.13H). In other words, the low C:N ratios, low lignin concentrations, and low tannin concentrations in litter, which are generated by high nutrient availability in soil, subsequently serve to increase the rate of nutrient cycling. These experiments with aspen provide an elegant demonstration that the phytochemical landscape, by participating in a positive feedback processes, is both a cause and a consequence of environmental nutrient dynamics. Linking this feedback loop with that already described between the phytochemical landscape and trophic interactions (figure 1.1, loop A, chapters 4 and 5) is the subject of chapter 8.

Linking Trophic Interactions with Ecosystem Nutrient Dynamics on the Phytochemical Landscape

8.1 PUTTING IT ALL TOGETHER: LINKING CYCLES AND GENERATING FEEDBACK

In chapters 4 and 5, I outlined a feedback loop by which (a) autotroph chemistry influences trophic interactions, and (b) trophic interactions influence autotroph chemistry (figure 1.1 loop A). In chapters 6 and 7, I described a second feedback loop by which (c) autotroph chemistry mediates the dynamics of nutrient flux in ecosystems and (d) nutrient availability influences the expression of autotroph chemistry (figure 1.1, loop B). Combining these together, the phytochemical landscape represents a nexus that links trophic interactions and nutrient dynamics in a larger feedback process (figure 1.1, loop C; figure 8.1).

If this approach has merit, we should be able to better understand the effects of trophic interactions on nutrient dynamics, and vice versa, by following associated changes on the phytochemical landscape. By extension, we should expect that trophic interactions that are too weak to change the phytochemical landscape would also have relatively weak effects on nutrient dynamics at the ecosystem level. In turn, under circumstances in which changes in nutrient availability have only weak effects on phytochemistry, they should have only weak effects on trophic interactions (chapter 9).

In this chapter, I describe the evidence that trophic interactions influence nutrient dynamics through changes on the phytochemical landscape (figure 8.1). I then review the evidence that nutrient dynamics in ecosystems cascade back up to influence trophic interactions by their effects on autotroph chemistry. I conclude the chapter by providing examples in which we can observe the entire feedback loop between trophic interactions and nutrient dynamics, as well as the associated changes on the phytochemical landscape.

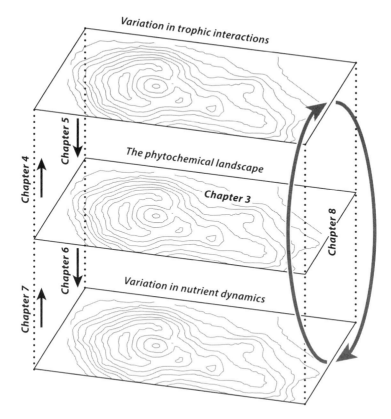

FIGURE 8.1. Variation on the phytochemical landscape is the nexus linking trophic interactions and nutrient dynamics. In this chapter, I focus on the entire feedback loop (large arrows) from trophic interactions to nutrient dynamics and back again. This large feedback loop arises from the combination of processes described in chapters 4–7.

8.2 FROM TROPHIC INTERACTIONS
TO ECOSYSTEM PROCESSES

In chapter 5, I described a number of different ways in which herbivores and their natural enemies can influence the phytochemical landscape. Mechanisms include selective foraging on the landscape of fear, phytochemical induction, and any concomitant evolutionary changes that occur as the result of herbivore preferences among autotroph genotypes or species. Then, in chapter 6, I illustrated how the chemistry of autotroph residues helps to determine rates of decomposition and nutrient flux in terrestrial and aquatic ecosystems. Here, I integrate these processes to describe when and how trophic interactions have effects on nutrient dynamics at the ecosystem level. I begin by describing the phytochemistry-mediated influences that herbivores are known to have on ecosystem nutrient fluxes, and then add natural enemies to the picture once herbivore effects have been firmly established. Perhaps a majority of the effects that enemies and herbivores have on nutrient cycling are trait-mediated indirect effects (Abrams 1995, Ohgushi et al. 2012), wherein enemies change the behavior of herbivores (Schmitz et al. 2010) and herbivore activity influences the phytochemical landscape that is available for subsequent decomposition (Hunter et al. 2012).

8.3 EFFECTS OF HERBIVORY ON NUTRIENT DYNAMICS

There are diverse mechanisms by which herbivores influence nutrient dynamics in ecosystems (Hunter 2001a, Hunter et al. 2012) (table 8.1). Most are relevant to both aquatic and terrestrial systems, although one (through-fall) may be limited to the soils of terrestrial environments. The mechanisms are (a) deposition of feces and urine, (b) inputs of cadavers, (c) herbivore-mediated changes in the chemistry of precipitation (canopy through-fall, terrestrial only), (d) sloppy feeding that releases nutrient-rich materials, (e) induced exudation from cells or roots, (f) reductions in residue quantity, (g) selective foraging among autotrophs, (h) induced residue recalcitrance, (i) increased residue quality, (j) changes in

TABLE 8.1. Pathways by which herbivores can influence ecosystem nutrient dynamics through the fast or slow decomposition cycles.

Pathway	Cycle	Expectation
Feces and Urine	Fast	Accelerate
Cadavers	Fast	Accelerate
Herbivore-Modified Through-Fall*	Fast	Accelerate
Sloppy Feeding	Fast	Accelerate
Induced C Exudation	Fast	Accelerate
Reductions in Residue Inputs	Slow	Decelerate
Selective Foraging	Slow	Decelerate or Accelerate
Induced Residue Recalcitrance	Slow	Decelerate
Increased Residue Quality	Slow	Accelerate
Changes in Microclimate	N/A	Decelerate or Accelerate
Changes in Resource Uptake	N/A	Decelerate or Accelerate

Note: Expectations are for acceleration or deceleration of biogeochemical cycling. The overall effects of herbivores in any particular ecosystem will depend on the relative importance of each of these pathways.
 *Terrestrial systems only

microclimate, and (k) changes in resource uptake. Because these processes can have opposing effects on rates of nutrient flux (Hunter et al. 2012), their relative importance in a given ecosystem will determine the overall effects of herbivores on decomposition and nutrient dynamics (Bardgett and Wardle 2003). Critically, the relative strength of these pathways varies with a suite of other biotic and abiotic variables, generating strong context-dependence in the effects of herbivores on nutrient cycling (McSherry and Ritchie 2013).

To understand how herbivores influence nutrient dynamics through the nexus of phytochemistry, it is important to distinguish between fast-cycle and slow-cycle effects on decomposition (McNaughton et al. 1988) and whether herbivores accelerate or decelerate nutrient cycling (Ritchie et al. 1998). Fast-cycle effects are those mediated by labile inputs derived from herbivore activity (table 8.1). For example, inputs such as herbivore feces, cadavers, nutrients released by sloppy feeding, and induced C exudates all provide sources of C and nutrients that are readily utilized by decomposers. They short-circuit the typical pathways by which autotroph residues are recycled in ecosystems and form part of the fast cycle (figure 8.2). Because these types of herbivore inputs are typically more labile than senesced autotroph residues, they generally cause rapid increases in microbial activity and accelerate rates of nutrient cycling (Hamilton and Frank 2001, Yang 2004, Madritch et al. 2007).

However, herbivores also have slow-cycle effects on nutrient dynamics by influencing long-term decomposition rates of senesced primary production (figure 8.2). In this case, herbivores can either accelerate or decelerate nutrient cycling, depending upon their impact on the chemistry of senesced autotroph residues on the phytochemical landscape. Herbivores can accelerate nutrient cycling when their activities make senesced residues less recalcitrant (Classen et al. 2006) or decelerate nutrient cycling when they induce increases in residue recalcitrance (Pastor and Naiman 1992, Findlay et al. 1996).

So, do herbivores increase or decrease overall rates of nutrient cycling? The answer in any given ecosystem will arise from a dynamic balance among the alternative pathways listed in table 8.1 (Bardgett and Wardle 2003, Pineiro et al. 2010). For example, browsing by herbivores on woody plants in the African savanna tends to decrease litter quality by inducing higher levels of N resorption during leaf senescence. As a result, litter in heavily browsed areas has a higher C:N ratio and should decompose more slowly. However, high inputs of herbivore dung and feces at these heavily browsed sites appears to offset the effects of low litter quality on nutrient cycling (Fornara and Du Toit 2008), such that fast-cycle effects of herbivory overcome slow-cycle effects. In other words, we need to understand the relative strengths of the mechanisms linking herbivores to nutrient fluxes before we can predict how nutrient dynamics will respond to herbivore activity; we need

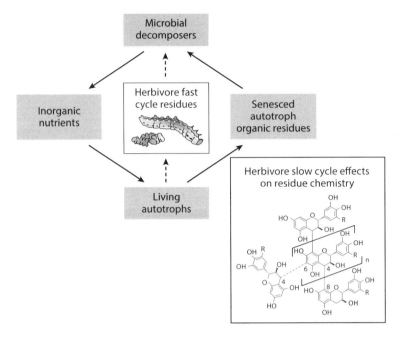

FIGURE 8.2. Labile residues that result from herbivore activities can short-circuit (dashed arrows) the typical recycling of nutrients (solid arrows) from senesced autotroph residues. The fast cycle produced by this short-circuit will increase rates of nutrient cycling. Herbivores can also change the chemistry of senesced autotroph residues, thereby influencing the rate of the typical slow cycle. See table 8.1 for examples.

to understand which matters most to variation on the phytochemical landscape. I describe these mechanisms in more detail below.

8.3.1 Shit Happens and Then You Die: Fast-Cycle Effects of Herbivores on Nutrient Cycling

Feces and Urine. All of the fast-cycle effects of herbivores on biogeochemical cycles should act to accelerate rates of nutrient cycling (table 8.1) as herbivores convert portions of the phytochemical landscape to less recalcitrant forms. For example, herbivore feces generally have lower ratios of C:N and C:P than do senesced autotroph tissues (Hunter et al. 2003b), which will favor rapid cycling of fecal material (chapter 6).

The effects of herbivore egestion and excretion on nutrient dynamics is particularly important in terrestrial communities of perennial plants, in which leaf senescence is normally preceded by nutrient resorption and storage. Because herbivore feces contain nutrients that would otherwise have been resorbed by plants before leaf senescence, herbivore feces can double rates of nutrient return from plants to the environment (Hollinger 1986). Although many plant second-ary metabolites (PSMs) persist in herbivore feces (Madritch et al. 2007), fecal C is often more labile than the C in plant litter (Christenson et al. 2010), allow-ing C-limited microbes to immobilize rapidly the nutrients returned in feces (Christenson et al. 2002). In turn, grazing on microbial decomposers and feces by heterotrophs then remobilizes nutrients (Zimmer and Topp 2002, Reynolds et al. 2003) for assimilation by plants (Frost and Hunter 2007), storage in min-eral soil (Christenson et al. 2010), or leaching losses (Townsend et al. 2004). During periods of high herbivore activity, the export in streams of feces-derived N can represent substantial losses of N from forested ecosystems (Eshleman et al. 1998, Reynolds et al. 2000). In soils, effects of herbivore waste on nutri-ent fluxes can persist for several years (Hudson et al. 2009) and can change the balance of bacterial and fungal decomposers in the environment (Wardle et al. 2010).

The chemistry of herbivore feces reflects both the nutrient and PSM concen-trations of the original phytochemical landscape. Herbivores, and their gut micro-biomes, process autotroph cells and tissues to varying degrees, extracting only a proportion of available nutrients and catabolizing a subset of PSMs. For example, genetic variation in the nutrient and tannin concentration of aspen leaves is main-tained in the feces of caterpillars that feed on those trees (Madritch et al. 2007). In turn, fecal chemistry influences the decomposition of the feces itself, and of the associated leaf litter in the soil (see priming, chapter 6). A more extensive study of 130 caterpillar species confirms that fecal chemistry is determined in large part by plant chemistry (Kagata and Ohgushi 2012), although differences among her-bivores in their abilities to digest nutrients and PSMs from plants also influence fecal chemistry (Madritch et al. 2007, Couture and Lindroth 2014). In aquatic sys-tems, algal secondary metabolites can become concentrated in the feces of zoo-plankton. For example, dimethylsulfoniopropionate (DMSP) from algae becomes concentrated in the feces of copepods or fish that feed on those algae (Tang 2001, Hill and Dacey 2006). Subsequent microbial breakdown of DMSP in feces may accelerate rates of sulfur cycling and lead to the release of gaseous DMS into the atmosphere, with potential consequences for climate change. Unfortunately, relatively few studies have explored how PSMs in the feces of aquatic herbivores influence subsequent nutrient dynamics.

Effects of labile feces inputs on nutrient dynamics can be dramatic. For exam-ple, in subarctic mountain birch forests, climate change has caused an expansion in the ranges of two geometrid moth species that defoliate birch trees (Jepsen

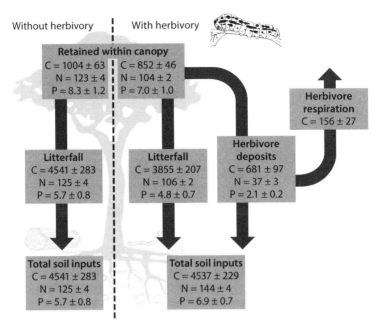

FIGURE 8.3. Herbivory in the canopy of lowland Amazon tropical forest increases inputs of N and P from canopy to soil by 15% and 21%, respectively. Modified from Metcalfe et al. (2013).

et al. 2011). In areas with increasing levels of defoliation, soil biogeochemical processes have changed markedly. Higher levels of defoliation are associated with lower fungal:bacterial ratios in soils, higher densities of enchytraeids, lower soil C:N ratios, and higher concentrations of soil ammonium (Kaukonen et al. 2013). Under the highest levels of defoliation, grasses have replaced cowberry as the dominant understory vegetation. All of these results are consistent with acceleration in the rate of nutrient cycling, whereby insect feces replace leaf litter and carbon exudation as the dominant flux of carbon from aboveground to below. In other words, herbivore activity is engineering the phytochemical landscape into more labile chemical forms. At the opposite climatic extreme, insect defoliation in the canopies of tropical forests also affects nutrient flux (Fonte and Schowalter 2005), increasing inputs of both N and P from trees into soil (Metcalfe et al. 2013) (figure 8.3).

Urinary inputs (excretion) by herbivores also influence patterns of N cycling in ecosystems (Steinauer and Collins 1995). For example, white-tailed deer in eastern North America deposit urine in discrete patches, increasing the heterogeneity of nutrient availability on the forest floor. In turn, discrete patches of urine-derived N influence the spatial pattern of herb-layer cover (Murray et al. 2013).

Urinary inputs by large mammal grazers can even change the relative abundance of microbial genes responsible for soil N transformations (Yang et al. 2013). In other words, herbivores convert the N of the phytochemical landscape into forms that alter microbial gene expression and subsequent nutrient dynamics.

In terrestrial ecosystems, root herbivores also influence nutrient cycling by causing fluxes of C and N from plants to soil. For example, cicada nymphs are root xylem feeders, tapping in to the water and nutrient transport systems of plant roots. It is ironic that cicadas probably have their greatest sustained impact on ecosystem structure and function when they are hidden belowground. Research is unfortunately biased toward the adult stage, which is much more easily studied (Williams and Simon 1995). However, cicada nymphs can reach very high densities and prodigious levels of secondary production. As Dybas and Davis (1962) point out, "Calculated on an annual basis, the net average annual productivity of cicada protoplasm per acre . . . compares favorably with man's best efforts at beef production under the most carefully controlled forage conditions."

At such high densities, it is no surprise that feeding by cicada nymphs influences the cycling of nutrients from plants to soil. Cicada nymphs consume an average of 1.3 liters of xylem per nymph per year (Ausmus et al. 1978). Using published data on the amino acid and ureide content of the xylem of 60 eastern North American tree species (Barnes 1963), we can estimate N flux through cicada nymphs in temperate forest ecosystems (Hunter 2008). Xylem contains between 17 and 80 µg of amino acid and/or ureide per mL of sap, with N representing an average of 18.82 percent of compound mass. Estimating from these numbers, each cicada consumes an average of 4–20 mg N per year in 1.3 liters of xylem. Given that *Magicicada* densities can reach 3.75 million ha^{-1} (Dybas and Davis 1962), we can scale up. At their highest densities, cicadas could cost their tree hosts between 15 and 75 kg of N ha^{-1} y^{-1} in 4,875,000 L of water. In other words, cicadas are transferring huge quantities of N from plant to insect and then to soil. Cicadas do not assimilate all of the xylem that they take in, and much of the N will be excreted and available to soil microbes and plant roots quite rapidly.

In aquatic systems, feces and urine deposition by consumers also plays a major role in nutrient dynamics (Wotton and Malmqvist 2001, Vanni 2002). In combination with algal exudates (chapter 6), the byproducts of herbivore ingestion and digestion contribute substantially to powering the microbial loop and associated fluxes of C and N (Jumars et al. 1989, Saba et al. 2011). Moreover, because of their diel vertical migration to avoid predation pressure, zooplankton translocate a portion of their ingested C and N from the photic zone to darker, deeper waters through respiration, egestion, and excretion (Longhurst and Harrison 1988, Longhurst et al. 1990).

What happens to the fecal and excretory products of zooplankton? The fate of zooplankton fecal pellets likely depends upon their size, with smaller pellets being degraded and recycled rapidly within the water column, whereas larger pellets contribute to the flux of materials (such as marine snow, chapter 6) from the photic zone to the benthos (Turner 2002). Whether they are recycled in place or subsidize benthic phytochemical landscapes, aquatic herbivore feces are important mediators of nutrient dynamics. For example, the fecal pellets of copepods are significant sources of dissolved organic carbon (DOC) (Thor et al. 2003) and N (Vincent et al. 2007) in aquatic ecosystems, serving as hot spots of microbial activity in the water column. Likewise, the fecal pellets of sea urchins can degrade rapidly, contributing to local recycling of nutrients (Sauchyn and Scheibling 2009). However, during large defoliation events, when urchins reduce kelp beds to bare substrate, rates of fecal production exceed rates of physical and microbial degradation such that fecal material becomes a temporary storage pool for kelp-derived C and N. Pellets that are subsequently translocated to deeper waters likely serve as substantial subsidies for benthic ecosystems (Sauchyn et al. 2011).

In addition to feces, aquatic herbivores excrete sufficient nitrogenous waste to influence ecosystem N dynamics. For example, *Daphnia* excrete enough ammonium from consuming phytoplankton to stimulate algal production and compensate in part for population losses due to *Daphnia* grazing (Sterner 1986). Similarly, grazing by snails on palatable marine algae can double the ammonium concentration in water, thereby supporting primary production by less palatable algae in the same habitat (Bracken et al. 2014). Omnivores like mussels, which consume zooplankton, bacteria and detritus as well as phytoplankton, excrete enough ammonium to increase the abundance of algal epibionts on their shells by up to 10-fold in comparison to the abundance of algae on artificial mussels (Aquilino et al. 2009). Zooplankton grazers may also release substantial quantities of urea into the water column (Saba et al. 2011). In short, excretion and egestion by herbivores represent a ubiquitous pathway of C and N flux through aquatic phytochemical landscapes.

At a vastly finer scale, marine viruses are increasingly implicated in the collapse of algal blooms, the lysis of algal cells, and the release of organic C and nutrients into the aquatic environment. Although we are at the early stages of understanding virus-phytoplankton ecology, marine viruses may have effects comparable to marine herbivores on the dynamics of nutrients (Fuhrman 1999, Suttle 2007). For example, *Prochlorococcus* is one of the most abundant cyanobacteria in the oceans, contributing substantially to C fixation in the photic zone. *Prochlorococcus* is attacked by viruses that lyse cells and therefore release DOM into the water column (Avrani et al. 2011). This cyanobacteria-virus interaction presumably supplies resources for subsequent use by heterotrophic bacteria.

Likewise, the cosmopolitan marine coccolithophore *Emiliania huxleyi* generates some of the largest algal blooms in the ocean and accounts for an astonishing one-third of all marine calcium carbonate production (Iglesias-Rodriguez et al. 2008). Viral infection of *E. huxleyi* reprograms fatty acid metabolism to support viral assembly; as they contribute to bloom collapse, viruses have significant impacts on the cycling of both C and Ca in the ocean (Rosenwasser et al. 2014). Below the photic zone, sulfur (S)-oxidizing bacteria are abundant and widespread chemolithoautotrophs, which are important contributors to C fixation in the dark waters of the oceans. Viruses that attack these bacteria influence S cycling by killing and lysing cells. Remarkably, some viral strains also contain bacterial genes that encode subunits of the enzyme that oxidizes sulfur (Anantharaman et al. 2014). As a consequence, the marine viruses may serve as a reservoir of genes that mediate the marine sulfur cycle, facilitating the S cycle through horizontal transfer of the appropriate genetic machinery. In short, herbivores and viral agents of disease appear to be major drivers of nutrient flux from the phytochemical landscapes of aquatic ecosystems.

Cadavers. Herbivore biomass is processed and repackaged plant chemistry (chapter 3) with the distribution and abundance of herbivores determined in part by variation on the phytochemical landscape (chapter 4). In turn, herbivore cadavers are well known to influence nutrient cycling in terrestrial and aquatic ecosystems (Beasley et al. 2012). In prairies, carcasses of bison, deer, and cattle introduce concentrated pulses of nutrients into soil that generate landscape patches of high nutrient availability and high plant production. In other words, there is spatial correlation among nutrients, autotroph chemistry, and herbivore activity (figure 2.4), the signature of links between nutrient dynamics and trophic interactions on the phytochemical landscape (chapter 9). Carcass-mediated changes in prairie vegetation dynamics can persist for at least five years, introducing spatial complexity that promotes plant diversity (Towne 2000). Similarly, in temperate forests, the cadavers of large mammal herbivores such as deer generate biogeochemical hot spots of nutrient cycling, and introduce substantial spatial heterogeneity in nutrient availability (Bump et al. 2009b) (figure 8.4 A–F). The N introduced into soil by deer cadavers is recycled rapidly back into plants, elevating their foliar quality and reducing their C:N ratios (figure 8.4 G–I). As I noted in chapters 4 and 6, these plant foliar traits generally increase herbivore performance and increase rates of litter decomposition. In other words, deer carcasses are both a cause and a consequence of variation on the phytochemical landscape, so completing a fundamental feedback loop from herbivore populations to soil nutrient dynamics and back to herbivores (figure 8.1).

Because they can reach such high densities (Ausmus et al. 1978, Andersen 1987), we might expect that the cadavers of root-feeding organisms would also

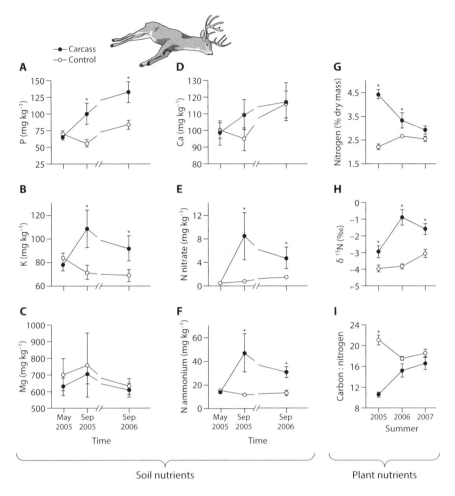

FIGURE 8.4. Carcasses of white-tailed deer are hot spots of biogeochemistry that increase nutrient availability in soil (A–F) and subsequent N concentrations in plant foliage (G–I). Modified from Bump et al. (2009b).

influence ecosystem processes. However, depressingly few data are available on this topic. In one fascinating study, a fungus that is both pathogenic to insects and mutualistic with plant roots has been shown to transfer N from insect cadavers to the roots of bean and switchgrass plants (Behie et al. 2012). The fungal transfer of cadaver N to plants may be an important but understudied pathway of N dynamics in some terrestrial soils (Klironomos and Hart 2001). Additionally, data are available for some root herbivores that emerge from a root-feeding nymphal stage into the adult stage aboveground. For example, cicadas cause substantial levels of nutrient transport out of the soil system as they emerge as adults aboveground. Using available data on cicada mass (Dybas and Davis, 1962), water, and nitrogen content (Yang 2004, 2006), we can estimate N and water flux from soil to the terrestrial environment during a 17-year emergence of northern periodical cicadas (Hunter 2008). A peak emergence will move between 4.25 and 8.5 kg N ha^{-1} and between 1,200 and 2,400 L water ha^{-1} from below- to aboveground. My estimates agree well with those from a riparian forest in Kansas, where nitrogen flux during cicada emergence averages 6.3 kg N ha^{-1}, peaking at 30 kg N ha^{-1} at the highest densities (Whiles et al. 2001).

These levels of N addition to aboveground ecosystems caused by cicadas are comparable with those of chronic N deposition caused by anthropogenic activities (Vitousek et al. 1997) and have measurable effects on ecosystem structure and function. For example, cicada cadavers directly increase microbial biomass and nitrogen availability in forest soils, and influence the growth and reproduction of forest plants (Yang 2004). Pulses of periodical cicada cadavers create bottom-up cascades that influence a suite of detritivores, omnivores, and predators on the forest floor (Yang 2006). However, as I stressed in chapter 1, top-down and bottom-up process are fundamentally linked through the nexus of phytochemistry, and this link is apparent during cicada emergence too. The increase in N availability caused by the decomposition of cicadas increases the foliar N concentration of plants on the forest floor. In turn, these plants represent higher quality food for herbivores and suffer increasing levels of defoliation from mammalian herbivores (Yang 2008). In other words, root-feeding cicadas induce a broad suite of changes in the structure and function of ecosystems by transforming the phytochemical landscape (figure 8.5).

In aquatic ecosystems, the three-dimensional nature of the environment and the important influence of currents means that herbivore cadavers may have their greatest impacts on nutrient availability at some distance from the location at which herbivore production occurred (Beasley et al. 2012). For example, in marine ecosystems, whale carcasses (or whale falls) serve as local hot spots of biological diversity and nutrient cycling in the deep ocean (Smith and Baco 2003, Rouse et al. 2004). Although extremely patchy, cetacean carcasses can play an important role over extended time periods in local C and nutrient flux. However, over a larger spatial scale, the carcasses of zooplankton may play a more general role in the fluxes of nutrients from herbivores back to the abiotic environment. In an oligotrophic coastal area of the Mediterranean, average fluxes of C and N in copepod carcasses from shallow to deeper waters are almost twice as high as the fluxes in copepod fecal pellets. Both carcasses and fecal pellets represent significant inputs of energy and nutrients to the benthic ecosystem (figure 8.6) (Frangoulis et al. 2011).

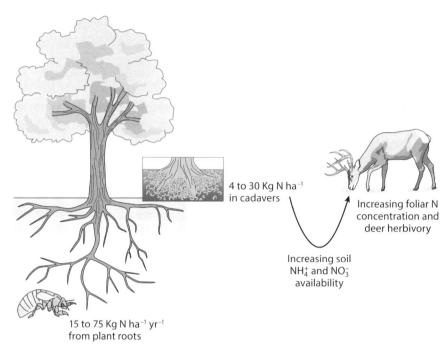

FIGURE 8.5. Periodical cicadas mediate important nutrient fluxes in terrestrial ecosystems. During the long nymphal stage, they transfer substantial quantities of N from plants to soil. During adult emergence, N is transferred to the soil surface where microbial decomposition of cicada cadavers makes N available to plants and their herbivores. Adapted from data in Whiles et al. (2001), Yang (2004, 2006, 2008), and Hunter (2008).

Through-fall. In terrestrial systems, rainfall that passes through plant canopies damaged by herbivores can leach a surprisingly high concentration of nutrients from damaged foliage and from herbivore feces trapped in the canopy (Tukey and Morgan 1963, Stadler et al. 2006). For example, a nominal 8 percent level of defoliation to black locust and red maple in the southern Appalachians results in a surprising 70 percent increase in the leaching of potassium from foliage (Seastedt et al. 1983). While this mechanism may be limited to terrestrial plants, it is possible that nutrients leached in this way can enter aquatic systems through runoff or ground water (Reynolds et al. 2000). I know of no studies that have explored increases in nutrient leaching from aquatic plants attacked by herbivores (I consider sloppy feeding on phytoplankton separately below). However, in terrestrial systems, this canopy through-fall can increase rates of N cycling in soils (Orwig et al. 2008), and increase fluxes of P from canopy to soil (Reynolds and Hunter

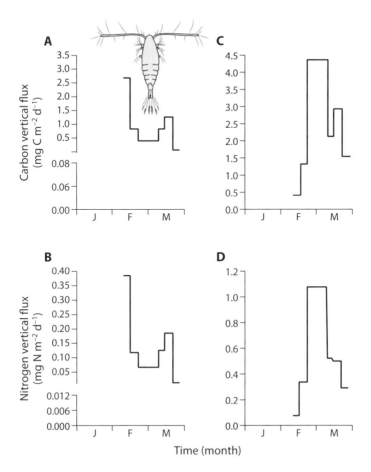

FIGURE 8.6. Variation during February and March in the vertical fluxes of carbon (A, C) and nitrogen (B, D) from shallow to deeper water in the fecal pellets (A, B) and carcasses (C, D) of zooplankton. Modified from Frangoulis et al. (2011).

2001). In short, herbivore activity generates significant nutrient flux in through-fall from the phytochemical landscape to the soil.

Green-fall. Many herbivores are messy eaters, dropping a significant proportion of the primary production that they are trying to consume (Risley 1986). In terrestrial systems, herbivore outbreaks sometimes generate green carpets of partially eaten plant material on the soil surface. Even when they don't drop the food that they are trying to eat, herbivores can induce premature abscission of plant organs, before sugars and nutrients have been completely resorbed (Faeth et al. 1981, Chapman et al. 2006). In both cases, the quality of plant tissue available to

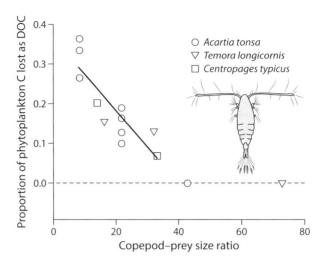

FIGURE 8.7. Copepods are particularly messy eaters when consuming larger phytoplankton, releasing up to 40% of phytoplankton C as dissolved organic carbon (DOC). Symbols represent three common copepod species. Modified from Moller (2007).

decomposers is generally higher than that of naturally senesced tissue (Fonte and Schowalter 2004), with the result that rates of decomposition and nutrient cycling increase (Risley and Crossley 1992, Reynolds and Hunter 2001).

There are also messy eaters in aquatic systems (Lampert 1978). For example, zooplankton may not ingest all of the phytoplankton cells that they trap in their appendages, and the cellular contents of ruptured phytoplankton may be released into the water as a source of DOC, ammonium, and urea. Copepods appear to be notably sloppy feeders, particularly when they consume phytoplankton that are large relative to their own body size (Moller 2007). Handling large phytoplankton causes up to 40 percent of algal C to be released as DOC, which is then available to support the microbial loop (figure 8.7). When the ratio of copepod size to phytoplankton size approaches 40, the C lost to sloppy feeding drops to zero. We might visualize this by imagining a human eating by hand a single walnut (no crumbs produced) versus an entire walnut loaf (an unfortunate source of detritus). The key point is that sloppy feeding links herbivory with subsequent nutrient dynamics by the effects of that feeding on the phytochemical landscape.

C Exudation. The last fast-cycle mechanism by which herbivores influence nutrient dynamics is by altering the exudation of labile C from autotrophs (Holland et al. 1996). In terrestrial ecosystems, C exudation from roots is called rhizodeposition, and can account for a remarkable 10–40 percent of all the C that is allocated by plants belowground (Gershenson and Cheng 2010). C exudation may support symbiotic microorganisms that mineralize organic matter and therefore help to replace nutrients lost to herbivores (Frost and Hunter 2008a). In some grasses, higher exudation of labile C following defoliation by herbivores increases daily N mineralization rates in the soil by 5-fold, thereby increasing rates of N uptake (Hamilton and Frank 2001, Hamilton et al. 2008). Similarly, intact *Phleum pratense* plants, raised in soil that has previously hosted defoliated plants, have shoot N concentrations that are 10 percent higher than those of plants grown in clean soils (Mikola et al. 2005). Presumably, higher C exudation by defoliated plants increases the availability of N in soils. In both of these cases, herbivore-induced increases in the availability of labile C on the phytochemical landscape then increase the rate of nutrient cycling and subsequent plant uptake.

Likewise, defoliation aboveground can stimulate the allocation of resources to arbuscular mycorrhizal fungi, an important source of plant P. Defoliation of *Plantago lanceolata* aboveground stimulates colonization by AMF, even though root growth declines with defoliation (Pietikainen et al. 2009). This suggests that defoliated plants are investing more C in their microbial mutualists to replace nutrients lost to defoliators. We have found similar results in our work with milkweed, in which defoliation of *Asclepias syriaca* by monarch caterpillars increases colonization of roots by mycorrhizal fungi (Vannette and Hunter 2014). However, chronic long-term defoliation can reduce overall levels of AMF in soils, while changing the structure of AMF communities belowground (Murray et al. 2010). The effects of herbivore-mediated changes in AMF abundance and community composition for the cycling of nutrients are not yet well characterized, and should prove an interesting area for future study.

In aquatic systems, heterotrophic bacteria certainly use C exudates from aquatic primary producers as a fundamental C source (Williams et al. 2009). However, I can find no studies examining defoliation-induced changes in C exudation in aquatic primary producers beyond those that signal to enemies of herbivores (chapter 5). For technical reasons, herbivore-induced exudation remains the least well studied of the fast-cycle mechanisms and its relative importance in terrestrial and aquatic systems cannot be assessed at this time.

In summary, herbivores can accelerate rates of decomposition and nutrient cycling through their fast-cycle effects on the phytochemical landscape. In all cases, these effects represent herbivore-mediated changes in the chemistry of autotroph production that enters the detrital food web. By decreasing the chemical recalcitrance of primary producers, their residues, and their exudates, herbivores can stimulate bacterial and fungal decomposers and accelerate nutrient cycling.

8.3.2 Slow-Cycle Effects of Herbivores on Nutrient Cycling

As we saw in chapter 6, variation in the chemistry of autotroph detritus is a fundamental driver of decomposition rate (Hättenschwiler and Vitousek 2000, Cornwell et al. 2008). Critically, herbivore activity can influence the relative expression of chemical traits on the phytochemical landscape (concentrations of nutrients, toxins, lignin, tannin) that affect decomposition rates (chapter 5). Whenever (a) herbivores influence the chemical traits of living autotrophs and (b) those traits persist in senesced autotroph residues, herbivores will influence slow-cycle processes during residue decomposition (figure 8.2). In chapter 5, I described a number of important effects of herbivores on phytochemistry. Three are particularly important in understanding the slow-cycle effects of herbivores on rates of nutrient cycling: phytochemical defense induction, changes in nutrient concentration, and selective foraging.

Herbivore-Induced Effects on PSMs and Nutrients. That herbivores influence the chemistry of the autotrophs on which they feed has become a cornerstone of chemical ecology (Green and Ryan 1972, Cronin and Hay 1996b). When chemical changes in primary producers persist into senesced organs or cells, and influence rates of decomposition, they are referred to as "afterlife effects" (Choudhury 1988, Findlay et al. 1996). These "ghosts of induction past" can either accelerate or decelerate nutrient cycling, depending on their impacts on residue recalcitrance.

When residues from damaged primary producers have higher concentrations of recalcitrant molecules (e.g., lignin and tannin, a typical response of deciduous woody plants to herbivore attack), the residues usually decompose more slowly than do residues from unattacked autotrophs. For example, lace bugs induce increases in the lignin content of oak litter, with subsequent declines in rates of litter decomposition (Kay et al. 2008). Likewise, litter from defoliated or gall-infested cottonwood trees contains higher condensed tannin concentrations and decomposes more slowly than does litter from herbivore-free trees (Schweitzer et al. 2005a, Schweitzer et al. 2005b). Finally, declines in the quality of birch litter induced by red deer browsing reduces the rate of litter decomposition (Harrison and Bardgett 2003), with concomitant reductions in dissolved organic C, nitrate, ammonium, and N mineralization in soil (Harrison and Bardgett 2004). In all of these cases, herbivore activity increases the recalcitrance of autotroph residues on the phytochemical landscape, reducing rates of nutrient cycling.

Unfortunately, we know essentially nothing about the impacts of induced chemical defenses in aquatic phytochemical landscapes on subsequent decomposition of algal residues. While induction of chemical defense in seaweeds is now well established (Cronin and Hay 1996b, Pavia and Toth 2000, Coleman et al.

2007b), I can find no studies that have investigated subsequent decomposition and nutrient release from previously induced seaweed residues. Likewise, while herbivores can induce increases in phytoplankton toxin production (Ianora et al. 2003), subsequent effects on nutrient dynamics are unknown. Given that algal residues drive much of the bacterial production in the water column, and can also accumulate in lake and marine sediments (chapter 6), effects of induction on residue decomposition should be an important area for future research.

Consequently, our knowledge of the effects of chemical induction on nutrient dynamics in aquatic ecosystems is limited to those caused by terrestrial plant residues that fall into aquatic systems. In forested streams, where light levels are often low and rates of photosynthesis are also low, leaf litter from forest vegetation provides up to 99 percent of the energy entering the streams (Tuchman et al. 2002). Accordingly, afterlife effects of litter chemistry from trees attacked by herbivores can influence rates of decomposition and nutrient cycling in stream ecosystems. For example, mite damage to cottonwood plants causes increases in leaf phenolics that persist in litter and reduce rates of decomposition in stream water by 50 percent (Findlay et al. 1996) (figure 8.8). Interestingly, exposure of cottonwood trees to elevated ozone concentrations also induces foliar phenolics and reduces rates of litter decomposition in stream water.

In chapter 5, I described how herbivores induce changes in the distribution of elements such as N and C among autotroph tissues (Babst et al. 2005). Reallocation of nutrients away from sites of damage following herbivore attack may change how those nutrients are subsequently recycled. For example, both milkweeds and oak preferentially allocate new N and C to storage (stems, tap roots) following attack by herbivores (Frost and Hunter 2008a, Tao and Hunter 2012a). Because fine roots and leaves generally turn over more rapidly than do tap roots and stem tissue, herbivore attack may reduce the rate at which N and C are recycled. Similarly, when aphids feed on the leaves of soybean plants, they cause reductions in foliar N concentrations that are maintained after leaf senescence. As a consequence, N mineralization rates from aphid-damaged litter are 40 percent lower than are those from undamaged soybean litter (Katayama et al. 2013).

Herbivores can also accelerate rates of nutrient cycling in those systems in which herbivore attack decreases the recalcitrance of residues on the phytochemical landscape. Again, our knowledge of such effects is limited to terrestrial ecosystems. In some nutrient-rich habitats, grazing may limit plant biomass more than does nutrient availability. As a consequence, plant tissues that regrow after grazing may contain higher nutrient concentrations and decompose more readily after they senesce (Hobbie 1992). This appears to be the case in the Serengeti, where a combination of higher plant quality and herbivore feces (above) result in grazing-induced increases in nutrient cycling at

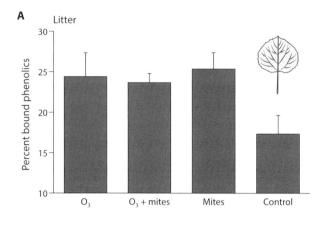

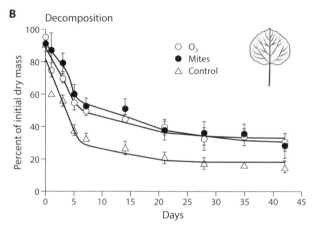

FIGURE 8.8. Damage to cottonwood leaves by mites or ozone (A) induces increases in litter phenolic concentrations that (B) reduce subsequent rates of litter decomposition in streams. Modified from Findlay et al. (1996).

landscape scales (McNaughton et al. 1988). Likewise, long-term grazing of the grass *Paspalum dilatatum* by herbivores in Argentina increases rates of nutrient cycling. In this case, herbivores induce increases in grass lamina production over sheath production, so reducing lignin:N ratios on the phytochemical land-scape and accelerating decomposition (Semmartin and Ghersa 2006). Whenever grazing causes a relative decline in recalcitrant molecules in comparison to pri-mary nutrients, we might expect the rate of litter decomposition to increase. For example, reindeer browsing on birch trees in Finland reduces the condensed tan-nin concentration of birch litter such that rates of decomposition and N cycling increase (Stark et al. 2007).

In many terrestrial systems, herbivores induce premature leaf abscission in their host plants, causing foliage to drop before complete resorption of nutrients (Faeth et al. 1981, Chabot and Hicks 1982). Herbivore-induced needle abscission prior to complete resorption of nutrients appears to be a common response of conifer trees (Chapman et al. 2006). For example, after the beetle *Ips typographus* attacks oriental spruce trees in Turkey, needle litter has lower C:N and lignin:N ratios and decomposes more rapidly (Sariyildiz et al. 2008). Similarly, defoliation of piñon pine by a stem-boring moth increases the N content of needle litter by 16 percent and increases the subsequent rate of litter decomposition (Classen et al. 2007b). Hemlock needle litter also contains higher N concentrations under trees that are attacked by hemlock woolly adelgid than under unattacked trees (Stadler et al. 2006). Finally, beetle damage to *Tamarix* trees increases the N and P concentrations of litter and decreases the C:N, C:P, and lignin:N ratios. As a consequence, litter from attacked trees decomposes 23 percent faster, releasing 16 percent more N and 60 percent more P than litter from unattacked trees (Uselman et al. 2011).

Although evergreen and deciduous trees often respond differently to herbivore attack, whereby litter of the former becomes less recalcitrant and litter of the latter becomes more recalcitrant (Chapman et al. 2006), there are enough counterexamples to make generalization difficult at present (Kielland et al. 1997, Hunter et al. 2012). However, the key point remains: Induced chemical defenses and incomplete nutrient resorption are powerful mechanisms by which herbivores cause changes on the phytochemical landscape. The result can be important afterlife effects on decomposition and nutrient cycling.

Effects on Residue Recalcitrance Mediated by Selective Foraging. In chapters 4 and 5, I described how the interactive forces of phytochemistry and predation pressure drive selective foraging by herbivores. In turn, selective foraging causes substantial changes on the phytochemical landscape. Here, I consider how those chemical changes influence nutrient dynamics and biogeochemical cycles.

Selective foraging results in a change in the relative representation of chemical forms on the phytochemical landscape. Preferred autotroph species are consumed at a higher rate than are less preferred species, such that the latter increase in relative abundance at the expense of the former. Because autotroph taxa vary widely in their chemistry (chapter 3), selective foraging results in a relative change in the chemistry of autotrophs in the environment, with subsequent effects on residue quality and nutrient dynamics (Pastor and Naiman 1992, Wardle et al. 2002).

Typically, autotrophs that are palatable to generalist herbivores are also those whose residues decompose most rapidly. In contrast, tissues that are unpalatable to generalist herbivores often share chemical traits (high lignin or tannin concentrations; high C:N, C:P, or lignin:N ratios) that reduce rates of decomposition (Cornelissen et al. 2004, Kurokawa et al. 2010). As a result, selective foraging often acts to decelerate rates of decomposition and nutrient cycling. For example, selective foraging by moose in North America tends to reduce the relative representation of (more palatable) deciduous trees in favor of (less palatable) conifers. Because conifer needles generally decompose more slowly than do abscised leaves of deciduous trees (Cornwell et al. 2008), heavy moose browsing reduces rates of organic matter turnover and nutrient cycling (Pastor et al. 1993, Pastor and Cohen 1997). Similarly, in an extensive study, Wardle et al. (2002) collected litter from 59 understory and 18 canopy tree species from 28 forested sites throughout New Zealand. They reported a significant positive correlation between the deleterious effect of mammal browsing on the density of a given plant species and the rate at which its litter decomposed. Again, this kind of slow-cycle effect of herbivory will act to decrease the rate of decomposition in the plant community as a whole. Finally, in oak savanna in the United States, insect and mammal herbivores selectively remove plants with high tissue N, including N-fixing legumes. Above- and belowground tissues, and the litter that they produce, become lower in N, and soil nitrate concentrations decline. Overall, rates of N cycling decline and productivity is halved in the presence of herbivores (Ritchie et al. 1998, Knops et al. 2000). In short, the evidence is now compelling that selective foraging by generalist terrestrial herbivores can reduce rates of nutrient cycling because selective foraging increases the average recalcitrance of the phytochemical landscape.

Because the dominant primary producers in large lakes and oceans lack the structural tissues of terrestrial plants, their residues are also more generally labile to decomposers than are those of most land plants (chapter 6). As a consequence, slow-cycle effects of herbivores on nutrient dynamics, mediated by selective foraging, may be generally weaker in aquatic than in terrestrial systems. Of course, selective foraging by zooplankton can influence phytoplankton community structure and therefore rates of nutrient uptake (chapter 5), but herbivore-mediated effects on the decomposition of phytoplankton residues are not well characterized. However, one notable exception may be the effects of copepods on the silica cycle. Grazing pressure by copepods appears to contribute to selection on the thickness of diatom frustules (shells). Because frustules are constructed from silica, the life and death of diatom blooms can drive silica cycling over substantial areas of the ocean. The sinking of thick-shelled (grazer resistant) diatoms in the Southern Ocean sequesters silica in sediments that might otherwise be recycled in surface waters. Selection by grazers may thereby limit silica availability, and diatom production, in other oceans of the world (Assmy et al. 2013). Additionally, zooplankton foraging on palatable phytoplankton can favor higher densities of toxin-producing species (Hay and Kubanek 2002). Because some algal toxins contain N and may resist decomposition (e.g., some forms of microcystin, chapter 6), selective foraging by zooplankton may facilitate the sequestration in sediments of N in algal toxins.

Of course, some herbivores are extreme specialists, including those that specialize on well-defended and low-nutrient autotrophs. Obviously, herbivory by such species may reduce the representation of recalcitrant residues in ecosystems (Kagata and Ohgushi 2011). The high degree of specialization among temperate insect herbivores (Novotny et al. 2002) may mean that their effects on nutrient cycling are determined more by the relative dominance of their particular host plant species than by community-wide effects of generalist herbivores across multiple plant taxa. Importantly, even specialist insect herbivores with very narrow dietary ranges can have a dramatic impact on watershed scale nutrient dynamics if their host plant species is dominant in the plant community. For example, high densities of the specialist oak sawfly, *Periclista*, can substantially increase the rate of N cycling in soils and N export in streams at the watershed scale (Reynolds et al. 2000), because its preferred host species contributes disproportionately to the phytochemical landscape.

Selective foraging by specialists can also result in the complete loss of autotrophs from the phytochemical landscape, with substantial effects on subsequent nutrient dynamics. For example, historically, eastern hemlock trees, *Tsuga canadensis*, have been a significant component of the riparian community in the southern Appalachian mountains of the United States. However, they are being lost from these communities because of attack by an exotic insect pest, the specialist hemlock woolly adelgid, *Adelges tsugae*. As a consequence, hemlock-rich areas are being replaced by hardwood stands, with significant impacts on nutrient cycling. Specifically, hemlock stands are generally cooler and support higher soil C, N, and P pools than do adjacent hardwood stands (Knoepp et al. 2011). In this case, the loss of a dominant tree species as a result of herbivore activity is causing local increases in soil and stream temperature and reductions in soil nutrient storage as rates of nutrient cycling increase.

The example of hemlock woolly adelgid illustrates that herbivores can actually increase decomposition rates through slow-cycle effects when they preferentially consume plant species with lower-than-average litter quality. For example, grazing by reindeer in northern Europe appears to promote the spread of plant species with litter that decomposes rapidly (Olofsson and Oksanen 2002). Similarly, browsing by moose in the Alaskan arctic increases the rate of nutrient cycling, at least in part by selective foraging and a gradual increase in the representation of nitrogen-fixing alder over willow (Kielland et al. 1997, Kielland and Bryant 1998). Finally, grasshoppers in prairie ecosystems can increase the quality of plant litter that reaches the soil, increasing rates of N cycling and overall productivity (Belovsky and Slade 2000).

While it is straightforward to understand how the evolution of resistance to herbivores might also select for autotroph tissues that decompose slowly (Uriarte 2000, Schweitzer et al. 2008), when might herbivore-mediated natural selection favor tissues that decompose rapidly? In some environments, herbivores may select for strategies of tolerance to herbivory rather autotroph resistance (Van der Meijden et al. 1988) (chapter 7). Tolerant autotrophs invest in rapid regrowth following damage while resistant species invest in chemical defense. If tolerance and regrowth demand high tissue nutrient concentrations, then selection for tolerance could increase rates of nutrient cycling (Hunter 2001a). Moreover, if autotroph tolerance is favored as an evolutionary strategy in high-resource environments (Coley et al. 1985), this may explain why herbivores appear to accelerate nutrient cycling in productive environments and decelerate nutrient cycling in unproductive environments (Bardgett and Wardle 2003). While the effects of herbivores on nutrient cycling may vary among systems, one key point remains: Whenever selective foraging by herbivores influences residue chemistry on the phytochemical landscape, nutrient dynamics are affected. Residue-mediated impacts of herbivores on nutrient cycling occur whether herbivores forage selectively within (Uriarte 2000) or among autotroph species (Pastor and Naiman 1992), and whether they increase (Yelenik and Levine 2010) or decrease (Belovsky and Slade 2000) the chemical recalcitrance of residues.

Additional Effects of Selective Foraging on Nutrient Dynamics. While changing the relative representation of recalcitrant residues is a dominant mechanism by which selective foraging influences nutrient dynamics, there are two other mechanisms that are worthy of brief mention (table 8.1). First, by altering the representation of species in autotroph communities, selective foraging can influence the physical structure of those communities with subsequent effects on microclimate (Classen et al. 2005). For example, the loss of hemlock trees to outbreaks of hemlock woolly adelgid in the riparian zones of eastern North America causes increases soil temperature and stream temperature that influence nutrient dynamics (Orwig et al. 2008, Knoepp et al. 2011). Soil moisture levels decline with the increase in insolation, causing declines in decomposition rates (Cobb et al. 2006). Changes in the levels of light reaching subcanopy plants can influence rates of carbon gain and subsequent resource allocation to chemical defense (Dudt and Shure 1994, Hunter and Forkner 1999). Similar effects may occur in aquatic communities that are dominated by macrophytes. Extensive defoliation of kelp forests by sea urchins influences the availability of light and the action of waves in some coastal marine communities (Estes et al. 2009).

Second, primary producers take up and utilize resources in a species-specific fashion (Tilman 1977, Hobbie 1992). When selective foraging by herbivores influences the relative abundance of autotroph species, it will also influence the overall uptake of nutrient resources by the primary producer community. If slow-growing, well-defended autotrophs are favored under high herbivore pressure, it can reduce the rate at which nutrients are taken up from the environment and recycled back to the environment (Pastor and Naiman 1992). Reduced rates of nutrient cycling may further reinforce dominance by slow-growing autotroph species (Hobbie 1992). Conversely, if species with recalcitrant residues are lost from communities as a result of herbivore activity, overall rates of nutrient uptake and nutrient cycling can increase (Knoepp et al. 2011).

Differential nutrient uptake is particularly important in aquatic phytoplankton communities. For example, in some temperate lakes, the relatively high palatability to zooplankton of the spring diatom bloom contributes to temporal replacement of diatoms by cyanobacteria. Cyanobacteria have a lower demand for silica and a higher demand for P than do most diatoms. Zooplankton grazing therefore mediates changes in phytoplankton community structure; phytoplankton nutrient stoichiometry; and the cycling of silica, N, and P in aquatic environments (Wetzel 1983, Sterner and Elser 2002). Similar effects of herbivore grazing on algal nutrient dynamics may occur in streams and rivers. In the Eel River in California, midge larvae construct retreats on the filamentous green alga, *Cladophora glomerata*, and consume the epiphytic algae that colonize the *Cladophora*. By changing the relative abundance of diatoms and cyanobacteria, midge larvae may change rates of N fixation and stream N cycling at much larger spatial scales (Furey et al. 2012).

8.3.3 The Relative Importance of Fast- and Slow-Cycle Effects of Herbivores on Nutrient Dynamics in Ecosystems

As I mentioned above, the relative importance in any given ecosystem of the multiple mechanisms by which herbivores influence nutrient dynamics (table 8.1) will determine the balance between fast- and slow-cycle processes and the overall impact of herbivores on nutrient dynamics. Given that the strengths of these herbivore-mediated pathways vary with other biotic and abiotic processes, we should expect substantial contingency in the effects of herbivores on nutrient dynamics (McSherry and Ritchie 2013). Can we make any broad generalizations about the relative importance of fast-cycle and slow-cycle effects of herbivores on nutrient dynamics?

I expect that slow-cycle effects will dominate in many terrestrial ecosystems, particularly in those such as forests in which most of the NPP remains unconsumed. Although slow cycle effects can either increase or decrease the rate of nutrient cycling (depending upon how herbivores influence residue quality, above), herbivore-induced reductions in the rate of nutrient cycling may dominate when rates of herbivory are modest. The impacts of natural selection on tissue palatability (Uriarte 2000, Cornelissen et al. 2004), impacts of selective foraging on plant community structure (Pastor and Naiman 1992), and afterlife effects of litter from damaged plants (Choudhury 1988, Findlay et al. 1996) likely combine to decrease rates of nutrient cycling in many systems. Clear counterexamples emerge when high densities of herbivore species result in a high proportion of plant material being consumed and converted into labile forms (McNaughton et al. 1988). Under such circumstances, fast-cycle effects resulting from feces deposition (Hunter et al. 2003b), sloppy feeding (Risley and Crossley 1988), and premature leaf abscission (Fonte and Schowalter 2004, 2005) can combine to accelerate rates of nutrient cycling (Reynolds and Hunter 2001, Frost and Hunter 2007).

In lake and marine systems, where phytoplankton dominate productivity and most autotrophs lack extensive structural tissues, herbivores should mediate primarily fast cycle effects and so increase the rate of nutrient cycling. The upper waters may be particularly strongly driven by fast-cycle effects in which zooplankton feces, N excretion, and sloppy feeding provide heterotrophic microbes with substantial sources of DOM (Wotton and Malmqvist 2001, Saba et al. 2011). Of course, marine snow and phytodetritus link pelagic and benthic phytochemical landscapes, and selective foraging by zooplankton may influence the relative abundance of algal groups and therefore the chemistry of dead cells that sink to deeper waters. While most dead algal cells are degraded above or at the surface of sediments (chapter 6), at least some enter long-term storage (Efting et al. 2011). Nonetheless, the typical effect of herbivores in aquatic ecosystems is probably to speed up the rate of nutrient cycling (Vanni 2002) by increasing microbial access to labile components of the phytochemical landscape.

A Case Study with Oaks. Assessing the relative impacts of herbivores on nutrient dynamics via fast- and slow-cycle processes requires long-term studies of multiple mechanisms simultaneously. We have been studying the effects of oak canopy herbivores on the phytochemical landscape and nutrient dynamics for over 25 years, comparing the strength of fast- and slow-cycle processes. The ecology of oak forests is important; there are about 600 oak species worldwide that occur from the tropics to northern latitudes and from the Americas through Europe to Asia. Oaks are foundation species in many forest communities in the northern hemisphere and support diverse communities of consumers. To date, our results suggest that the net effects of insect herbivores on oak chemistry will accelerate rates of nutrient cycling (figure 8.9).

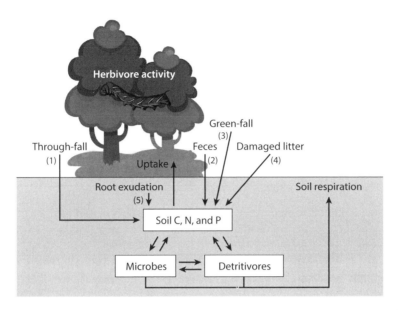

FIGURE 8.9. Pathways by which oak canopy herbivores influence nutrient dynamics. (1) Through-fall increases soil P by up to 16-fold, Collembola biomass up to 1.5-fold, and nematode biomass up to 7.5-fold; (2) feces increase soil N pools over 2-fold, Collembola biomass up to 3-fold, and nematode biomass up to 14-fold; (3) green-fall increases soil N by 1.2-fold and soil respiration by 1.27-fold; (4) herbivore-induced changes in litter quality have negligible effects on soil processes; (5) C exudates from roots are maintained at constant levels despite herbivore-induced declines in C allocation to roots. During outbreaks, fast-cycle mechanisms combine to increase soil N by 7-fold and N export in streams by 2-fold in oak-dominated watersheds, while trees recover about 16% of the N initially lost in herbivore feces. Modified from Hunter et al. (2012).

Diverse caterpillar species live in the canopies of oak trees. Sawflies in the genus *Periclista*, and Lepidoptera such as the gypsy moth, the winter moth, and the green oak leaf roller moth can reach outbreak densities, consuming a large proportion of oak leaf area. Because they short-circuit the typical pathway of nutrient resorption prior to foliar senescence and abscission (figure 8.2), caterpillar feces represent significant fluxes of labile C and N from the canopy to the soil. We have measured 5-fold increases in through-fall N concentration and 7-fold increases in soil N availability following defoliation events. When defoliation levels in the canopy reach 40 percent of leaf area removed, export of N doubles in streams draining forested watersheds (Reynolds et al. 2000).

Herbivore feces, modified through-fall, and sloppy feeding all contribute to the effects of oak herbivores on nutrient dynamics (Reynolds and Hunter 2001, Frost and Hunter 2004). For example, leaf fragments falling at rates typical of moderate herbivore densities increase rates of soil respiration by over 25 percent and soil nitrate availability by 20 percent. Through-fall additions typical of watersheds under attack by herbivores can increase soil P levels by a remarkable 16-fold. Even at endemic herbivore densities, feces from oak forest canopies can have an important effect on nutrient dynamics. At one of our field sites, variation in the input of herbivore feces explained 62 percent of the variation in soil nitrate availability (Hunter et al. 2003b).

Because a majority of decomposers are limited by the availability of labile C (chapter 6), herbivore feces can stimulate rapid immobilization of nutrients by bacteria and fungi (Christenson et al. 2002), and we have observed increases in microbial biomass following feces deposition (Frost and Hunter 2004). In some cases, immobilized N can enter the large pool of N stored in soils, and our labeling studies demonstrate that a substantial proportion of N in herbivore feces becomes bound in mineral soil (Frost and Hunter 2007). However, members of the brown food web can graze on microbial decomposers (see figure 6.2B), remobilizing mineral nutrients from herbivore feces. For example, substantial increases in the densities of fungal-feeding Collembola, and both fungal- and bacterial-feeding nematodes occur following the deposition of herbivore feces and through-fall in soils (Reynolds et al. 2003). Nutrients remobilized by trophic interactions within the brown food web are then available for uptake by plants, soil storage, or leaching losses.

The fast-cycle effects of oak herbivores mediated by inputs of feces, through-fall, and sloppy feeding should all act to accelerate the rate of nutrient cycling. To test this hypothesis, we applied ^{15}N-enriched insect feces to soil in oak meso-cosms (Frost and Hunter 2007). We applied the labeled feces during spring, when a majority of the defoliation to oaks takes place. Our expectation was that we might detect some of the labeled N in trees by the following growing season. However, N cycling and uptake by trees was much faster than we had anticipated. We observed some mineralization and export of N from feces within a week of deposition, and some uptake in the foliage of trees within a month. We then recovered some labeled N from the body tissue of insect herbivores feeding on those trees during late summer. Remarkably, N had cycled from trees, through herbivore feces, into soil, back into trees, and into a second generation of insect herbivores, all within a single growing season (Frost and Hunter 2007) (figure 8.10). This rapid rate of nutrient cycling provides strong support for the acceleration hypothesis (Ritchie et al. 1998). The key point here is that herbivore activity transforms the recalcitrant phytochemical landscape of the oak forest into chemically labile feces, so short-circuiting slow-cycle processes and increasing rates of nutrient cycling.

Oak herbivores can also change nutrient dynamics in soils by influencing rhizo-deposition (Frost and Hunter 2008a), a process by which plants feed C to microbial

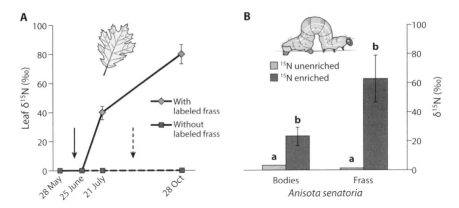

FIGURE 8.10. (A) Insect feces (frass) labeled with ^{15}N was applied to oak mesocosms in mid June (solid arrow). N from the labeled feces appeared rapidly in the foliage of oak trees. Caterpillars that were fed on those leaves in late August (dashed arrow) also incorporated labeled N from feces into their own tissues (B), illustrating rapid N cycling from spring-feeding herbivores to fall feeding herbivores within the same growing season. Based on Frost and Hunter (2007).

decomposers associated with their roots. Again, because many soil decomposers are C limited (Bardgett and Wardle 2010), rhizodeposition can stimulate rates of decomposition and subsequent nutrient release from organic matter. In short, plants may be able to compensate in part for the loss of N and other nutrients to herbivores by increasing rhizodeposition and nutrient uptake (Hamilton and Frank 2001, Hamilton et al. 2008). In oaks, the effects of herbivores on rhizodeposition appear to be a bit more complicated. Defoliation aboveground causes a dramatic reduction in the total amount of new C that is allocated to roots (Frost and Hunter 2008a). However, rates of C transfer to soil remain unchanged. This argues that rhizodeposition is maintained, despite lower concentrations of C available in roots. In other words, the proportion of root C allocated for rhizodeposition increases in order to maintain C flow to decomposers in the soil. The newly acquired N from soil is then allocated preferentially to storage, perhaps to minimize the consequences of any further herbivore attack (Tao and Hunter 2012a).

Are there any slow-cycle effects of herbivores on oak trees that could act to oppose the fast-cycle effects of feces, through-fall, sloppy feeding, and rhizodeposition on nutrient dynamics? In oak forests, rates of N and C cycling in litter and soil decline with increasing tannin concentrations in litter (Madritch and Hunter 2002, 2004). Because foliar tannin concentrations increase in oaks following defoliation by herbivores (Rossiter et al. 1988, Hunter and Schultz 1993), herbivores have the potential to influence nutrient cycling in soils through afterlife effects of the quality of oak litter on the phytochemical landscape.

However, although afterlife effects induced by herbivores appear to be strong in some systems (Findlay et al. 1996, Chapman et al. 2006), they are weak in oak forests. Experiments with red oak, *Quercus rubra*, suggest that, by the time leaves senesce, damaged litter is actually lower in tannin than is undamaged litter, and effects of insect damage on litter decomposition rates are minor and transient (Frost and Hunter 2008b). In other words, there is no simple link in red oak between the defensive chemistry induced by herbivores and the subsequent chemistry of litter. Because herbivore feces can serve as a rapid source of N for defoliated oaks (Frost and Hunter 2007), and N recovery from soil can reduce the magnitude of tannin induction (Hunter and Schultz 1995), we investigated whether the presence or absence of feces might mediate any afterlife effects of damaged oak litter on nutrient dynamics. Surprisingly, feces deposition resulted in a slight increase in the lignin:N ratio of litter. As with defoliation, there were weak afterlife effects of feces deposition on litter quality and nutrient dynamics that were short-lived and minor in comparison to fast-cycle effects (Frost and Hunter 2008b).

The weak slow-cycle effects of oak herbivores on nutrient dynamics are not limited to red oak trees in temperate forests. We have found similar results with three sub-tropical species of scrub oak, *Q. geminata*, *Q. myrtifolia*, and *Q. chapmanii*, in Florida. As with red oaks, litter from trees receiving herbivore damage is lower in tannin than is litter from undamaged trees. However, herbivore-induced increases in the lignin:N ratios of leaves do persist into the litter. Importantly, changes in litter chemistry caused by herbivores are minor in comparison to chemical variation from year to year (Hall et al. 2005a). This highlights the importance of multiyear studies to gain perspective on the relative strength of different ecological processes in driving foliar and litter chemistry. As with red oak trees, the weak afterlife effects of herbivore damage on litter chemistry in scrub oaks do not translate into meaningful effects on litter decomposition. Rather, the location in which litter decomposes has a greater impact on decomposition and nutrient dynamics than does slow-cycle effects of herbivores (Hall et al. 2006).

A weakness of our studies of herbivores in oak forests is that we have not assessed any impacts of natural selection or selective foraging by herbivores on leaf and litter quality. We know from other systems that natural selection on the palatability of plants for herbivores may translate into variation in litter quality and nutrient dynamics (Uriarte 2000, Cornelissen et al. 2004). There is certainly substantial genetic variation in foliar chemistry within and among oak species (Forkner and Hunter 2000, Klaper et al. 2001), which may facilitate evolution of chemically based resistance against herbivores (Feeny 1970, Forkner et al. 2004). However, we are as yet unable to assess the relative magnitude of historical evolutionary processes and contemporary ecological processes, mediated by herbivores, in determining the chemical recalcitrance of oak-derived inputs to the detrital system.

The effects of selective foraging by herbivores on foliar and litter chemistry are well established in some systems (Pastor and Naiman 1992, Mason et al. 2010). Tree species in our study sites vary substantially in their rates of decomposition (Ball et al. 2008), and changes in their relative abundance could cause dramatic shifts in nutrient dynamics in both soils and streams (Kominoski et al. 2007, Ball et al. 2009b). In some temperate forests in North America, deer browsing causes significant shifts in plant community structure (Horsley et al. 2003), although subsequent effects on nutrient dynamics are not well established. At present, we are unable to assess the importance of selective foraging by herbivores on the chemistry of litter in our study sites. However, our experiments demonstrate that insect herbivore activity in oak forests significantly accelerates nutrient cycling through a broad combination of fast-cycle mechanisms that change the phytochemical landscape (Hunter et al. 2012) (figure 8.9).

8.4 EFFECTS OF PREDATORS ON NUTRIENT DYNAMICS

Let's be clear. Every fast-cycle and slow-cycle mechanism that I have described above, by which herbivores influence nutrient dynamics in ecosystems, can be mediated by the effects of predators and parasites on herbivore distribution, abundance, and behavior. The distribution of herbivore feces, urine, cadavers, and sloppy feeding (fast-cycle effects), as well as the differential consumption of autotroph species by herbivores (slow-cycle effects) are in part a function of the landscape of fear upon which herbivores forage. Wherever and whenever natural enemies influence significantly herbivore foraging or herbivore density, they will transmit those effects to the phytochemical landscape, and so on to the ecosystem processes that that those landscapes dictate (see figure 2.4). As a consequence, the widespread loss of top predators and the trophic downgrading of planet Earth is having significant consequences for ecosystem nutrient dynamics (Estes et al. 2011, Ripple et al. 2014b).

Predators exert both direct and indirect effects on nutrient cycling (Werner and Peacor 2003, Majdi et al. 2014). Indirect effects, mediated through changes in the distribution, abundance, and behavior of herbivores, may be stronger than direct effects. For example, herbivores experiencing predation risk may favor autotroph diets that contain labile carbon sources, thereby mitigating in part the energy costs associated with predation stress. As a consequence, predators change the relative abundance of labile and recalcitrant residues that enter the detrital pool (Hawlena and Schmitz 2010b, Leroux et al. 2012). Additionally, because all herbivores are selective feeders, consuming primary producers out of proportion to their representation in the community, predator-mediated changes in the density of generalist herbivores alter autotroph community structure and the phytochemical landscape (chapter 5). However, direct effects of predators on nutrient dynamics may operate also, as predators add their own feces, urine, and cadavers into ecosystems. Because predator production is linked fundamentally to the chemical quality of primary production (chapter 4), predator-derived inputs contribute to closing the loop between trophic interactions and nutrient dynamics in ecosystems (Allgeier et al. 2013).

8.4.1 Indirect Effects of Predators on Nutrient Dynamics Mediated by Herbivores

Foraging Behavior and Autotroph Community Structure. By their impacts on herbivores, predators generate pervasive trait-mediated and density-mediated effects on autotroph community structure and subsequent nutrient cycling (Schmitz et al. 2010). In one of the best-studied systems to date, Oswald Schmitz and colleagues have illustrated how spider-induced changes in the foraging behavior of herbivores in old fields influence nutrient cycling. As described in chapter 5 (section 5.2.2), risk imposed by a sit-and-wait spider predator, *Pisaurina mira*, causes a generalist grasshopper, *Melanoplus femurrubrum*, to alter its feeding preference from a grass, *Poa pratensis*, to a dominant herb, *Solidago rugosa*. *P. mira* therefore maintains evenness in plant community structure by preventing *S. rugosa* from outcompeting other members of the plant community. In contrast, the hunting spider *Phidippus rimator* directly reduces the density of its grasshopper prey. As a consequence, *P. rimator* favors overall decreases in plant species diversity, and increases by 168 percent the relative abundance of *S. rugosa*.

Critically, the increase in *S. rugosa* performance facilitated by *P. rimator* results in higher annual net primary production (ANPP) and higher rates of N mineralization (Schmitz 2008a) (figure 8.11). N mineralization rates increase because the changes in plant community composition mediated by *P. rimator* decrease the average C:N ratio of the phytochemical landscape; litter C:N ratio is a fundamental determinant of decomposition rate (chapter 6). In contrast, the sit-and-wait spider, *P. mira*, increases the diversity of the plant community, decreases ANPP, and reduces the rate of N mineralization (figure 8.11). The reduced rate of N mineralization that results from risk of *P. mira* attack likely emerges from increased demand by grasshoppers for labile carbon in the diet, to meet the energetic demands of a higher metabolic rate under predation stress. By preferentially consuming diets with labile carbon stores, predation-stressed grasshoppers

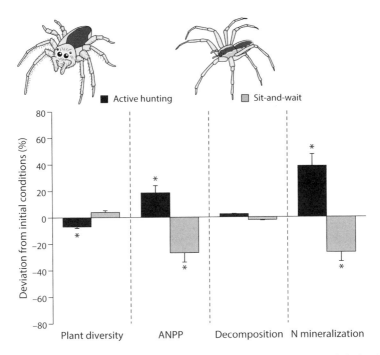

FIGURE 8.11. Active hunting spiders (black bars) and sit-and-wait spiders (light bars) have important but opposite effects on ecosystem processes. Effects are mediated by changes in the foraging behavior of their grasshopper prey that translate to changes on the phytochemical landscape. ANPP is aboveground net primary production. Modified from Schmitz (2008a).

should leave behind plants with an average 10 percent higher C:N ratio to enter the detrital pool (Hawlena and Schmitz 2010a). The key point is that both species of spider are capable of changing the phytochemical landscape of old fields, with significant consequences for ecosystem nutrient dynamics. In addition to changes in N cycling, fear of predation results in net increases in plant carbon storage and allocation to belowground tissues (Strickland et al. 2013).

The work by Schmitz and colleagues provides an excellent example of how the phytochemical landscape both generates, and is generated by, trophic interactions, thereby linking both autotroph chemistry and trophic interactions to nutrient dynamics (figure 8.1). For example, in the absence of inherent chemical differences in foliage between *S. rugosa* and other old-field plant species, spider-mediated changes in grasshopper foraging behavior could have no impact on the average recalcitrance of leaf litter. In this system, predators cannot influence nutrient cycling without preexisting standing variation in foliar chemistry on the phytochemical landscape. At the same time, variation in risk imposed by spiders helps to determine the relative abundance of recalcitrant organic tissues on the soil surface, and the rate at which soil N is cycled. However, there is no primary driver (nutrients, plants, herbivores, or predators) in this integrated system; the effects of any one are both causes and consequences of the effects of all others. Variation in the phytochemical landscape is simply the nexus through which these links of mutual causality must flow, and arguments about the dominance of top-down and bottom-up forces dissolve. The landscape of fear, the phytochemical landscape, and the cycling of nutrients cannot be understood as independent units.

So, by a combination of trait-mediated (e.g., foraging behavior) and density-mediated indirect effects, predators can change the structure of autotroph communities, the phytochemical landscape, and the cycling of nutrients (Schmitz 2008a). The most well-known impacts of predators on autotroph community structure are those generated by trophic cascades (Carpenter et al. 1985), which have now been documented in all of the world's major biomes (Estes et al. 2011). The chemical diversity and identity of autotroph residues that remain following trophic cascades then determine the rate of nutrient cycling in many ecosystems (chapter 6). In section 5.2.2, I described how trophic cascades influence autotroph biomass and community composition, and I will not revisit those processes here. Rather, I will describe a few examples in which the effects of predators have been followed all the way through to explicit changes in nutrient dynamics.

Previously I noted that, when fast-cycle effects of herbivores on nutrient dynamics dominate over slow-cycle effects, we should expect herbivores to increase the rate of nutrient cycling in ecosystems. By the same logic, increasing predation pressure in those same systems should serve to slow down the rate at which nutrients cycle. This appears to be the case following the reintroduction of wolves into Yellowstone National Park, wherein grazing ungulates tend to increase the rate of N mineralization in soils. Following the reintroduction of wolves, N mineralization rates have declined, apparently caused by changes in the spatial patterns of grazing by ungulates on the new landscape of fear (Frank 2008).

Likewise, density-mediated trophic cascades generated by insectivorous birds and mammals can influence P cycling in the soil of tropical forests. Predators reduce the density of spiders, thereby increasing the density of microbivores, and facilitating the turnover of P that is otherwise immobilized in microbial biomass (Dunham 2008). This study supports the growing recognition that generalist predators that are nourished in part by green food webs can influence rates of decomposition and nutrient cycling by feeding (also in part) on members of detrital food webs (Hunter et al. 2003a, Hines and Gessner 2012) (figure 6.2). Because the density and diversity of detritivores influences significantly the rate of plant decomposition (Srivastava et al. 2009), trophic cascades that impact detritivores will influence nutrient dynamics in ecosystems (Gessner et al. 2010).

The introduction of arctic foxes onto islands in the Aleutian archipelago provides an interesting example of predator-mediated changes in nutrient dynamics, although not by a traditional trophic cascade. Rather, foxes reduce the transfer of nutrients across ecosystem boundaries, with subsequent effects on plant community structure. Specifically, on islands upon which foxes have been introduced, the foxes prey on seabirds and severely reduce their abundance. Consequently, inputs of seabird guano, derived from marine productivity, are negligible on islands with foxes in comparison to those on fox-free islands. The overall result is a reduction in soil nutrient availability and a transition from grassland to tundra habitat on islands inhabited by foxes (Croll et al. 2005). I consider the role of ecosystem subsidies in mediating links between trophic interactions and nutrient dynamics in more detail in chapter 9.

The importance of trophic cascades in aquatic ecosystems is well established (Shurin et al. 2006), and I will not review that literature again here. I would simply point out that the indirect effects of predators on nutrient cycling are pervasive in diverse aquatic ecosystems, from small to large spatial scales. For example, in freshwater ecosystems ranging from bromeliads, through ponds, streams, and lakes, predators change the rate of C cycling and CO_2 efflux by cascading effects on primary producer biomass. In larger lakes, the presence of piscivorous fish reduces densities of planktivorous fish, increases densities of zooplankton, and decreases densities of phytoplankton in a classic trophic cascade. The result is a low rate of C fixation such that lakes are net sources rather than net sinks of CO_2 (Schindler et al. 1997). In comparison, fish and insect predators in bromeliads, ponds, and streams generally reduce zooplankton densities, facilitate higher densities of phytoplankton, and generate aquatic ecosystems that are net sinks for CO_2 (Atwood et al. 2013). Because of the inherent coupling of C cycling with that of other nutrients, predator-induced changes in C fixation will also induce changes in mineral nutrient uptake and dynamics (Sterner and Elser 2002).

As in terrestrial systems, aquatic predators can change the relative abundance of herbivores and primary producers, with consequences for the phytochemical landscape (chapter 5) and subsequent nutrient dynamics (chapter 6). Size-selective foraging by fish predators on larger zooplankton can shift the zooplankton assemblage to individuals of smaller size. In turn, smaller zooplankton species feed more efficiently on smaller phytoplankton (e.g., figure 8.7), and increase the rate at which P is recycled (Bartell 1981). Indeed, the size structure of phytoplankton communities is key to understanding the biogeochemistry of aquatic ecosystems (Marañón et al. 2013) (section 7.2.1), and trophic cascades influence both the size structure and palatability of phytoplankton assemblages (Weis and Post 2013).

Cadaver Distribution. In section 8.3.1, I described how herbivore cadavers represent labile sources of C and nutrients on landscapes that accelerate nutrient cycling through fast-cycle effects. Importantly, predators influence the distribution and abundance of herbivore cadavers on the landscape, with consequences for spatial variation in nutrient flux, microbial activity, and subsequent primary production (Carter et al. 2008, Barton et al. 2013b). For example, in a semiarid shrub-steppe ecosystem, carrion decomposition accounts for less than 1 percent of the ecosystem's nutrient budget, but generates important hot spots of biogeochemical activity (Parmenter and MacMahon 2009). Within these hot spots, cadavers release their nutrients in a predictable sequence, starting with K and Na, and ending with Ca (figure 8.12A); they elevate substantially the availability of limiting nutrients in soils below cadavers. In at least one system, the fear of predation can actually change the nutrient stoichiometry of herbivore cadavers. Carcasses of grasshoppers that have been stressed by the presence of spiders exhibit higher C:N ratios than do cadavers of unstressed grasshoppers. While the cadavers themselves decompose at equivalent rates, those from stressed grasshoppers inhibit the decomposition of associated plant litter, thereby linking the fear of predation with decomposition and nutrient dynamics in soil (Hawlena et al. 2012).

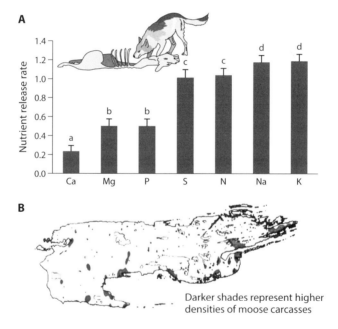

FIGURE 8.12. Carrion left behind by predators releases nutrients (A) in a predictable sequence, with rapid cycling of Na and K and slower cycling of Ca (the regression slope illustrates the rate of nutrient loss over time). On Isle Royale, the spatial distribution of moose carrion (B) is influenced strongly by the hunting patterns of wolves. Modified from Parmenter and MacMahon (2009) and Bump et al. (2009a).

In one of the best systems studied to date, wolves influence both spatial and temporal heterogeneity in soil nutrients by their effects on the distribution of ungulate carcasses (Bump et al. 2009a). Using a 50-year record of more than 3,600 moose cadavers in Isle Royale National Park, Bump and colleagues report that wolves generate strong spatial clustering of soil nutrients, soil microbes, and plant nutrient content. The spatial clustering of cadavers arises from wolf hunting behavior, and may be self-reinforcing (figure 8.12B). Because wolf-generated carcasses also generate zones of high-nutrient plant foliage, those areas may attract subsequent mammal grazers and therefore additional predation pressure. In short, the wolves of Isle Royale are exemplars of the feedback loop from predators, through soil nutrient dynamics, and back to predators, with high-nutrient plant foliage providing the nexus through which the feedback takes place (Bump et al. 2009a). The spatial correlation that results among nutrient dynamics, the phytochemical landscape, and the strength of trophic interactions is the hallmark of the phytochemical landscape concept (figure 2.4, chapter 9).

8.4.2 Direct Effects of Predators on Nutrient Dynamics

Because of energy losses during trophic transfer (Lindeman 1942), and the pyramids of energy that result, we might expect that predators should impose their greatest impact on nutrient dynamics by changing the behavior of their more productive herbivore prey. However, the standing biomass of predators can exceed that of herbivores if predation rates are sufficiently high. Whenever the turnover of prey biomass into predator biomass is substantial, we might expect to observe significant direct effects of predators on nutrient dynamics (Vanni 2002). As I stressed in chapter 4, the phytochemical landscape plays a fundamental role in mediating the transfer of energy and nutrients from plants, through herbivores, to natural enemies. In turn, natural enemies recycle nutrients (directly and indirectly) back into the environment (figure 8.1).

Feces and Urine. As with herbivores, the products of predator egestion and excretion facilitate rapid recycling of nutrients in ecosystems. For example, the recycling of P from fish excretion can provide anywhere from 5 percent to 36 percent of the P demand of lake phytoplankton (Schindler et al. 1993). When piscivorous fish are abundant, and therefore zooplankton densities are high, zooplankton (i.e., herbivore) excretion is a dominant source of P for phytoplankton. However, when piscivorous fish are rare, and planktivorous fish limit the density of zooplankton, excretion by planktivorous fish can be a major source of P for phytoplankton. In other words, the size structure of fish populations helps to determine the direct effects of predators on nutrient dynamics in lakes.

Elegant experimental work has illustrated how the excretory products of predators stimulate nutrient availability and subsequent production in seagrass ecosystems. When artificial reefs are introduced into seagrass habitats, the reefs serve as areas of aggregation for herbivorous and predatory fish. Both contribute nutrients in their excreta to the seagrass ecosystem, stimulating seagrass production (Allgeier et al. 2013). Notably, the nutrients derived from predator excretion differ in elemental stoichiometry from those derived from herbivore excretion, with predators adding back substantially more P per unit N (figure 8.13). P availability may be particularly important for stimulating seagrass production.

Critically for the topic of this volume, nutrient inputs from predators on artificial reefs can radically alter algal community structure and the phytochemical landscape. For example, predator excretion on artificial reefs in the Florida Bay favors rapidly growing microalgae and periphyton, while facilitating replacement of the slower growing *Thalassia testudinum* with the nutrient demanding seagrass *Halodule wrightii* (Dewsbury and Fourqurean 2010). Seabird communities also contribute substantial nutrient subsidies through their feces into seagrass ecosystems. Seabird inputs can change the relative abundance of seagrass species and therefore the phytochemical landscape available to marine herbivores (Powell et al. 1991). This series of examples from seagrass habitats illustrates nicely the feedback loop from predators, to environmental nutrient availability, to variation in phytochemistry, and back to trophic interactions (figure 8.1). Similar effects likely occur on coral reefs, where the structure and diversity of fish communities determine how N and P are stored and recycled from fish biomass to influence primary production and coral health (Allgeier et al. 2014).

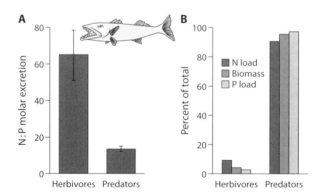

FIGURE 8.13. On artificial reefs in seagrass ecosystems, the excretory products of predators contribute relatively more phosphorus (A) and a greater proportion of the overall nutrients returned by excretion (B) to primary production than do the excretory products of herbivores. Modified from Allgeier et al. (2013).

Predation also serves to change the relative availability of limiting nutrients in freshwater ecosystems. For example, the excretory products of freshwater mussels (omnivores) can alter the relative abundance of N and P in streams, with consequences for subsequent primary production. Mussel excretion has a high N:P ratio that alleviates strict N limitation, and imposes N and P colimitation on primary producers. As a consequence, stream sections without mussels are N-limited and characterized by N-fixing cyanobacteria whereas reaches with mussels are dominated by diatoms (Atkinson et al. 2013). Again, by changing the relative availability of limiting nutrients, omnivores in streams influence autotroph community structure and the subsequent phytochemical landscape.

By recycling nutrients rapidly in their feces and urine, predators may change the rate at which nutrients cross ecosystem boundaries. For example, damselfly larvae in bromeliad tanks are important predators of midge larvae that feed on detritus in the bromeliads. The midge larvae act as sinks for N as they pupate and leave the bromeliad tanks. By consuming midge larvae within the tanks, damselfly larvae recycle a substantial portion of midge N through their feces, which is subsequently available for uptake by the bromeliads themselves (Ngai and Srivastava 2006). Of course, many natural enemies actually stimulate cross-ecosystem movement by their prey, so generating subsidies in one ecosystem at the expense of losses in another. I discuss cross-ecosystem subsidies further in section 8.5.3 and in chapter 9. Although fear-induced movement of prey is one obvious mechanism generating translocation of nutrients, there are some fascinating additional pathways. For example, nematomorph parasites of crickets induce their hosts to enter streams where they attract fish predators, release benthic invertebrates from predation pressure, and so increase rates of litter decomposition (Sato et al. 2012).

Back in the terrestrial environment, ant nests in forest canopies release a consistent stream of excreta that influences nutrient dynamics on the forest floor. Arboreal ants that feed on insect herbivores (or their sugar-rich excretory products) increase litter decomposition on the forest floor because their excreta is 7-fold higher in P, 23-fold higher in K, and 3-fold higher in N than is leaf litter (Clay et al. 2013). Given the typically high biomass of ants in tropical forest ecosystems, their impacts on nutrient dynamics may be substantial.

Predator Cadavers. Just like herbivores, predators die and decompose, and their carcasses can represent substantial inputs into the detrital pool. Although we might hypothesize that differences in the stoichiometry (Fagan and Denno 2004) or toxin concentration (Rothschild 1964) of herbivores and predators should generate differences in the rate at which their cadavers are recycled, I can find limited evidence in the literature to support that view. In one fascinating study, jellyfish species were shown to vary in their cadaver nutrient ratios, with consequences both for nutrient release during decomposition and for the structure of associated microbial populations (Tinta et al. 2012). We need many more studies of this kind before we can understand the factors that mediate the differential release of nutrients from the cadavers of most consumers.

Whatever their relative chemical quality, predator cadavers represent important energy and nutrient inputs in certain ecosystems. In some streams, salmon carcasses translocate significant subsidies of nutrients and energy from the marine environment to freshwater habitats. Translocation of carcasses (and their nutrients) from sites of production to sites of decomposition is a common feature of aquatic ecosystems (Beasley et al. 2012). Moreover, as they decompose, salmon carcasses accentuate differences in the rates of decomposition among leaf litter species in those streams. Salmon carcasses generally decrease rates of stream litter breakdown by attracting detritivores away from litter and onto the labile fish cadavers (Bretherton et al. 2011). In other words, differences in the chemical quality of litter and carcasses as food for detritivores determines in part the rates of decomposition and nutrient release from both substrates.

Overall, there is overwhelming evidence that predators have cascading effects on the dynamics of nutrients in ecosystems. While those effects may vary with predator size or age (Werner and Gilliam 1984, Rudolf and Rasmussen 2013) and with the relative importance of the direct and indirect pathways described above (Schmitz 2008a), there is little doubt that the loss of predators from ecosystems will influence how nutrients cycle through those ecosystems (Estes et al. 2011). Because a majority of the mechanisms by which predators influence ecosystem processes are mediated by the chemical quality of living or dead autotroph tissues and cells (above), changes on the phytochemical landscape should provide substantial predictive power in understanding the effects of predators on nutrient dynamics.

8.5 EFFECTS OF NUTRIENT DYNAMICS ON TROPHIC INTERACTIONS

In the previous section, I described how the effects of trophic interactions can cascade down from predators, through herbivores, and through the phytochemical landscape, to influence the availability of nutrients at the ecosystem level. However, variation in nutrient availability also cascades up (Hunter and Price 1992b) to influence trophic interactions, mediated in part by the effects of nutrients on autotroph chemistry (figure 8.1). In chapter 7, I described the wide diversity of terrestrial and aquatic ecosystems in which the availability of nutrients such as N, P, Si, and Fe influence the chemistry of individual primary producers, the structure of autotroph communities, and the phytochemical landscape upon which trophic interactions take place. In chapter 4, I also described how the phytochemical landscape mediates interactions between herbivores and autotrophs, and between herbivores and their natural enemies. Integrating these two bodies of work together, it should be clear how the dynamics of nutrients in ecosystems mediate trophic interactions through the nexus of phytochemistry (figure 1.1 loop C, figure 8.1).

A key concept here is that high nutrient availability does not simply increase the *amount* of food available for herbivores, but also changes *the nutritional and defensive chemistry* of that food. In other words, it is not sufficient to relate environmental nutrient availability to the productivity of terrestrial and aquatic ecosystems; such relationships have been recognized for decades (Schlesinger and Bernhardt 2013). Critical here is the idea that trophic interactions depend more on the nutritional and defensive chemistry of what herbivores eat than on simple primary production (chapter 4). Effects of nutrient availability on phytochemistry and subsequent trophic interactions can arise from changes in the phenotype of individual autotrophs (Forkner and Hunter 2000) and from changes in autotroph community structure (de Sassi et al. 2012, Gough et al. 2012).

In addition to improving ecological theory, there are some very practical reasons for seeking to understand how nutrient availability influences trophic interactions. Increasing anthropogenic inputs of nutrients into natural ecosystems make it increasingly urgent that we understand the effects of nutrient availability on species interactions. Over the past 150 years, human activity has increased atmospheric N deposition by an order of magnitude across large areas of the Northern Hemisphere (Galloway et al. 2004), with significant impacts in both terrestrial and aquatic ecosystems (Vitousek et al. 1997). One simple prediction is that atmospheric N deposition should increase autotroph N concentrations, and thereby benefit N-limited herbivores, increasing their abundance (Erelli et al. 1998). As expected, N deposition often leads to increases in the N concentrations of primary producers (Liu et al. 2013a), with subsequent increases in the abundances, per capita rates of increase, and carrying capacities of herbivores (Hartley et al. 2003, Zehnder and Hunter 2008). However, atmospheric N deposition may also change the quality or quantity of PSMs or alter amino acid profiles, both of which can reduce herbivore performance (Throop and Lerdau 2004). Additionally, N deposition can change autotroph community composition (Stevens et al. 2004, Isbell et al. 2013a) and thereby the phytochemical landscape upon which trophic interactions take place (chapter 4). Moreover, N deposition influences the abundance and activity of natural enemies (Strengbom et al. 2005). In short, we need detailed mechanistic studies to help understand how and when anthropogenic nutrient deposition will alter trophic interactions (Tylianakis et al. 2008).

8.5.1 Terrestrial Systems

We have studied the effects of nutrient availability on trophic interactions on oak trees over a number of years (Hunter and Schultz 1995, Forkner and Hunter 2000, Zehnder et al. 2010) and have found the following general patterns. First, experimental fertilization of oak trees increases foliar nutrient concentrations and decreases foliar chemical defenses. Second, the higher quality foliage of fertilized trees increases the densities of both insect herbivores and invertebrate predators on oak trees in a bottom-up cascade (Hunter and Price 1992b) that is based on the chemistry, not quantity, of available foliage. Third, while bird predation can significantly reduce the densities of insects on oak trees (Marquis and Whelan 1994, Zehnder et al. 2010), we generally observe strong effects of bird predation only where soil nutrient availability is high (Forkner and Hunter 2000). In other words, nutrient availability in the soil influences the presence and magnitude of the top-down trophic cascade. As I described earlier, trophic interactions on oak trees have important effects that cascade down to influence the availability of mineral nutrients in soil (section 8.3.3). In combination, these bottom-up and top-down cascades generate a negative feedback loop from soil nutrients up to predators and back down to soil nutrients that is mediated by phytochemistry (figure 8.14). Negative feedback loops are a prerequisite for equilibrium dynamics (Royama 1992), and may contribute to the stability of the ecological processes at all points in figure 8.14.

Two notable results emerge from our work on oaks. First, when nutrient availability is low, densities of natural enemies may simply track those of their prey. Second, high nutrient availability increases the power of predators to suppress the populations of their prey. These results are consistent with the hypothesis that herbivore production reflects the nutrient stoichiometry of autotrophs (Sterner and Elser 2002), or the effects of plant defenses on access to nutrients (Coley et al. 1985), rather than levels of primary production per se. In turn, the suppression of herbivore populations by predators increases with herbivore production (Arab and Wimp 2013) and, sometimes, the quality of herbivore tissues (Malzahn et al. 2007, Sarfraz et al. 2009). These hypotheses are similar in some ways to predictions that emerge from the classic exploitation ecosystem hypothesis (Oksanen et al. 1981), but based on access by herbivores to high-quality phytochemistry rather than high rates of autotroph production.

How common is to observe donor control or weak prey suppression in low-nutrient environments, but increases in prey suppression in high-nutrient

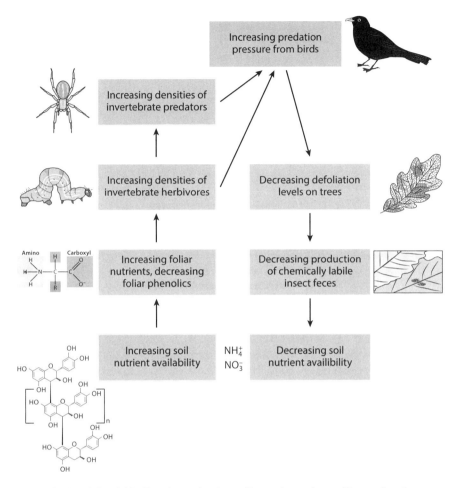

FIGURE 8.14. Left-hand side: Experimental and sampling work on oak trees illustrate how increases in soil nutrient availability cascade up to influence plant chemistry, herbivore density, invertebrate predator density, and the activity of birds on trees. Effects of bird predation also cascade down to influence defoliation levels, but only on those trees growing in high-nutrient soils. Right-hand side: Trophic interactions cascade back down to influence soil nutrient availability (see section 8.3.3). In combination, the two cascades generate a negative feedback loop, mediated by the phytochemical landscape, which may contribute to the stability of all processes in the two cascades.

environments? At least in terrestrial systems, the ability of predators to suppress herbivore populations has been shown repeatedly to increase with nutrient availability. For example, soil nutrient availability influences the density of galling insects on sea-oxeye daisy and the subsequent efficacy of parasitoids that attack the galling insects (Stiling and Rossi 1997). As on oaks, it is the quality of sea-oxeye daisy tissues, not their available biomass, which mediates the trophic cascade. Likewise, nutrient availability determines whether or not ladybeetles induce trophic cascades by consuming aphids on grasses—cascades occur only in plots where soil nutrient availability is high (Fraser and Grime 1998). Interestingly, stronger prey suppression by invertebrate predators on high nutrient plants may also result from elevated production of volatile organic signals when soil nutrient availability is high (Low et al. 2014) (chapter 5). Finally, as in our studies on oak, the ability of bird predators to suppress caterpillar populations on *Vaccinium* increases with plant fertilization treatments (Strengbom et al. 2005). All of these studies are consistent with the high-performance/high-mortality hypothesis (Singer et al. 2012), which predicts increases in the relative suppression of prey by predators when those prey feed on high-quality primary production on the phytochemical landscape.

Based on what we learned in chapters 4 and 5, it should be no surprise that some of the effects of nutrient availability on trophic interactions depend on how nutrients influence the expression of PSMs on the phytochemical landscape. For example, high nutrient availability can reduce or eliminate the induced chemical changes that occur in autotroph tissues following attack by herbivores. In turn, the mitigation of chemical induction influences subsequent trophic interactions. Specifically, fertilization of Alaska paper birch trees prevents the development of delayed inducible resistance (DIR), a suite of changes in foliar chemistry that includes decreases in foliar N content and increases in leaf phenolic concentrations (Bryant et al. 1993). In turn, DIR in birch trees reduces the quality of foliage for herbivores for up to three subsequent years (Haukioja and Niemela 1977). Similarly, fertilization of oak trees prevents the increases in foliar condensed tannin concentration and foliar astringency that generally follow defoliation by gypsy moth larvae. In this case, the effects of nutrient availability on subsequent trophic interactions are clear—distributions of gall-forming and phloem-feeding insects are skewed away from gypsy moth damage on unfertilized (= induced) trees while distributions are unaffected by gypsy moth damage on fertilized (= uninduced) trees (Hunter and Schultz 1995). In both of these cases, nutrient availability cascades up to influence trophic interactions by changing the phytochemical landscape.

8.5.2 Aquatic Systems

In aquatic ecosystems, the availability of nutrients (both naturally and anthropogenically derived) influences the chemistry of individual primary producers and the relative abundance of algal species that differ in their nutritional and toxin chemistry (chapter 7). Given that algal chemistry mediates many aquatic trophic interactions (chapter 4), we should be able to link nutrient dynamics in aquatic ecosystems with subsequent rates of herbivory and predator-prey interactions. For example, low P availability in freshwater ecosystems promotes high algal C:P ratios that can limit the performance of zooplankton herbivores. Over the short-term, phenotypic plasticity in the nutrient stoichiometry of phytoplankton cells generates variation in the C:P content and subsequent performance of zooplankton (Malzahn et al. 2007). Additionally, over the last 120 years, some *Daphnia* species have adapted evolutionarily to increased anthropogenic P inputs and eutrophication, including changes in the efficiency with which they utilize P from their phytoplankton diets (Frisch et al. 2014). The evidence is now strong that nutrient availability influences the phytochemical landscape, and therefore interactions between phytoplankton and zooplankton.

But are the effects of nutrient availability on phytoplankton chemistry and herbivore performance transmitted to the third trophic level? The answer appears to be sometimes yes and sometimes no. For example, in an experiment exploring the effects of P availability on the performance of phytoplankton, *Daphnia*, and fish, individual *Daphnia* exhibited higher C:P ratios when fed on low-P algae than when fed on high-P algae. However, the effect was not transferred across another trophic level to fish predators of the *Daphnia* (figure 8.15A). In other words, P availability influenced the chemistry of algal cells and their interactions with herbivores, but not interactions between fish and *Daphnia* (Boersma et al. 2009); in this example, the cascade up stopped at the herbivore level. However, nutrient availability, through its impact on phytoplankton stoichiometry, can influence natural enemy performance in some cases. For example, P limitation of the cryptophyte, *Rhodomonas salina*, increases the C:P ratio of their copepod herbivores. In turn, planktivorous fish that feed on the P-limited copepods grow more slowly and are in poorer physiological condition than are fish that feed on P-sufficient copepods (Malzahn et al. 2007). In other words, the availability of inorganic P in water influences the phytochemical landscape, with effects that extend upward to the third trophic level; note that such effects are in addition to any effects of nutrient limitation on prey biomass.

Similar effects of P limitation, transmitted from phytoplankton (*Scenedesmus acutus*) to *Daphnia magna*, reduce the performance of a bacterial pathogen, *Pasteuria ramose*, that infects *Daphnia* hosts (Frost et al. 2008). When C:P ratios are high in *Daphnia* hosts, rates of infection by the bacterium decline along with reductions in the within-host multiplication rate of the pathogen. In other words, the effects of P availability on host-pathogen dynamics depend at least in part on the individual chemistry of zooplankton hosts, which depend in turn on the nutrient stoichiometry of their phytoplankton diet (figure 8.15B, C). P is not the

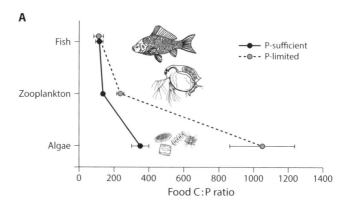

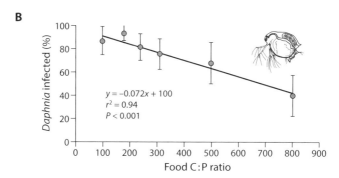

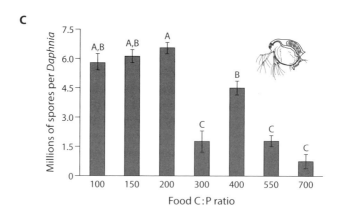

FIGURE 8.15. (A) The availability of inorganic P influences the C:P ratio of algae and zooplankton, but not the C:P ratio of planktivorous fish. However, algal C:P ratio influences (B) the proportion of *Daphnia* infected with a bacterial pathogen, *Pasteuria ramose*, and (C) the number of bacterial spores produced per infected *Daphnia*. Modified from Frost et al. (2008) and Boersma et al. (2009).

only nutrient that mediates the strength of trophic interactions in lakes. The availability of potassium (K) can limit the spread of a fungal pathogen that attacks *Daphnia* populations (Civitello et al. 2013). Using a powerful combination of lake surveys, laboratory bioassays, modeling, and lake mesocosm experiments, the authors illustrate that the spread of the pathogen, *Metschnikowia bicuspidata*, among *Daphnia dentifera* hosts increases with the availability of K in lake water. Experimental addition of K appears to increase the quality of *Daphnia* individuals as hosts for the fungus, thereby facilitating disease outbreaks.

Environmental nutrient availability also influences trophic interactions in streams, sometimes by unusual mechanisms. For example, leaf litter that falls into streams supports higher quality bacteria and algae on its surface (periphyton) when that litter is nutrient-rich. In turn, the availability of high-quality periphyton increases the rate of development and alters the physical phenotype of amphibian grazers (Stoler and Relyea 2013). What makes this example particularly interesting is that variation in phytochemistry generated in the terrestrial environment causes differential rates of nutrient release in the aquatic environment. The effects of these nutrients then cascade up to influence periphyton chemistry and the phenotypic plasticity of amphibian herbivores. Because amphibian phenotype also affects amphibian-predator interactions (Relyea 2002), the effects of nutrient availability may be transferred to the third trophic level.

In addition to influencing the stoichiometry of phytoplankton cells, environmental nutrient availability influences per capita toxin production by algae. The production of saxitoxin, domoic acid, microcystin and many other potent algal toxins on the phytochemical landscape varies with the local availability of N, P, Si, and other nutrients in water (section 7.2.2). In turn, per capita toxin production by phytoplankton influences the behavior, performance, and relative abundance of zooplankton herbivores (chapter 4). For example, in Swedish lakes, high concentrations of microcystin on the phytochemical landscape tend to favor cyclopoid copepods and rotifers over *Daphnia* and calanoid copepods, thereby changing the structure of lake food webs (Hansson et al. 2007). Additionally, some *Daphnia* are able to control the proportion of toxic and nontoxic phytoplankton cells that they consume, with the degree of selective foraging increasing with experience (Tillmanns et al. 2011).

Published effects of nutrient availability on the PSMs of macroalgae are much less dramatic than are those reported among toxin-producing phytoplankton. However, there have been far fewer experimental studies of the relationships between nutrient availability and macroalgal chemical defense than studies of phytoplankton toxins or terrestrial plant defenses. Nonetheless, at least some macroalgae alter their allocation to PSMs based on nutrient levels in water. For example, phlorotannin concentrations in brown algae generally decline in response to

N and P enrichment (Yates and Peckol 1993, Van Alstyne and Pelletreau 2000). While the role of phlorotannins in algal defense remains controversial (Kubanek et al. 2004, Amsler and Fairhead 2006), they appear to deter herbivore grazing in some cases (Jormalainen and Ramsay 2009). As a consequence, phlorotannins may sometimes serve as a phytochemical link between environmental nutrient availability and herbivore-macroalgal interactions.

In contrast to phlorotannins, terpenoid production by macroalgae appears largely insensitive to local variation in nutrient availability (Cronin and Hay 1996a). For example, in the red alga, *Portieria hornemannii*, monoterpene chemical defenses are unaffected by the availability of N and P in the water. Rather, monoterpene production appears to vary more with light than with nutrient availability (Puglisi and Paul 1997). Interestingly, the synthesis of defensive compounds by red algae may be more related to concentrations of dissolved inorganic C than to concentrations of N or P (Mata et al. 2012). In short, with the limited sample size available at this time, it appears that the largest effects of nutrient availability on the herbivores of macroalgae may be mediated by algal nutritional chemistry rather than algal chemical defense (Hemmi and Jormalainen 2002, Barile et al. 2004).

Perhaps the most well known effects of nutrient availability on trophic interactions in aquatic ecosystems are those resulting from harmful algal blooms. Eutrophication increases the frequency of both microalgal and macroalgal blooms, which usually act to reduce overall algal diversity and the diversity of consumers at higher trophic levels (Anderson et al. 2012, Lyons et al. 2014). By changing the relative abundance of algal species, nutrient-induced algal blooms reorganize the entire phytochemical landscape of aquatic ecosystems (chapter 7). Blooms can release high concentrations of toxic compounds into the water that have both direct and indirect effects on higher-level consumers (Hay and Kubanek 2002). Phycotoxins are often broadly active against metazoans, including zooplankton, other invertebrates, fish, mammals, and birds (Hackett et al. 2004, Lewitus et al. 2012). In other words, the effects of phycotoxins can both propagate up food chains through trophic channels and also act directly against predators as environmental toxins. The result can be a complete restructuring of aquatic food webs. As I described in chapter 4, dinoflagellate brevetoxins can kill millions of marine fish when released by cell lysis into the environment (Van Dolah 2000). Alternatively, the accumulation through trophic transfer of toxins such as nodularin (from cyanobacteria) and domoic acid (from diatoms) can kill higher-level predators including copepods, fish, birds, seals, and even whales (Sopanen et al. 2009, Sieg et al. 2011).

These studies illustrate a fundamental principle—that the ecosystem-level availability of nutrients, by altering the phytochemical landscape, determines in part the strength of subsequent trophic interactions. As I described earlier in this chapter, those same trophic interactions cascade back down to influence nutrient dynamics at the ecosystem level. In combination, they generate an overall feed-back loop that links nutrient dynamics and trophic interactions through the nexus of phytochemistry (figure 8.1).

8.5.3. Resource Subsidies

As I noted previously, a substantial proportion of algal blooms in coastal waters results from nutrient pollution arising from inputs in the terrestrial environment. More generally, this pollution can be viewed as a nutrient subsidy across ecosystem boundaries; such subsidies can illustrate the power of nutrient availability to modify trophic interactions (Polis et al. 1997a). Resource subsidies introduce autotroph residues, animal waste products, or animal cadavers into ecosystems, all of which serve as resources for detritivores and decomposers. Critically, the nutrients released by the degradation of resource subsidies are available to influence the phytochemical landscape and subsequent trophic interactions.

For example, nutrient-poor islands in the Gulf of California generally depend on nutrient and energy subsidies from the marine environment (Polis et al. 1997b). On these islands, bird guano increases the concentrations of N and P in soils up to 6-fold in comparison to islands without birds. In turn, primary producers on guano islands take up these nutrient subsidies during years with sufficient rainfall to promote primary production (Anderson and Polis 1999). Similarly, plants on islands near New Zealand also depend upon nutrients provided in bird guano. On islands invaded by rats, seabird populations have declined, thereby reducing soil nutrient availability, reducing plant nutrient concentrations, and altering the structure of soil food webs. In this system, nutrient flow arising from seabird guano cascades up through the phytochemical landscape to at least the level of secondary consumers (Fukami et al. 2006) (figure 8.16).

Likewise, chronic inputs of seaweed onto the shores of islands in the Bahamas decompose to provide nutrients that increase the nitrogen content of the phytochemical landscape. In turn, higher foliar nutrient levels increase defoliation of native island plants by herbivores that respond positively to increases in plant nutrient content (Piovia-Scott et al. 2013). Predators (lizards) also respond numerically to higher herbivore densities, and have an increasing impact on herbivore populations in a fashion analogous to those of birds in our own work on oak trees (figure 8.14). On the Bahamas, the effects of nutrient subsidies on increasing herbivore densities trump those of increasing predation pressure so that, overall, levels of herbivory increase with nutrient subsidy (Piovia-Scott et al. 2013).

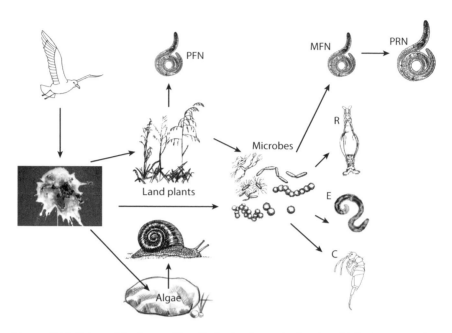

Figure 8.16. Rats on islands near New Zealand reduce colonies of seabirds that, on rat-free islands, import high concentrations of N and P in their guano. The nutrient subsidies provided by seabird guano cascade up to influence the phytochemical landscape, herbivores, and predators in terrestrial ecosystems. PFN = plant feeding nematodes, MFN = microbivorous nematodes, PRN = predatory nematodes, R = rotifers, E = enchytraeids, C = Collembola. Based on Fukami et al. (2006).

Note that this result, derived from sampling efforts under chronic natural levels of herbivory, differs from short-term (pulse) experiments, which illustrate a shift in lizard diet in response to seaweed inputs (Spiller et al. 2010). With short-term pulses of seaweed inputs, lizards shift from autochthonous plant-based food chains to allochthonous seaweed-based food chains, freeing herbivores on native plants from predation pressure and increasing levels of herbivory. In other words, the strength and duration of nutrient inputs influence their impacts on trophic interactions.

Insects that emerge as adults after completing aquatic larval stages can also transfer nutrients from aquatic to terrestrial ecosystems. In northeast Iceland, the emergence of midges from lakes transfers N to the terrestrial ecosystem, increasing foliar N concentrations of nearby willow trees by around 10 percent. In turn, the higher quality of willow foliage causes caterpillar densities on trees to increase by 4- to 6-fold, and caterpillar mass to increase by 72 percent (Bultman et al. 2014). Overall, midge-derived N induces a bottom-up cascade in the terrestrial

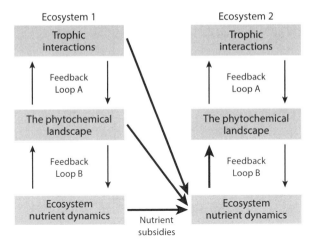

FIGURE 8.17. Resource subsidies link the trophic interactions and ecosystem processes in one ecosystem with those in a second ecosystem. In this example, inputs of inorganic nutrients, or of organic tissues from autotrophs, herbivores, and predators, serve as nutrient subsidies that can cascade up in the second ecosystem. Subsidies of a variety of forms are considered in more detail in chapter 9.

environment that increases herbivore density and performance. More generally, then, resource subsidies act to link the trophic interactions and ecosystem processes in one system with the trophic interactions and ecosystem processes in a second system (figure 8.17) (chapter 9).

In summary, nutrient availability has pervasive effects on the chemistry of individual primary producers and on the structure of their communities. Nutrient-mediated effects on the phytochemical landscape then cascade up to influence the production and tissue quality of herbivores. In turn, herbivore production and quality influences the preference, performance, and abundance of higher trophic levels, the effects of which cascade back down to influence nutrient dynamics at the ecosystem level (figure 8.1).

8.6. FINAL THOUGHTS ON FEEDBACK LOOPS

Trophic interactions are both a cause and consequence of nutrient dynamics in ecosystems. The pathways that I have described so far in this chapter include those that cascade down from predator activity to nutrient availability, and those that cascade up from nutrient availability to predators. Critically, in both cases, changes in the nutritional and defensive chemistry of primary producers serve as a nexus through which we can study and understand the links between trophic interactions and ecosystem processes (figure 1.1). While consumer-mediated effects on nutrient dynamics have been studied for many years (Kitchell et al. 1979), I have argued here that studying associated changes on the phytochemical landscape provides a mechanistic understanding of the feedback process linking trophic interactions and nutrient dynamics. I conclude this chapter with just a few short examples where complete feedback loops, from trophic interactions to nutrients dynamics and back again, have been reported in the literature. These examples serve to illustrate that there may be important structural and dynamical consequences in both terrestrial and aquatic systems that arise from feedback processes linking trophic interactions with nutrient dynamics (Elser et al. 2012). Feedback processes play a central role in ecological theories of population dynamics (Royama 1992), the structure of food webs (McCann 2011), and ecosystem processes (Schlesinger and Bernhardt 2013).

Earlier in this chapter, I described some of our own work that links herbivore activity in temperate forest canopies with the availability of mineral nutrients in forest soil, and the overall effect of canopy herbivores in increasing the rate of N cycling (Hunter et al. 2012). During this work, we also established the existence of a feedback loop by which herbivore-mediated changes in soil N dynamics could influence subsequent levels of herbivory. Specifically, using ^{15}N-enriched feces from oak-feeding caterpillars, we observed that trees that were damaged by caterpillars in a given year recovered less of the N mineralized from feces by microbes than did undamaged trees. Over a two-year period, defoliated trees recovered 20 percent less N from soil than did undamaged trees, resulting in lower foliar N concentrations than those measured in undamaged trees (Frost and Hunter 2007). Critically, the lower foliar N concentrations resulted in subsequent reductions in herbivore density, subsequent reductions in leaf damage, and therefore lower N losses to herbivores and their feces. In this case, the feedback loop between trophic interactions and nutrient dynamics is a negative feedback process (figure 8.14) that could serve to constrain both the populations of herbivores on trees and the losses of N from forest trees to long-term storage in soils or export in streams (Reynolds et al. 2000). Herbivore-generated cycles in N availability have been implicated previously in cycling of herbivore populations (Haukioja et al. 1985, Hunter 1994) and represent a form of predator-prey dynamics, in which the prey is N and the predators are caterpillars.

In our studies to date, we have not measured soil nutrient dynamics in the field over long enough time periods to assess the presence of any herbivore-mediated cycles in soil nutrient dynamics that match the periodicity of herbivore activity in the canopy. Additionally, in forest ecosystems, a wide variety of natural enemies, including vertebrates (Singer et al. 2012), invertebrate parasites (Turchin et al. 2003), and pathogens (Elderd et al. 2013), also influence the dynamics of insect herbivore populations and the damage that they inflict on their host trees. As a consequence, there are important potential feedback processes to explore between the dynamics of natural enemies and the dynamics of soil nutrients in forested ecosystems. For example, given the strong links between nutrient availability and the chemistry of oak foliage (Hunter and Schultz 1995, Forkner and Hunter 2000), and the strong links between oak foliar chemistry and gypsy moth-virus interactions (Hunter and Schultz 1993, Elderd et al. 2013), it seems reasonable to predict that the gypsy moth virus may both generate and respond to spatial and temporal variation in soil nutrient dynamics. Testing such a prediction will require simultaneous long-term monitoring of soil nutrients, phytochemistry, herbivore populations, and pathogen dynamics.

Moving from forests to savanna ecosystems, termites (*Macrotermes*) in Lake Mburo National Park in Uganda concentrate nutrients into biogeochemical hot spots around their large termitaria. In turn, termitaria support diverse and nutrient rich plant communities that attract large mammal grazers. In other words, effects of termites on soil nutrient availability cascade up to influence heterogeneity on the phytochemical landscape and subsequent herbivory by ungulates. Moreover, in this intimately linked system, feces deposition during ungulate grazing also enriches nutrient availability around termitaria, which further magnifies the effects of termites on plant productivity and diversity (Okullo and Moe 2012). In other words, feedback processes link termites and ungulates through their interactive effects on nutrient dynamics and the phytochemical landscape.

Feedback loops from consumers to nutrient dynamics and back to consumers are also common in seminatural and managed ecosystems. For example, in the north Asian steppe, overgrazing by livestock reduces N availability soil. In turn, low-N soils promote the growth of low-N plants with high carbohydrate to protein ratios. Such plants increase the survival, growth rate, and adult mass of locusts, which may then exhibit outbreak dynamics (Cease et al. 2012). In this example, the population dynamics of locusts at large spatial scales may result from the combination of (a) the cascade down from large herbivores to soil N dynamics, and (b) the cascade up from soil N dynamics, through changes in plant protein chemistry, to invertebrate herbivores. It is interesting to consider that locust outbreaks in this region may derive in part from the loss of apex predators that may once have kept natural ungulate populations at much lower densities.

While overgrazing by livestock may reduce soil N in the north Asian steppe, corralling of livestock overnight in thorn enclosures (bomas) in East Africa concentrates ungulate fecal material into nutrient-rich glades that can persist for more than a century after the bomas are abandoned. Glades are nutrient-rich islands on the phytochemical landscape that attract native and domestic grazers. Grazers continue to subsidize these hot spots of biogeochemical activity with nutrient inputs in feces and urine. In turn, nutrients derived from egestion and excretion increase the nutritional quality of surrounding vegetation, causing higher densities of both insect herbivores and predatory lizards on trees (Donihue et al. 2013). In this case, the legacy effects of grazing on soil nutrient dynamics cascade back up to influence positively both herbivores and predators through the nexus of plant nutritional chemistry.

In aquatic ecosystems, enormous fluxes in nutrients, mediated by trophic interactions, can occur over rather short time periods. The rise and fall of algal blooms, with their specific nutrient demands and differential algal stoichiometry (chapter 7), set the stage for linking trophic interactions with nutrient dynamics through the phytochemical landscape. The seasonal turnover and diversity of algal species in lakes and oceans depend, at least in part, on the differential effects of grazers and pathogens on phytoplankton with different cell nutrient concentrations (Prowe et al. 2012). For example, coccolithophores such as *Emiliana huxleyi* cause algal blooms in the ocean where silica levels are low relative to the availability of N and P (Yunev et al. 2007). As blooms develop, they represent substantial sinks for C, N, and P in ocean waters. However, rapid recycling of these nutrients back into the environment occurs as a result of trophic interactions. In a recent study, a species-specific viral infection was shown to cause the rapid collapse of an *E. huxleyi* bloom in the North Atlantic Ocean, whereby two thirds of the 24,000 tons of carbon (and associated nutrients) fixed by the bloom were returned to the environment within just one week (Lehahn et al. 2014). In short, trophic interactions mediate the release of nutrients that are then available for decomposition by heterotrophic bacteria and subsequent algal uptake.

Likewise, the nutrient chemistry of macroalgae in intertidal environments mediates interactions between grazers and nutrient dynamics. Environmental N availability limits the ability of brown algae such as *Fucus vesiculosus* to take up inorganic P from the environment (Perini and Bracken 2014). This nutrient colimitation causes seasonal variation in the quality of algae as food for consumers— temporal variation on the phytochemical landscape. In turn, herbivores vary seasonally in their negative (consumptive) and positive (nutrient recycling) effects on algae (Perini 2013). In other words, the feedback loop between nutrient dynamics in the intertidal zone and grazing herbivores is mediated in part by the differential nutrient chemistry of brown algae.

In ecosystems in which the ratio of consumer biomass to producer biomass is relatively high, the standing stock of animal tissues represents a significant pool of nutrients. Like the large mammals and large predators supported by grasslands, some lake and marine ecosystems are characterized by relatively high consumer:producer biomass ratios, where the biomass of fish and higher predators can reach substantial levels. In such systems, the links between trophic interactions and nutrient dynamics should be particularly strong (Kitchell et al. 1979) as nutrients flow into and out of the predator compartment of the food web. While there now seems little doubt that predators in aquatic systems can act as both sinks and sources of key nutrients (Vanni et al. 2013), the role of primary producer chemistry in mediating the flows of nutrients between large predators and the abiotic environment remains poorly understood in most aquatic ecosystems, except when harmful algal blooms completely restructure food webs (above). As Vanni et al. (2013) point out, it should be a priority to understand the role of variation in the quality, rather than just the quantity, of primary and secondary production for generating links between large predators and nutrient cycling in our lakes and oceans.

In summary, there is strong evidence from both aquatic and terrestrial ecosystems of complete feedback loops, from trophic interactions to nutrients dynamics and back again, that are mediated by variation on the phytochemical landscape. These feedback loops combine ecological pathways that have traditionally been considered as either bottom-up or top-down into a single circular causal system (Hutchinson 1948). In the presence of strong feedback, it makes no sense to argue that predators are more or less important in determining nutrient dynamics than are herbivores or autotroph chemistry. Likewise, primary producers have no more primacy in this circular causal system than does any other trophic level. However, their variable chemistry provides us with a critical mechanistic understanding of how trophic interactions and nutrient dynamics are linked. In the final chapter of this book, I explore opportunities for future work and some testable predictions that arise from studying ecological and evolutionary processes on the phytochemical landscape.

CHAPTER 9

Synthesis and Prospects
for Future Work

9.1 INTRODUCTION

The phytochemical landscape, that complex spatial and temporal combination of elemental nutrients, toxins, and recalcitrant molecules, is the nexus that links trophic interactions with nutrient dynamics in ecosystems. Over ecological and evolutionary timescales, the chemistry of primary producers is both a cause and a consequence of variation in trophic interactions, including herbivory, predation, parasitism, and disease. Simultaneously, autotroph chemistry is both a cause and a consequence of variation in nutrient pools and fluxes at the ecosystem scale. The combination of these powerful processes generates large-scale feedback loops in ecological and evolutionary time. These feedback loops help us to understand how apex predators can influence ecosystem processes and how nutrient fluxes mediate the power of predation (figure 9.1). I have argued here that the feedback processes between nutrient dynamics and trophic interactions have important consequences for the spatial and temporal dynamics of diverse living systems. As we continue to alter Earth's ecosystems, our ability to predict the consequences of environmental change is becoming ever more important. Primary producers play a fundamental role in supporting complex food webs, and in supporting the decomposers that recycle organic materials. Understanding how the phytochemical landscape mediates the links between trophic interactions and nutrient dynamics will help us prepare for the novel ecosystems of tomorrow.

And given the current pace of environmental change, the stakes are simply too high to delay. Although we are making progress in understanding how phytochemistry links trophic interactions with nutrient dynamics (Burghardt and Schmitz 2015) there remains much work left to do. In this chapter, I describe some important questions that remain to be answered, and offer a series of testable predictions about how the phytochemical landscape should mediate links between trophic interactions and ecosystem processes.

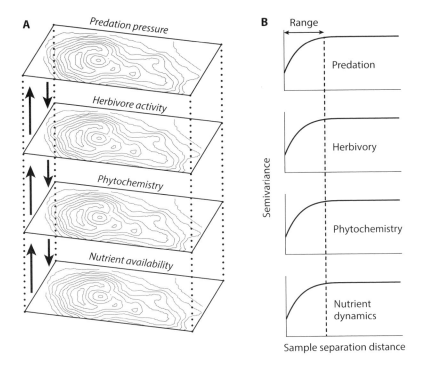

FIGURE 9.1. (A) Contour lines on maps link points of equivalent predation pressure, herbivory, phytochemistry, and nutrient flux. The spatial correlation among map layers illustrated in (A) results in matching ranges of spatial dependence (B), in which samples taken close together are more similar to one another than are samples taken at increasing distance from each other.

9.2 PRIORITY 1: LET'S MAKE SOME MAPS

Reductionist experimental studies are essential to the discipline of ecology, because they provide us with cause-and-effect evidence of the mechanisms underlying ecological interactions. For example, controlled laboratory and mesocosm studies of induced responses to herbivory have been central to understanding how autotroph chemistry in terrestrial and aquatic systems responds to consumer activity (Schultz and Baldwin 1982, Cronin and Hay 1996b). However, our interest in mechanistic studies should be driven primarily by a desire to explain patterns that we observe in nature. Frankly, much of the value of experimental studies is lost if we don't have available the observational data with which to describe ecological pattern at relevant spatial and temporal scales.

In chapter 2, I noted how few spatially explicit studies of the phytochemical landscape have been published to date. Such maps of chemical heterogeneity are a prerequisite for superimposing maps of ecological (Gallardo and Covelo 2005, Moore et al. 2010) and evolutionary (Andrew et al. 2007) interactions onto the phytochemical landscape. With an increasing number of tools available to obtain multivariate estimates of autotroph chemistry at landscape scales (Foley et al. 1998, Kokaly et al. 2009, Richards et al. 2015), and the availability of cheap and accurate GPS units, we should begin to prioritize the collection of spatially explicit phytochemistry data. We can then use those data to explore the spatial structure of ecological and evolutionary processes.

Moreover, just as all ecologists understand the need for repeated counts of organisms to analyze their population dynamics, so we should accept the need to make repeated measurements of our phytochemical landscapes to understand their chemical dynamics. We are in desperate need of phytochemical time-series data from multiple ecosystem types. In an era in which we can repeatedly sample genomes at landscape scales, we can surely also make repeated measures of chemical phenotypes at landscape scales. The benefits of doing so will be substantial, particularly when those data are associated with estimates of trophic interactions and nutrient dynamics on the same landscapes (below).

9.3 PRIORITY 2: ASSESS THE FREQUENCY AND STRENGTH OF SPATIAL CORRELATION

A fundamental and testable prediction of the approach described here is that there exists spatial correlation between pattern and process on the phytochemical landscape (figure 9.1). Specifically, we should expect to see spatial patterns of phytochemistry correlate significantly with spatial patterns of both trophic interactions and nutrient dynamics. We might estimate the strength of trophic interactions by the densities of herbivores and their natural enemies, or more precisely, with estimates of the interaction strengths that we typically use in ecological models of trophic interactions (Speight et al. 2008). Likewise, nutrient dynamics can be estimated by a variety of techniques at fine spatial and temporal scales (Robertson et al. 1999). The key is to have available high-quality spatial and temporal data of phytochemistry, trophic interactions, and nutrient dynamics.

The statistics required to assess the strength of spatial correlation among such data are quite well established. They have been used successfully to identify spatial correlation between the phytochemical landscape and nutrient dynamics (Gallardo and Covelo 2005, Rodriguez et al. 2011), and between the phytochemical landscape and trophic interactions (Moore et al. 2010, Marsh et al. 2014). To my knowledge, similar analyses have not yet been completed that include spatial correlation simultaneously among trophic interactions, the phytochemical landscape, and nutrient dynamics.

Prediction 1: A central and testable prediction arising from the phytochemical landscape concept is the existence of cascading spatial correlation among trophic interactions, the phytochemical landscape, and nutrient dynamics (figure 9.1).

In the example illustrated in figure 9.1, mapping of consumer activity, phytochemistry, and nutrient dynamics suggests strong spatial correlation among these variables (Fig 9.1A), with hot spots of phytochemistry associated with hot spots of trophic interactions and nutrient flux. Recall that these correlations can be positive or negative in sign (chapter 2, see also below). The strong spatial correlation can be expressed on a series of semivariograms (Cressie 1993) (figure 9.1B). In these plots, there is spatial autocorrelation apparent within each level, so that samples taken close to one another are more similar to each other (exhibit less variance) than are samples taken farther apart. There is a characteristic distance, the range, over which spatial autocorrelation occurs. Beyond that distance, samples are no more similar to one another than they are to the rest of the population. In this hypothetical example, the fact that trophic interactions, phytochemistry, and nutrient dynamics share a similar range (figure 9.1B) is consistent with the hypothesis that linked biological processes are generating their common spatial patterns. By the same logic, a breakdown in spatial correlation (and a significantly different range) between lower and higher levels in the interaction web might indicate that one of the feedback loops in figure 9.1A is weak or compromised. Analyses of the kind illustrated in figure 9.1B could be used as an initial screen for the identity of the predators and the herbivores with the greatest potential links to variation on the phytochemical landscape and therefore spatial variation in nutrient dynamics.

9.4 PRIORITY 3: UNDERSTANDING TIME LAGS AND THE TEMPORAL SCALE OF SPATIAL CORRELATION ON THE PHYTOCHEMICAL LANDSCAPE

9.4.1 Interactions between Time and Space

Feedback processes play a central role in population, community, and ecosystem ecology (Odum 1953). However, the temporal scale at which feedback occurs has important effects on the dynamics of systems (May 1974, Turchin 1990). The feedback processes I have described in this book operate at a wide variety of temporal scales, from minutes to millennia (chapter 1). For example, the arrival of a predator, and the concomitant change in how herbivores forage on the landscape of fear (chapter 5), may occur much more rapidly than subsequent changes in autotroph chemistry and nutrient cycling. On the phytochemical landscape, the chemistry of individual autotrophs may change rapidly (seconds, hours, days) in response to herbivore damage (Hu et al. 2013). However, the phytochemical landscape may also respond much more slowly to the effects of selective foraging by herbivores (Pastor and Naiman 1992), or to evolution of plant secondary metabolism by the forces of natural selection (Fine et al. 2013).

It should be a research priority to understand and reconcile the temporal scales at which feedback processes occur on the phytochemical landscape. Time lags in the responses of primary producer chemistry to trophic interactions or to nutrient dynamics could generate complex temporal dynamics, including phytochemical cycles (figure 9.2). The potential for phytochemical induction to generate cyclic dynamics in consumer populations and in nutrient dynamics has been recognized previously (Haukioja 1980, Underwood 1999, Reynolds et al. 2012). Additionally, consumer outbreaks can change both the relative abundance of autotrophs on the landscape (chapter 5) and the subsequent chemistry of herbivore-derived inputs (chapter 8), leading to stochastic temporal patterns on the phytochemical landscape (figure 9.2). While there is strong empirical support linking the temporal dynamics of consumer populations with the temporal dynamics of nutrients in both aquatic (Kitchell et al. 1979, Schindler et al. 1993) and terrestrial (Reynolds et al. 2000) ecosystems, such studies are only rarely conducted in a spatial context (Hunter et al. 2003b, Reynolds et al. 2003).

Prediction 2: The form of temporal synchrony among consumer activity, phytochemistry, and nutrient dynamics will show spatial structure on the phytochemical landscape. Specifically, patches of similar average phytochemistry in space will exhibit more similar temporal dynamics than will patches of different average phytochemistry (figure 9.2).

Moreover, just as populations of predators are correlated on delay with populations of their prey in classic predator-prey cycles (Varley et al. 1973), so we should expect time lags in the correlations among consumer activity, autotroph chemistry, and nutrient dynamics on the phytochemical landscape. This is why it is so critical that we collect time-series data of phytochemistry (above). Cross-correlation of data from multiple time series provides a valuable tool for exploring temporal synchrony in ecological processes (Bjornstad et al. 1999), and can detect both positive and negative associations among ecological variables that are measured concurrently. Cross-correlation analyses, incorporating systematic analyses of time lags of different length, should be used to explore temporal synchrony among consumer activity, phytochemistry, and nutrient

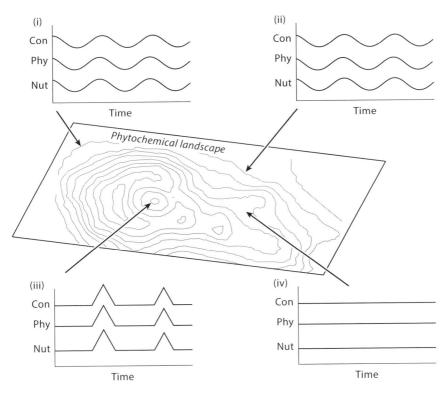

FIGURE 9.2. Temporal patterns of association among consumers (Con), phytochemistry (Phy), and nutrient dynamics (Nut) will vary in space on the phytochemical landscape. Locations of equivalent average phytochemistry (i and ii) will show more similar temporal patterns of association than will locations of different average phytochemistry (compare ii, iii, iv).

dynamics. Importantly, when cross-correlation analyses indicate the presence of important time lags, those time lags should reflect the ecological or evolutionary processes connecting phytochemistry with trophic interactions and nutrient dynamics. In short, in the same way that detecting time lags of one, two, or three generations in population time series can be used to infer the types of direct and delayed density-dependent processes that are operating (Royama 1992), so too we can use the length of time lags in the correlations among consumers, phytochemistry, and nutrient dynamics to infer the action of ecological processes.

Prediction 3: There will exist time lags in cross-correlations among the time-series of consumers, phytochemistry, and nutrient dynamics. The length of these time lags will themselves vary on the phytochemical landscape,

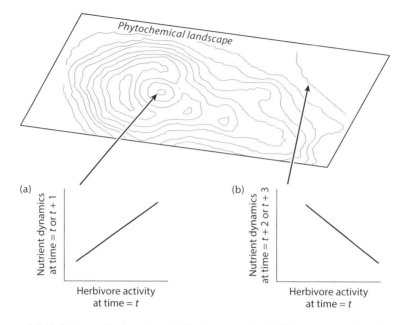

FIGURE 9.3. Variation on the phytochemical landscape results in different temporal lags between herbivore activity and nutrient cycling. On this hypothetical map, contour lines join points in space with equal N to lignin ratios in autotrophs. In (A), a region of high N to lignin ratio, herbivore activity favors fast-cycle effects and accelerated nutrient cycling, so that levels of herbivory at time t are correlated positively with nutrient dynamics at time t or shortly thereafter (e.g., time t+1). In (B), a region of low N to lignin ratio, herbivory favors slow-cycle effects, increases in phytochemical recalcitrance, and deceleration of nutrient cycling. Slow-cycle effects become apparent after longer time lags (e.g., time t+2 or t+3).

depending upon the ecological and evolutionary processes that are linking trophic interactions with nutrient dynamics (figure 9.3).

Prediction 4: Short time lags between trophic interactions and nutrient dynamics will indicate the dominance of fast-cycle effects (chapter 8) in which the feces, cadavers, and sloppy feeding of consumers act rapidly to accelerate nutrient cycling (figure 9.3 arrow a). In contrast, longer time lags between trophic interactions and nutrient dynamics will indicate the dominance of slow-cycle effects (chapter 8), whereby consumers influence the relative recalcitrance of autotroph residues that enters the detrital pathway (figure 9.3, arrow b).

In the hypothetical example illustrated in figure 9.3, we can envisage the contour lines as joining points on the phytochemical landscape with the same N to lignin ratio. Regions on the phytochemical landscape characterized by high N to lignin ratios (arrow a) may map to high rates of herbivore population growth, high rates of autotroph consumption, and high inputs of feces and urine. These fast-cycle inputs (McNaughton et al. 1988) may accelerate nutrient cycling (Ritchie et al. 1998) with almost no time delay (i.e., in time t or time t+1) (Frost and Hunter 2007). In contrast, regions on the phytochemical landscape characterized by low N to lignin ratios (arrow b) may map to lower overall rates of autotroph consumption. In such regions of the phytochemical landscape, inputs of feces and urine may be of less importance than are the slow-cycle effects of herbivores on nutrient cycling (McNaughton et al. 1988), notably increases in autotroph residue recalcitrance that can arise from phytochemical induction (Findlay et al. 1996) or selective foraging (Pastor et al. 1993). Effects of herbivory at time t would not become apparent until more time had passed (e.g., t+2 or t+3). The units of time would depend on the system of study, and would likely be much shorter for phytoplankton-dominated communities in oceans than for tree-dominated communities in boreal forests.

Recently, it has been suggested that the relative dominance of fast- and slow-cycle effects of herbivores on nutrient dynamics in terrestrial ecosystems might vary with the defense syndrome of plants, with fast-cycle effects more typical of tolerant plants and slow-cycle effects more typical of resistant plants (Burghardt and Schmitz 2015). To the extent that these syndromes are differentially clustered on the phytochemical landscape, with more tolerant plant species distributed in regions of high nutrient availability, they could produce patterns similar to those illustrated in figure 9.3.

We can also use the sign (positive or negative) of the cross-correlation between the time series of herbivore activity and nutrient dynamics to assess the frequency of acceleration or deceleration of nutrient dynamics caused by consumer activity. A negative cross-correlation between herbivore abundance and nutrient cycling would indicate deceleration whereas a positive cross-correlation would indicate acceleration. In addition to analyzing temporal cross-correlation of time-series data, we can use spatial correlation in a similar way (figure 2.4B). Negative spatial correlation between herbivore activity and rates of nutrient flux would indicate support for the deceleration hypothesis.

Prediction 5: Negative spatial or temporal correlation between herbivore activity and nutrient cycling rate, in support of the deceleration hypothesis, will occur on regions of the phytochemical landscape characterized by low autotroph nutrient concentration and high concentrations of recalcitrant organic molecules (lignin, cellulose). Positive spatial or temporal correlation between herbivore activity and nutrient cycling rate, in support of the acceleration hypothesis, will occur on regions of the phytochemical landscape characterized by high autotroph nutrient concentration and low concentrations of recalcitrant organic molecules (figure 9.3, figure 2.4B).

Moreover, given the lower prevalence of recalcitrant cells or structures synthesized by aquatic than by terrestrial autotrophs (Shurin et al. 2006), we might generally expect herbivores to accelerate nutrient cycling in aquatic ecosystems with greater frequency than they do in terrestrial ecosystems.

Prediction 6: Spatial and temporal cross-correlation between herbivore activity and the rate of nutrient cycling will be positive more often in phytoplankton-dominated ecosystems than in macrophyte-dominated ecosystems.

9.4.2 A Few Thoughts on Feedback Processes
under Evolutionary Time Scales

In this book, I have focused primarily on linking the ecological processes of trophic interactions with those of nutrient dynamics through the nexus of phytochemistry. It would take a second book of at least comparable size to consider in detail the implications of these interaction pathways for evolutionary processes, and the reciprocal effects of evolution on the ecological interactions that I describe here. Although there is not space to consider them, evolutionary processes are fundamental to generating the phytochemical landscapes that link population-level and ecosystem-level interactions.

There is growing interest in the interface between evolutionary and ecosystem processes, and the role of phytochemistry in defining that interface (Elser et al. 2000, Bailey et al. 2009). At the most basic level, evolution by natural selection plays a key role in determining the expression of traits in autotrophs, microbes and animals that determine how they interact with one another and how they influence the pools and fluxes of elemental nutrients. For example, in chapter 3, I described the strong genetic component underlying the expression of nutritional and defensive chemistry of terrestrial and aquatic primary producers (Agrawal et al. 2012). Similarly, evolutionary processes influence the physiological and behavioral traits of higher trophic levels that mediate the ecological processes of herbivory, predation, and mutualism (Fryxell and Sinclair 1988, Karban and Agrawal 2002), as well as the consequences of those interactions for ecosystem processes (Chislock et al. 2013). In short, the nutritional and defensive chemistry of primary producers, and the defensive and offensive traits of consumers, are in large part the products of evolution by natural selection.

In figure 9.4, I illustrate a general framework within which genotype and environment interact to influence the traits of organisms that, through their impacts on nutrient dynamics, feed back to influence both subsequent genetic change and the environment in which that change takes place. The term *trait*

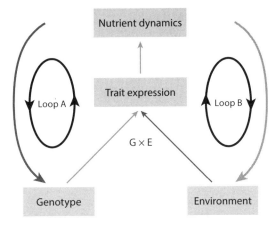

FIGURE 9.4. An evolutionary perspective on links between trophic interactions and nutrient dynamics on the phytochemical landscape. Here, trait expression refers to traits of both primary producers *and* consumers, including the chemistry of autotrophs, as well as the physiology, morphology, and behavior of higher trophic levels. Trait expression has genotypic and environmental components, and trait expression influences nutrient dynamics at the ecosystem level. In turn, nutrient dynamics influence the relative fitness of genotypes (thereby generating loop A), and the environmental factors that influence trait expression (thereby generating loop B). A template of this type can encompass a community genetics viewpoint (pale arrows) in which ecosystem processes are seen as part of the extended phenotype of organisms.

expression in figure 9.4 is meant to encompass both the chemistry of primary producers *and* those physiological, morphological, and behavioral responses of other biota (decomposers, herbivores, predators, mutualists) that mediate the effects of autotroph chemistry on nutrient dynamics. Because of the feedback loops in figure 9.4, trait evolution is both a cause and a consequence of nutrient dynamics at the ecosystem level (loop A). Similarly, nutrient dynamics are a component of the environment that influences subsequent trait expression (loop B). Additionally, the figure illustrates a pathway (palest arrows) by which a nutrient flux, or the environmental consequences of that flux, can be considered part of the extended phenotype of an organism in a community genetics framework (Whitham et al. 2003). A research agenda for linking ecological and evolutionary processes should include estimating the relative strengths of these pathways within and among ecosystems and exploring feedback between evolutionary change and ecosystem function.

The evidence for links between evolutionary and ecosystem processes has been strengthening in recent years (Zuppinger-Dingley et al. 2014). For example, the expression of condensed tannins in the foliage and litter of cottonwood trees has a strong genetic basis, and influences litter decomposition and nutrient flux in both aquatic (Driebe and Whitham 2000) and terrestrial (Schweitzer et al. 2005a) ecosystems. Likewise, genetic variation in tannin and lignin concentration in oak litter influences carbon and nitrogen fluxes in oak sandhill ecosystems (Madritch and Hunter 2002). That the effects of plant genotype on nutrient dynamics can feed back to influence subsequent evolutionary change in plant traits is not as well established yet. Perhaps the best example to date comes from Hawai'i, where populations of the tree *Metrosideros polymorpha* appear to be locally adapted to soil nutrient availability (Treseder and Vitousek 2001). In common garden experiments, tree genotypes from low N soils express higher lignin concentrations and lower N concentrations in their litter than do genotypes from more fertile soils. Litter with high lignin and low N concentration decomposes more slowly in this ecosystem, reducing rates of nutrient flux, and perhaps selecting yet further for tree genotypes with foliar traits typical of low N soils. In other words there may be a positive feedback loop, similar to loop A in figure 9.4, by which low-nutrient environments select for high lignin/low N genotypes that, in turn, maintain low nutrient conditions in the soil (Treseder and Vitousek 2001). We will need many more studies of this type, at multiple trophic levels, before we can make accurate generalizations about the nature of feedback between evolutionary processes and nutrient dynamics.

9.5 PRIORITY 4: EXPLORING VARIATION IN THE STRENGTH OF FEEDBACK BETWEEN TROPHIC INTERACTIONS AND NUTRIENT DYNAMICS ON THE PHYTOCHEMICAL LANDSCAPE

Analyses of the kinds described in figures 9.1 and 9.3 provide tools with which we can explore variation in the strength of the feedback processes linking trophic interactions with nutrient dynamics on the phytochemical landscape. When we observe strong spatial correlation (figure 9.1A), strong temporal cross-correlation (figure 9.3), and equal range sizes on semivariograms (figure 9.1B), we have significant support for the hypothesis that all of the links in figure 9.1A are operating. However, we should always initiate experiments at appropriate spatial and temporal scales to confirm that the proposed mechanisms of feedback are actually operating.

Inevitably, we should expect to see substantial variation in the strength of these links in ecological systems. Not all predators, parasites and agents of disease have strong impacts on herbivore abundance or behavior. Not all herbivores influence equally the chemistry of their autotroph hosts. And autotroph chemistry is only one factor influencing the dynamics of nutrients in terrestrial and aquatic ecosystems (chapter 1). Likewise, primary producers, herbivores and predators may respond differentially to variation in local nutrient dynamics. When any of the spatial and temporal correlations among trophic interactions, phytochemistry, and nutrient dynamics are weak, when the relationships illustrated in figures 9.1 and 9.3 are compromised, we have the opportunity to learn about the relative importance of additional factors operating in our ecological systems. In essence, we get to ask why the feedback loops in figure 9.1 have become decoupled. The concept of the phytochemical landscape as the mediator of links between trophic interactions and nutrient dynamics has substantial support (chapters 3–8) but the concept is also highly instructive when we can understand how and when its links become weak. Below, I provide a few examples of factors that may weaken the feedback processes illustrated in figure 9.1, and what the consequences might be for trophic interactions and nutrient dynamics. The phytochemical landscape provides a guide, or a template, for understanding the consequences of variation in interaction strength in ecosystems.

9.5.1 Donor Control and Subsidies

By definition, donor control describes an ecological system in which feedback is effectively absent. The concept of donor control is applied most commonly to consumer-resource interactions and is commonly invoked when predator densities respond to temporal variation in densities of their prey, but do not in turn affect the renewal rate of their prey resources (Hawkins 1992) (figure 9.5A). Donor control represents a kind of ecological forcing—a decoupling of typical consumer-resource feedback processes—and the concept can be applied at any trophic level.

Resource subsidies that cross ecosystem boundaries can foster donor control, with significant consequences for ecological networks (Polis et al. 1997a). Importantly, the concept of the phytochemical landscape provides a **mechanistic basis** for understanding how subsidies and donor control influence ecosystems. Note that subsidies can occur at any level of the web illustrated in figure 9.1. Critically, the rate of renewal of subsidies is essentially independent of their cascading effects in ecosystems, thereby decoupling certain feedback processes within the web. For example, I have already described (in chapter 4) the dramatic consequences of P pollution for the phytochemical landscape generated by toxic algal blooms in the Great Lakes of the United States (Michalak et al. 2013). Blooms of toxic algae have profound consequences for trophic interactions (chapter 4) and for nutrient flux (chapter 6). However, because changes in the trophic interactions that result from algal toxins have no subsequent impact on rates of P pollution in the Great Lakes—there is no feedback loop between trophic interactions and rates of P renewal—the ecosystem exhibits some characteristics of donor control rather

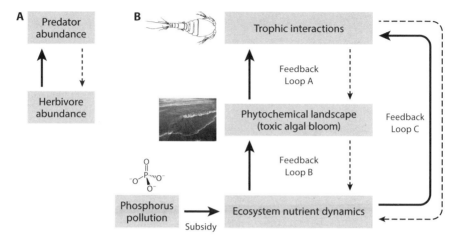

FIGURE 9.5. Donor control (A) is characterized by the strong dependence of a consumer on the availability of its resource without a reciprocal effect of the consumer on the rate of resource renewal. Powerful ecosystem subsidies (B) may foster donor control at the ecosystem level, and decouple the feedback loops (loops A and B) that generate reciprocal links between trophic interactions and nutrient dynamics (loop C).

than strong feedback (figure 9.5B). Note that the phytochemical landscape still mediates the interactions flowing from nutrient availability to trophic interactions, but not vice versa.

Subsidies are not limited to nutrient transfer. The spillover of predators from highly productive agricultural systems into adjacent natural landscapes (Cronin and Reeve 2005) can cause significant reductions in the densities of native herbivores (Rand et al. 2006). Predator spillover essentially serves as a top-down subsidy into the recipient habitat. As in bottom-up subsidies of nutrients, top-down subsidies of predators decouple the typical feedback processes inherent in consumer-resource interactions, and maintain stronger predator forcing of the system than would occur in the absence of the subsidy.

We can illustrate more clearly the causes and consequences of predator spillover with reference to the phytochemical landscape (figure 9.6). Our crop ecosystem represents a steep but flat plateau on the phytochemical landscape, wherein plant nutrient content is very high relative to that of the surrounding natural matrix. On the agricultural patch, a fertilizer subsidy (S1) maintains high nutrient availability and decouples the strong bottom-up forcing from any reciprocal top-down effects; high abundance of natural enemies is maintained largely by donor control. In turn, natural enemy spillover into the matrix acts as a second subsidy (S2), this time from the top down. Because natural enemy abundance depends largely on subsidy from the agricultural patch, it is decoupled from local feedback processes within

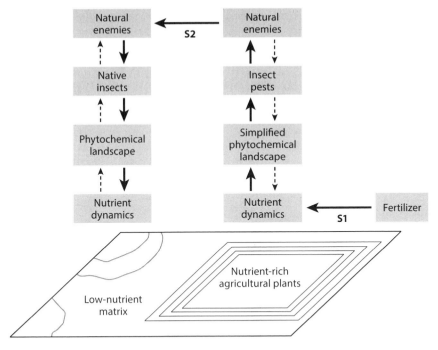

FIGURE 9.6. Fertilization of agricultural fields generates regular patches of high-nutrient plants (nutrient plateaus) on the phytochemical landscape. The fertilizer subsidy (S1) fosters donor control and the production of many natural enemies within the agricultural system. Natural enemies spill over as a top-down subsidy (S2) into adjacent parts of the phytochemical landscape, generating cascading effects in the matrix habitat. In both systems, subsidies decouple the typical feedback processes linking trophic interactions and nutrient dynamics.

the natural matrix habitat. As a consequence, natural enemies can exert substantial top-down forcing on the local herbivore community, with significant implications for the conservation of threatened herbivore species (Rand and Louda 2006). Our recent theoretical work has confirmed that spillover of predators from patches of high phytochemical quality can increase the ratios of predators to prey, and depress prey population growth, in lower quality patches on the phytochemical landscape (Riolo et al. 2015).

These examples of anthropogenic subsidies in aquatic and terrestrial ecosystems serve to emphasize a rather simple point; human activities can decouple trophic interactions from nutrient dynamics by establishing environmental forcing and disrupting natural feedback processes. An equally simple prediction, and one that will be of no surprise to those who manage water quality, is that we should be able to reestablish natural feedback processes by managing or terminating the artificial subsidies that we introduce into ecosystems (Michalak et al. 2013).

Perhaps less intuitively obvious is the idea that overharvesting of top predators in marine, lake, and terrestrial environments is the equivalent of a negative subsidy. We reduce or eliminate the feedback processes operating between carnivores and their prey, thereby imposing environmental forcing in the ecosystem, just as we do with positive subsidies. As a consequence, whenever we reduce the densities of top carnivores in ecosystems, we will weaken or eliminate the feedback processes by which predators influence the phytochemical landscape and, by extension, their vital role in nutrient cycling (Estes et al. 2011, Ripple et al. 2014b) (chapter 8).

9.5.2 Beyond the Effects of Nutrients and Herbivores on the Phytochemical Landscape

The major premise of this book is that feedback between trophic interactions and the phytochemical landscape (chapters 4 and 5) and between the phytochemical landscape and nutrient dynamics (chapters 6 and 7) result in fundamental linkage between trophic interactions and nutrient dynamics in ecosystems (chapter 8). But are there ecological systems in which variation on the phytochemical landscape derives largely from forces other than nutrient dynamics and herbivory? In such systems, trophic interactions and nutrient dynamics may be weakly related; phytochemistry will only function as the nexus linking trophic interactions with nutrient dynamics when phytochemical patterns derive from feedback between adjacent layers in the interaction web (figure 9.1).

> Prediction 7: When variation on the phytochemical landscape arises largely from processes other than nutrient availability or herbivory, trophic interactions and nutrient dynamics will be weakly coupled.

Throughout this book, I've described ecosystems from the tropics to the poles, and from streams to the ocean floor, in which nutrients, phytochemistry, and trophic interactions are coupled strongly over ecological and evolutionary time scales. However, there are other ecological forces capable of generating substantial variation on the phytochemical landscape.

Perhaps the most obvious of these is human activity. In our food and fiber production systems, in our urban and suburban landscapes, and when we overharvest natural resources, we impose radically new patterns on the phytochemical landscape and disrupt the links between trophic interactions and nutrient dynamics. Food and fiber production provide the clearest examples of human-induced changes to the phytochemical landscape. In fact, the whole point of these production systems is to generate a phytochemical landscape that fills our particular needs—for cellulose fibers (wood, cotton), for starch (potato, rice),

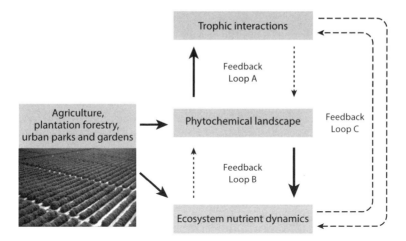

FIGURE 9.7. When we impose our own designs on the phytochemical landscape, we decouple the feedback processes (loops A–C) that link trophic interactions and nutrient dynamics. In our systems of managed phytochemistry, we overwhelm feedback processes with environmental forcing (solid arrows), with inevitable consequences for species interactions and nutrient cycling.

for oils (oil palm, rapeseed), or for vitamins and minerals (carrots, tomatoes, spinach, etc.). Because these phytochemical landscapes are not the products of natural interactions between nutrient dynamics and trophic interactions, they very obviously serve to break the links between ecosystem processes and trophic dynamics (figure 9.7).

Of course, conservation biologists have recognized that agricultural systems and urban environments hold great potential for the conservation of biological diversity and ecosystem processes. Essentially, managing agricultural (Vandermeer and Perfecto 2007) and urban environments (Hunter and Hunter 2008) for conservation purposes demands reestablishing some semblance of the natural phytochemical landscape (Haan et al. 2012). In doing so, we also reestablish fundamental links between trophic interactions and nutrient dynamics (Philpott et al. 2008, Gonthier et al. 2013). My personal view is that we could benefit from reframing many of the discussions in conservation biology as discussions about conserving, restoring, or engineering more natural phytochemical landscapes. Even when we are unable to restore all of the autotroph species that typify a particular habitat, we might be able to restore much of the spatial and temporal phytochemical diversity that supports the local ecology.

Figure 9.7 illustrates a very dramatic example of what we might call phytochemical forcing, in which factors external to trophic interactions and nutrient dynamics drive spatial and temporal patterns on the phytochemical landscape.

Phytochemical forcing might serve to foster "control from the middle out" (Trussell and Schmitz 2012, Burghardt and Schmitz 2015), in which trophic dynamics depend primarily on the quality or quantity of primary production. I have stressed throughout this book that autotrophs hold no particular primacy; variation in autotroph chemistry is connected by feedback to trophic interactions and nutrient dynamics, and phytochemistry is both a cause and a consequence of variation in these processes. Under this view of circular causal systems, no particular factor is in control of any other. However, control of trophic dynamics from the middle out might be manifest in systems in which forces other than trophic interactions and nutrient dynamics force strong patterns on the phytochemical landscape.

While humans are clearly powerful engineers of phytochemical landscapes, and may facilitate control from the middle out (figure 9.7), other ecological forces can also override the role of herbivores and nutrients in mediating spatial and temporal patterns of autotroph chemistry. Water availability provides a particularly compelling example. In chapter 4, I described interactions among a gall-forming sawfly, its willow host plants, and its natural enemies. In this system, long shoots on willow trees are more nutritious and less well defended than are short shoots, and are the preferred sites of sawfly oviposition (Craig et al. 1989). Critically, the availability of long shoots depends on the availability of water (Price and Clancy 1986), such that patterns of precipitation and drought drive variation in the dynamics of sawflies and their natural enemies (Clancy and Price 1986, Hunter and Price 1998) (figure 4.12). Because neither sawflies nor natural enemies can influence rates of precipitation, there is very little feedback between trophic interactions and the phytochemical landscape. As a consequence, sawfly population dynamics can be predicted accurately based upon relatively simple models of precipitation and drought (Price and Hunter 2005, Price and Hunter 2015). While the effects of willow sawflies on nutrient dynamics have not been investigated to date, it seems unlikely that they will have much impact; environmental forcing of the phytochemical landscape compromises the feedback loop between trophic interactions and nutrient dynamics. While water availability is the main driver of phytochemical variation in this system, the principle should hold whenever the availability of high-quality autotroph tissues is limited by other factors. This leads to an additional prediction that is strongly linked to prediction 7 above:

Prediction 8: The feedback processes between trophic interactions and nutrient dynamics will be weak or negligible when neither trophic interactions nor nutrient availability limit the production of high-quality chemistry on the phytochemical landscape. Such circumstances may favor trophic control from the middle out.

9.5.3 Effects of Disturbance on the Phytochemical Landscape

Disturbance events—fire, flooding, wind storms and so on—certainly have the potential to compromise feedback between trophic interactions and nutrient dynamics by forcing change on the phytochemical landscape. Excellent texts have been written on the multiple consequences of disturbance for ecological systems (Pickett and White 1987, Frelich 2008) and I can't do justice to the topic here. However, there are two brief points that I'd like to make that are particularly relevant to the topic of this book.

First, while disturbance may temporarily disrupt the links between trophic interactions and nutrient dynamics, the time period during which feedback processes are compromised may be quite short, even following severe disturbance. For example, following the 1980 eruption of Mount St. Helens in the western United States, we might assume that the loss of soil and the deposition of pyroclastic materials would minimize the strength of ecological feedback for decades. In reality, trophic relationships were reestablished soon after the first plants colonized. Lupine reinvasion was suppressed at the colonizing edge by insect herbivores that reduced plant spread, whereas herbivore populations in the center of lupine patches were held in check by a suite of natural enemies (Fagan and Bishop 2000). Lupines are early colonizers in part because of their ability to fix atmospheric N; they are critical mediators of the N cycle. In short, while severe disturbance radically changed the ecosystem type and the identity of the organisms participating in the processes, feedback between trophic interactions and nutrient cycling was reestablished very quickly, mediated by lupine ecology on the phytochemical landscape (Bishop 2002).

Nonetheless, we might imagine that frequent disturbance events that repeatedly modify the phytochemical landscape could prevent the reestablishment of strong links between trophic interactions and nutrient dynamics. For example, on Isle Royale, we might imagine that if the time between a sequence of severe disturbance events was shorter than the time it takes for selective foraging by moose to influence soil nutrient dynamics (Pastor et al. 1998), that particular link between trophic interactions and nutrient cycling would be compromised.

Prediction 9: Frequent disturbance to the phytochemical landscape will decouple nutrient dynamics from trophic interactions when the intervals between

disturbance events are shorter than the time scales over which feedback naturally operates between trophic interactions and nutrient dynamics.

Second, some disturbances are both a cause **and** a consequence of variation
on the phytochemical landscape. In other words, there can exist a feedback loop
between disturbance, succession of the phytochemical landscape, and the probability of subsequent disturbance. Whenever there exist cycles of disturbance and
succession, whereby ecosystems gradually return to the susceptible phytochemical state for additional disturbance, then the disturbance itself becomes an inte-
gral part of the process linking trophic interactions with nutrient dynamics. Fire
is the classical example of a disturbance category that initiates a successional
process that ultimately predisposes the ecosystem to additional fire. What we
call a high fuel load is simply an abundance of flammable phytochemistry—dry
accumulations of energy-rich cellulose and lignin, sometimes with flammable
hydrocarbon resins—that are particularly prone to burning. Fires in grasslands
and forests initiate successional trajectories in phytochemistry that ultimately
lead back to fire.

Fire-induced successional changes on the phytochemical landscape have
pervasive effects on trophic interactions and nutrient dynamics. For example,
fire frequency varies across the boreal forest landscapes of America, with the
greatest frequency occurring in the northwest (Stocks et al. 2002). This trans-
continental variation in fire frequency establishes a spatial mosaic on the phytochemical landscape, with early successional plant species, characterized by
high nutrient chemistry, available more frequently in the northwest. This spatial mosaic of nutrient-rich phytochemistry maps closely to spatial variation in
the densities of snowshoe hare (*Lepus americanus*) that exploit nutrient-rich
birch twigs (Paragi et al. 1997). Critically, spatial variation in snowshoe hare
abundance generates variation in the selection pressure for chemical defense on
the phytochemical landscape (Bryant et al. 2009). As a consequence, juvenile
birch trees secrete more of the toxic resin, papyriferic acid, from glands on their
twigs, in those regions with pervasive fire and abundant hare populations (figure
9.8). In this fascinating example, fire imposes variation in both nutrients and
toxins on the phytochemical landscape through a mixture of direct ecological
effects on plant communities and indirect evolutionary effects mediated by herbivores (Bryant et al. 2009).

That the frequency and severity of fire is also a consequence of the phytochemical landscape is aptly demonstrated when invasive plant species alter historical
fire regimes. This has occurred in the Great Basin region of the United States,
where invasion by the nonnative grass *Bromus tectorum* has increased the frequency of wildfires by 2- to 4-fold (Balch et al. 2013). Critically, because fire is

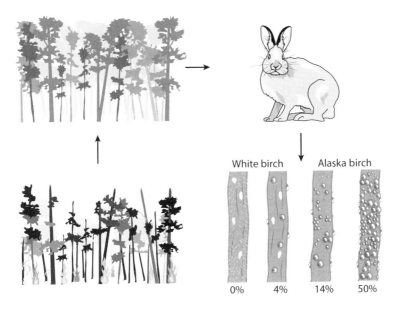

White birch   Alaska birch

0%   4%   14%   50%

FIGURE 9.8. A spatial mosaic of fire frequency generates a spatial mosaic of high-nutrient birch trees on the landscape. High-nutrient birches favor a high abundance of snowshoe hare that, in turn, impose selection for increased defensive resin on the twigs of juvenile birch trees. Based on data in Bryant et al. (2009).

both a cause and a consequence of variation on the phytochemical landscape, it means that fire frequency can respond to variation in trophic interactions. This is illustrated in figure 9.9, where the addition of a feedback loop between fire and the phytochemical landscape provides a pathway by which predators might influence fire frequency and intensity. This is exactly what happens when the loss of large predators from grassland and forest ecosystems intensifies grazing on a subset of the plant community, so altering the characteristic fire regime of the ecosystem (Estes et al. 2011). For example, reductions in wolf densities in Alaska cause corresponding increases in moose densities. In turn, moose selectively forage on nutritious deciduous plant species, thereby speeding up by more than 15 years the rate of succession to more flammable evergreen forest (Feng et al. 2012). Therefore, at least in part, wolves mediate the frequency of fire, and rates of forest succession, by their impact on the phytochemical landscape.

Moreover, by displacing or killing biota, and by volatilizing nutrients stored in organic matter, fire can also have direct impacts on trophic interactions and nutrient dynamics (lines (i) and (ii) in figure 9.9) with cascading effects throughout the interaction web. For example, fires in central Minnesota favor high densities of lace bugs on bur oak trees. Lace bugs induce high concentrations of lignin in bur

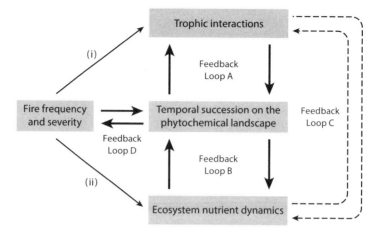

FIGURE 9.9. Because the frequency and severity of fire is both a cause and a consequence of variation on the phytochemical landscape (loop D), it can participate in important feedback processes with trophic interactions and nutrient dynamics. While fire can have direct effects on trophic interactions (i) and nutrient dynamics (ii), those same effects can feed back to influence fire dynamics through their impacts on the phytochemical landscape.

oak litter that subsequently reduce rates of litter decomposition (Kay et al. 2008). In other words, fire induces a top-down effect of trophic interactions on nutrient dynamics in addition to any direct effects of fire on soil nutrients. In contrast, from the bottom-up, fires in the Rocky Mountains of the western United States can cause temporary pulses of N availability in soil. Such soil N pulses can increase the ability of wild tobacco plants to induce their N-rich alkaloid defenses, and may reduce the subsequent palatability of tobacco plants to their native herbivores (Lou and Baldwin 2004).

In summary, some ecological disturbances have the potential to decouple variation on the phytochemical landscape from trophic interactions and nutrient dynamics. However, the links between trophic dynamics and nutrient dynamics can be reestablished very rapidly. Moreover, some kinds of disturbance events, like fire, become dynamic participants in the feedback processes linking trophic interactions and nutrient dynamics through the phytochemical landscape.

9.6 PRIORITY 5: COMPARING THE ROLE OF THE PHYTOCHEMICAL LANDSCAPE IN TERRESTRIAL AND AQUATIC ECOSYSTEMS

There exists enormous diversity in aquatic phytochemistry (chapter 3), with important consequences for the feedback processes linking trophic interactions and nutrient dynamics (chapters 4–8). Nonetheless, the effects of phytochemical variation on ecological and evolutionary processes remain better understood in terrestrial than in aquatic ecosystems. A research priority should be to explore in more detail the causes and consequences of variation in the phytochemical landscape of marine and lake ecosystems. I note just a couple of potential differences here, while stressing the importance of additional research effort.

First, a majority of marine and lake autotrophs are single-celled bacteria or eukaryotic algae (chapter 3). Because they are small and free-living, they are subject to transport by waves, currents, and tides. These advection processes might serve to decouple trophic interactions and nutrient dynamics in space. Moreover, dead aquatic autotrophs make up a significant proportion of marine snow (chapter 6), and can sink below the local ecological processes taking place in well-illuminated upper waters. Sinking or subduction (Omand et al. 2015) of phytodetritus again may disconnect in space trophic interactions from nutrient dynamics. Moreover, when dead autotrophs sink in aquatic ecosystems, a temporal lag is introduced; seasonal turnover of water (lakes) or nutrient upwelling (oceans) are then required to reestablish the links among phytochemistry, nutrient cycling, and subsequent trophic interactions (Wetzel 1983).

Second, in comparison to terrestrial autotrophs, most aquatic autotrophs do not invest in conspicuous structural tissues. They lack stems, roots, branches and petioles constructed of recalcitrant mixtures of organic molecules including complexes of lignin, cellulose, hemicellulose, and suberin (chapter 6). The consequences of this difference have been described previously; herbivores consume a greater proportion of aquatic than of terrestrial primary production (Duarte and Cebrian 1996), and trophic cascades are more common in aquatic than in terrestrial ecosystems (Shurin et al. 2006). Moreover, nutrients sequestered in aquatic primary production turn over faster than do nutrients in terrestrial primary production (Cebrian 2004).

In my view, these are issues of spatial and temporal scaling rather than fundamental ecological differences between aquatic and terrestrial ecosystems. Nutrients trapped in phytodetritus on the bottom of lakes, awaiting seasonal turnover, are not fundamentally different from nutrients trapped in the litter of terrestrial ecosystems during periods of cold temperature (temperate and boreal forest, tundra) or drought (deserts). Likewise, nutrient upwelling in coastal regions can be considered a form of ecosystem subsidy (figures 9.5B and 9.6); subsidies are well known in both aquatic and terrestrial ecosystems. Perhaps the biggest difference, then, is that the phytochemical landscape of terrestrial ecosystems is more recalcitrant to decomposition and nutrient cycling than is the phytochemical landscape of aquatic systems. I considered this in some detail in chapter

6, and I won't repeat that discussion here. It does, however, lead to a testable prediction:

> Prediction 10: In terrestrial systems, both trophic interactions and nutrient dynamics will be correlated in space with a broad mixture of phytochemical traits, including nutrient concentrations, toxins, and structural molecules. In aquatic systems, spatial correlations will be dominated by nutritional indices of the phytochemical landscape (e.g., N, P, Fe) associated with the small size, rapid growth, and short lifespans of phytoplankton.

Despite the importance of advection in aquatic ecosystems, spatial correlation does occur among phytoplankton nutrients, trophic interactions, and nutrient cycling. In other words, our rivers, lakes and oceans are still ecologically patchy, with major currents and smaller eddies contributing to the boundaries of those patches (Mahadevan 2014). At large spatial scales, the oceans contain provinces of nutrient cycling that map onto phytoplankton biomass and production (Capone 2014). For example, in the tropical Atlantic Ocean, there are provinces of both high and low N fixation, divided by the Intertropical Convergence Zone (ITCZ). North of the ITCZ, currents favors northward transport of iron-rich dust originating from African deserts. These iron-rich waters support high populations of N-fixing cyanobacteria that influence both N and P cycling—they add N from the atmosphere into ocean waters while depleting P concentrations (Schlosser et al. 2014). Productive and nutrient-rich phytoplankton then support production at higher trophic levels.

In summary, aquatic landscapes may contain well-defined spatial patches with strong coupling among nutrient dynamics, phytoplankton chemistry, and trophic interactions. However, we need much more research to understand variation in the size, location, and persistence of such patches on aquatic landscapes, and in particular to understand feedback processes with large consumers (Esselman et al. 2015). In my view, many of the apparent differences between aquatic and terrestrial ecosystems are really issues of spatial and temporal scale rather than fundamental differences in ecological processes. The concept of the phytochemical landscape, and its role in linking trophic interactions with nutrient dynamics, should apply in water as well as on land.

9.7 CONCLUDING REMARKS

In this chapter, I have suggested some priorities for future research. I have also provided a few testable predictions that arise from applying the concept of the phytochemical landscape to understanding how consumers and nutrient cycles are linked in ecological systems. These predictions emerge because we already know a substantial amount about the mechanisms that link trophic interactions with phytochemistry (chapters 4 and 5) and that link phytochemistry with nutrient dynamics (chapters 6 and 7). We have begun to put these mechanisms together to understand links between population processes and ecosystem processes (chapter 8) but we need to become more spatially explicit as we move forward (chapter 9).

There should be substantial benefit from doing so. For example, using maps like those illustrated in figure 9.1A, we will be able to test hypotheses regarding the relative roles of different kinds of phytochemistry in driving ecological processes in space. We can also explore whether or not their relative importance varies among ecosystems and over time. Moreover, we can continue the important task of understanding how interactions among molecular types on the phytochemical landscape, including interactions among nutrients, toxins, and structural molecules, influence ecological processes in space and time (Moore et al. 2010, Tao et al. 2014). Finally, the phytochemical landscape provides us with a mechanistic framework with which to explore the causes and consequences of heterogeneity and patchiness in our ecological systems.

Of course, humans remain a dominant cause of spatial and temporal heterogeneity at landscape scales. Understanding and mitigating our deleterious effects on ecological systems should benefit substantially from understanding and mitigating our deleterious effects on the phytochemical landscape. Our activities often compromise the chemical diversity and heterogeneity of phytochemical landscapes, with inevitable consequences for trophic interactions, biological diversity, and ecosystem processes. Conservation and restoration of the phytochemical landscape will help to mitigate some of the negative consequences of global environmental change. Moreover, we can better manage the phytochemical landscapes of our production systems and urban areas to sustain the goods and services that they provide. As I noted in chapter 3, all life in essence is complex chemistry, and it makes good sense to manage prudently the phytochemical landscapes that support us.

References Cited

Abbott, K. C. and G. Dwyer. 2007. Food limitation and insect outbreaks: complex dynamics in plant–herbivore models. *Journal of Animal Ecology* **76**:1004–1014.

Abbott, K. C., W. F. Morris, and K. Gross. 2008. Simultaneous effects of food limitation and inducible resistance on herbivore population dynamics. *Theoretical Population Biology* **73**:63–78.

Aber, J. D., J. M. Melillo, and C. A. McClaugherty. 1990. Predicting long-term patterns in mass loss, nitrogen dynamics, and soil organic matter formation from initial fine litter chemistry in temperate forest ecosystems. *Canadian Journal of Botany* **68**:2201–2208.

Aber, J. D., K. J. Nadelhoffer, P. Steudler, and J. M. Melillo. 1989. Nitrogen saturation in northern forest ecosystems—Hypotheses and implications. *Bioscience* **39**:378–386.

Abrahamson, W. G., M. D. Hunter, G. Melika, and P. W. Price. 2003. Cynipid gall-wasp communities correlate with oak chemistry. *Journal of Chemical Ecology* **29**:208–223.

Abrahamson, W. G. and A. E. Weis. 1997. *Evolutionary Ecology across Three Trophic Levels: Goldenrods, Gallmakers, and Natural Enemies* (MPB-29). Princeton University Press, Princeton, NJ.

Abrams, P. A. 1995. Implications of dynamically variable traits for identifying, classifying, and measuring direct and indirect effects in ecological communities. *American Naturalist* **146**:112–134.

Adams, L. G. 2003. Marrow fat deposition and skeletal growth in caribou calves. *Journal of Wildlife Management* **67**:20–24.

Agrawal, A. A. 2004. Plant defense and density dependence in the population growth of herbivores. *American Naturalist* **164**:113–120.

Agrawal, A. A. 2007. Macroevolution of plant defense strategies. *Trends in Ecology & Evolution* **22**:103–109.

Agrawal, A. A. and M. Fishbein. 2006. Plant defense syndromes. *Ecology* **87**:S132-S149.

Agrawal, A. A., G. Petschenka, R. A. Bingham, M. G. Weber, and S. Rasmann. 2012. Toxic cardenolides: chemical ecology and coevolution of specialized plant-herbivore interactions. *New Phytologist* **194**:28–45.

Aguilera, M. A. and S. A. Navarrete. 2012. Functional identity and functional structure change through succession in a rocky intertidal marine herbivore assemblage. *Ecology* **93**:75–89.

Akiyama, K. and H. Hayashi. 2002. Arbuscular mycorrhizal fungus-promoted accumulation of two new triterpenoids in cucumber roots. *Bioscience Biotechnology and Biochemistry* **66**:762–769.

Alamri, S. A. 2012. Biodegradation of microcystin-RR by *Bacillus flexus* isolated from a Saudi freshwater lake. *Saudi Journal of Biological Sciences* **19**:435–440.

Alborn, H. T., T. C. J. Turlings, T. H. Jones, G. Stenhagen, J. H. Loughrin, and J. H. Tumlinson. 1997. An elicitor of plant volatiles from beet armyworm oral secretion. *Science* **276**:945–949.

Ali, J. G., H. T. Alborn, R. Campos-Herrera, F. Kaplan, L. W. Duncan, C. Rodriguez-Saona, A. M. Koppenhofer, and L. L. Stelinski. 2012. Subterranean, herbivore-induced plant volatile increases biological control activity of multiple beneficial nematode species in distinct habitats. *PLOS ONE* **7**.

Alldredge, A. L. and M. W. Silver. 1988. Characteristics, dynamics and significance of marine snow. *Progress in Oceanography* **20**:41–82.

Allen, M. M. 1984. Cyanobacterial cell inclusions. *Annual Review of Microbiology* **38**:1–25.

Allgeier, J. E., C. A. Layman, P. J. Mumby, and A. D. Rosemond. 2014. Consistent nutrient storage and supply mediated by diverse fish communities in coral reef ecosystems. *Global Change Biology* **20**:2459–2472.

Allgeier, J. E., L. A. Yeager, and C. A. Layman. 2013. Consumers regulate nutrient limitation regimes and primary production in seagrass ecosystems. *Ecology* **94**:521–529.

Allison, S. D. 2006. Brown ground: A soil carbon analogue for the green world hypothesis? *American Naturalist* **167**:619–627.

Allison, S. D. 2012. A trait-based approach for modelling microbial litter decomposition. *Ecology Letters* **15**:1058–1070.

Alpermann, T. J., B. Beszteri, U. John, U. Tillmann, and A. D. Cembella. 2009. Implications of life-history transitions on the population genetic structure of the toxigenic marine dinoflagellate Alexandrium tamarense. *Molecular Ecology* **18**:2122–2133.

Alpermann, T. J., U. Tillmann, B. Beszteri, A. D. Cembella, and U. John. 2010. Phenotypic varation and genotypc diversity in a planktonic population of the toxigenic marine dinoflagellate *Alexandrium tamarense* (Dinophyceae). *Journal of Phycology* **46**:18–32.

Altieri, A. H., M. D. Bertness, T. C. Coverdale, N. C. Herrmann, and C. Angelini. 2012. A trophic cascade triggers collapse of a salt-marsh ecosystem with intensive recreational fishing. *Ecology* **93**:1402–1410.

Amin, S. A., D. H. Green, M. C. Hart, F. C. Kuepper, W. G. Sunda, and C. J. Carrano. 2009. Photolysis of iron-siderophore chelates promotes bacterial-algal mutualism. *Proceedings of the National Academy of Sciences* **106**:17071–17076.

Amsler, C. D. and V. A. Fairhead. 2006. Defensive and sensory chemical ecology of brown algae. Pages 1–91 *in* J. A. Callow, editor. *Advances in Botanical Research*. Vol. 43, *Incorporating Advances in Plant Pathology*. Academic Press Ltd-Elsevier Science Ltd, London.

Anantharaman, K., M. B. Duhaime, J. A. Breier, K. A. Wendt, B. M. Toner, and G. J. Dick. 2014. Sulfur oxidation genes in diverse deep-sea viruses. *Science* **344**:757–760.

Andersen, D. C. 1987. Below-ground herbivory in natural communities: a review emphasizing fossorial animals. *The Quarterly Review of Biology* **62**:261–286.

Anderson, D., P. Glibert, and J. Burkholder. 2002. Harmful algal blooms and eutrophication: Nutrient sources, composition, and consequences. *Estuaries and Coasts* **25**:704–726.

Anderson, D. M., A. D. Cembella, and G. M. Hallegraeff. 2012. Progress in understanding harmful algal blooms: Paradigm shifts and new technologies for research, monitoring, and management. Pages 143–176 *in* C. A. Carlson and S. J. Giovannoni, editors. *Annual Review of Marine Science*. Vol. 4. Annual Reviews, Palo Alto, CA.

Anderson, D. M., D. M. Kulis, J. J. Sullivan, and S. Hall. 1990. Toxin composition variations in one isolate of the dinoflagellate *Alexandrium fundyense*. *Toxicon* **28**:885–893.

Anderson, T. R., M. Boersma, and D. Raubenheimer. 2004. Stoichiometry: Linking elements to biochemicals. *Ecology* **85**:1193–1202.

Anderson, W. B. and G. A. Polis. 1999. Nutrient fluxes from water to land: seabirds affect plant nutrient status on Gulf of California islands. *Oecologia* **118**:324–332.

Andrew, R. L., R. Peakall, I. R. Wallis, and W. J. Foley. 2007. Spatial distribution of defense chemicals and markers and the maintenance of chemical variation. *Ecology* **88**:716–728.

Andrianasolo, E. H., L. Haramaty, R. Rosario-Passapera, C. Vetriani, P. Falkowski, E. White, and R. Lutz. 2012. Ammonificins C and D, hydroxyethylamine chromene derivatives from a cultured marine hydrothermal vent bacterium, *Thermovibrio ammonificans*. *Marine Drugs* **10**:2300–2311.

Antunes, J. T., P. N. Leao, and V. M. Vasconcelos. 2012. Influence of biotic and abiotic factors on the allelopathic activity of the cyanobacterium *Cylindrospermopsis raciborskii* strain LEGE 99043. *Microbial Ecology* **64**:584–592.

Aquilino, K. M., M. E. S. Bracken, M. N. Faubel, and J. J. Stachowicz. 2009. Local-scale nutrient regeneration facilitates seaweed growth on wave-exposed rocky shores in an upwelling system. *Limnology and Oceanography* **54**:309–317.

Aquilino, K. M. and J. J. Stachowicz. 2012. Seaweed richness and herbivory increase rate of community recovery from disturbance. *Ecology* **93**:879–890.

Arab, A. and G. M. Wimp. 2013. Plant production and alternate prey channels impact the abundance of top predators. *Oecologia* **173**:331–341.

Ardiles, V., J. Alcocer, G. Vilaclara, L. A. Oseguera, and L. Velasco. 2012. Diatom fluxes in a tropical, oligotrophic lake dominated by large-sized phytoplankton. *Hydrobiologia* **679**:77–90.

Arditi, R. and L. R. Ginzburg. 2012. *How Species Interact: Altering the Standard View on Trophic Ecology*. Oxford University Press, Oxford, UK.

Arim, M. and P. A. Marquet. 2004. Intraguild predation: a widespread interaction related to species biology. *Ecology Letters* **7**:557–564.

Arndt, S., B. B. Jorgensen, D. E. LaRowe, J. J. Middelburg, R. D. Pancost, and P. Regnier. 2013. Quantifying the degradation of organic matter in marine sediments: A review and synthesis. *Earth-Science Reviews* **123**:53–86.

Asner, G. P., R. E. Martin, R. Tupayachi, C. B. Anderson, F. Sinca, L. Carranza-Jiménez, and P. Martinez. 2014. Amazonian functional diversity from forest canopy chemical assembly. *Proceedings of the National Academy of Sciences* **111**:5604–5609.

Assmy, P., V. Smetacek, M. Montresor, C. Klaas, J. Henjes, V. H. Strass, J. M. Arrieta, U. Bathmann, G. M. Berg, E. Breitbarth, B. Cisewski, L. Friedrichs, N. Fuchs, G. J. Herndl, S. Jansen, S. Krägefsky, M. Latasa, I. Peeken, R. Röttgers, R. Scharek, S. E. Schüller, S. Steigenberger, A. Webb, and D. Wolf-Gladrow. 2013.

Thick-shelled, grazer-protected diatoms decouple ocean carbon and silicon cycles in the iron-limited Antarctic Circumpolar Current. *Proceedings of the National Academy of Sciences* **110**:20633–20638.

Atkinson, C. L., C. C. Vaughn, K. J. Forshay, and J. T. Cooper. 2013. Aggregated filter-feeding consumers alter nutrient limitation: consequences for ecosystem and community dynamics. *Ecology* **94**:1359–1369.

Atwood, T. B., E. Hammill, H. S. Greig, P. Kratina, J. B. Shurin, D. S. Srivastava, and J. S. Richardson. 2013. Predator-induced reduction of freshwater carbon dioxide emissions. *Nature Geoscience* **6**:191–194.

Ausmus, B. S., J. M. Ferris, D. E. Reichle, and E. C. Williams. 1978. Role of belowground herbivores in mesic forest root dynamics. *Pedobiologia* **18**:289–295.

Avise, J. C. 2000. *Phylogeography. The history and formation of species.* Harvard University Press, Cambridge, Massachusetts, USA.

Avrani, S., O. Wurtzel, I. Sharon, R. Sorek, and D. Lindell. 2011. Genomic island variability facilitates *Prochlorococcus*-virus coexistence. *Nature* **474**:604–608.

Awmack, C. S. and S. R. Leather. 2002. Host plant quality and fecundity in herbivorous insects. *Annual Review of Entomology* **47**:817–844.

Azam, F., T. Fenchel, J. G. Field, J. S. Gray, L. A. Meyer-Reil, and F. Thingstad. 1983. The ecological role of water-column microbes in the sea. *Marine Ecology Progress Series* **10**:257–263.

Babikova, Z., L. Gilbert, T. J. A. Bruce, M. Birkett, J. C. Caulfield, C. Woodcock, J. A. Pickett, and D. Johnson. 2013. Underground signals carried through common mycelial networks warn neighbouring plants of aphid attack. *Ecology Letters* **16**:835–843.

Babst, B. A., R. A. Ferrieri, D. W. Gray, M. Lerdau, D. J. Schlyer, M. Schueller, M. R. Thorpe, and C. M. Orians. 2005. Jasmonic acid induces rapid changes in carbon transport and partitioning in *Populus*. *New Phytologist* **167**:63–72.

Bailey, J. K., J. A. Schweitzer, F. Ubeda, J. Koricheva, C. J. LeRoy, M. D. Madritch, B. J. Rehill, R. K. Bangert, D. G. Fischer, G. J. Allan, and T. G. Whitham. 2009. From genes to ecosystems: a synthesis of the effects of plant genetic factors across levels of organization. *Philosophical Transactions of the Royal Society B: Biological Sciences* **364**:1607–1616.

Balch, J. K., B. A. Bradley, C. M. D'Antonio, and J. Gómez-Dans. 2013. Introduced annual grass increases regional fire activity across the arid western USA (1980–2009). *Global Change Biology* **19**:173–183.

Baldwin, I. T. 1989. The mechanism of damaged-induced alkaloids in wild tobacco. *Journal of Chemical Ecology* **15**:1661–1680.

Baldwin, I. T., A. S. Eric, and T. E. Ohnmeiss. 1994. Wound-induced changes in root and shoot jasmonic acid pools correlate with induced nicotine synthesis in *Nicotiana sylvestris* Spegazzini and Comes. *Journal of Chemical Ecology* **20**:2139–2157.

Baldwin, I. T. and J. C. Schultz. 1983. Rapid changes in tree leaf chemistry induced by damage: evidence for communication between plants. *Science* **221**:277–279.

Ball, B. A., M. A. Bradford, D. C. Coleman, and M. D. Hunter. 2009a. Linkages between below and aboveground communities: Decomposer responses to simulated tree species loss are largely additive. *Soil Biology & Biochemistry* **41**:1155–1163.

Ball, B. A., M. A. Bradford, and M. D. Hunter. 2009b. Nitrogen and phosphorus release from mixed litter layers is lower than predicted from single species decay. *Ecosystems* **12**:87–100.

Ball, B. A., M. D. Hunter, J. S. Kominoski, C. M. Swan, and M. A. Bradford. 2008. Consequences of non-random species loss for decomposition dynamics: Experimental evidence for additive and non-additive effects. *Journal of Ecology* **96**:303–313.

Baltensweiler, W., G. Benz, P. Bovey, and P. Delucchi. 1977. Dynamics of larch bud moth populations. *Annual Review of Entomology* **22**:79–100.

Baltensweiler, W. and A. Fischlin. 1988. The larch budmoth in the Alps. Pages 331–351 *in* A. A. Berryman, editor. *Dynamics of Forest Insect Populations: Patterns, Causes, Implications.* Plenum, New York.

Banks, J. A., T. Nishiyama, M. Hasebe, J. L. Bowman, M. Gribskov, C. dePamphilis, V. A. Albert, N. Aono, T. Aoyama, B. A. Ambrose, N. W. Ashton, M. J. Axtell, E. Barker, M. S. Barker, J. L. Bennetzen, N. D. Bonawitz, C. Chapple, C. Y. Cheng, L. G. G. Correa, M. Dacre, J. DeBarry, I. Dreyer, M. Elias, E. M. Engstrom, M. Estelle, L. Feng, C. Finet, S. K. Floyd, W. B. Frommer, T. Fujita, L. Gramzow, M. Gutensohn, J. Harholt, M. Hattori, A. Heyl, T. Hirai, Y. Hiwatashi, M. Ishikawa, M. Iwata, K. G. Karol, B. Koehler, U. Kolukisaoglu, M. Kubo, T. Kurata, S. Lalonde, K. J. Li, Y. Li, A. Litt, E. Lyons, G. Manning, T. Maruyama, T. P. Michael, K. Mikami, S. Miyazaki ,S. Morinaga, T. Murata, B. Mueller-Roeber, D. R. Nelson, M. Obara, Y. Oguri, R. G. Olmstead, N. Onodera, B. L. Petersen, B. Pils, M. Prigge, S. A. Rensing, D. M. Riano-Pachon, A. W. Roberts, Y. Sato, H. V. Scheller, B. Schulz, C. Schulz, E. V. Shakirov, N. Shibagaki, N. Shinohara, D. E. Shippen, I. Sorensen, R. Sotooka, N. Sugimoto, M. Sugita, N. Sumikawa, M. Tanurdzic, G. Theissen, P. Ulvskov, S. Wakazuki, J. K. Weng, W. Willats, D. Wipf, P. G. Wolf, L. X. Yang, A. D. Zimmer, Q. H. Zhu, T. Mitros, U. Hellsten, D. Loque, R. Otillar, A. Salamov, J. Schmutz, H. Shapiro, E. Lindquist, S. Lucas, D. Rokhsar, and I. V. Grigoriev. 2011. The *Selaginella* genome identifies genetic changes associated with the evolution of vascular plants. *Science* **332**:960–963.

Bar-Yosef, Y., A. Sukenik, O. Hadas, Y. Viner-Mozzini, and A. Kaplan. 2010. Enslavement in the water body by toxic *Aphanizomenon ovalisporum*, inducing alkaline phosphatase in phytoplanktons. *Current Biology* **20**:1557–1561.

Barantal, S., H. Schimann, N. Fromin, and S. Hättenschwiler. 2014. C, N and P fertilization in an Amazonian rainforest supports stoichiometric dissimilarity as a driver of litter diversity effects on decomposition. *Proceedings of the Royal Society B: Biological Sciences* **281**.

Barbehenn, R. V., S. Cheek, A. Gasperut, E. Lister, and R. Maben. 2005. Phenolic compounds in red oak and sugar maple leaves have prooxidant activities in the midgut fluids of *Malacosoma disstria* and *Orgyia leucostigma* caterpillars. *Journal of Chemical Ecology* **31**:969–988.

Barbehenn, R. V. and C. P. Constabel. 2011. Tannins in plant-herbivore interactions. *Phytochemistry* **72**:1551–1565.

Barbehenn, R. V., A. Jaros, G. Lee, C. Mozola, Q. Weir, and J. P. Salminen. 2009. Tree resistance to *Lymantria dispar* caterpillars: Importance and limitations of foliar tannin composition. *Oecologia* **159**:777–788.

Barbehenn, R. V., J. Niewiadomski, and J. Kochmanski. 2013. Importance of protein quality versus quantity in alternative host plants for a leaf-feeding insect. *Oecologia* **173**:1–12.

Barber, I. 2005. Parasites grow larger in faster growing fish hosts. *International Journal for Parasitology* **35**:137–143.

Barbosa, P., P. Gross, and J. Kemper. 1991. Influence of plant allelochemicals on the tobacco hornworm and its parasitoid, *Cotesia congregata. Ecology* **72**:1567–1575.

Barbosa, P., J. Hines, I. Kaplan, H. Martinson, A. Szczepaniec, and Z. Szendrei. 2009. Associational resistance and associational susceptibility: Having right or wrong neighbors. *Annual Review of Ecology Evolution and Systematics* **40**:1–20.

Bardgett, R. D. and D. A. Wardle. 2003. Herbivore-mediated linkages between aboveground and belowground communities. *Ecology* **84**:2258–2268.

Bardgett, R. D. and D. A. Wardle. 2010. *Aboveground-Belowground Linkages: Biotic Interactions, Ecosystem Processes, and Global Change.* Oxford University Press, Oxford, UK.

Barile, P. J., B. E. Lapointe, and T. R. Capo. 2004. Dietary nitrogen availability in macroalgae enhances growth of the sea hare *Aplysia californica* (Opisthobranchia : Anaspidea). *Journal of Experimental Marine Biology and Ecology* **303**:65–78.

Barnes, B. V., D. R. Zak, S. R. Denton, and S. H. Spurr. 1998. *Forest ecology.* Wiley, New York.

Barnes, R. L. 1963. Organic nitrogen compounds in tree xylem sap. *Forest Science* **9**:98–102.

Barreiro, A., C. Guisande, M. Frangopulos, A. Gonzalez-Fernandez, S. Munoz, D. Perez, S. Magadan, I. Maneiro, I. Riveiro, and P. Iglesias. 2006. Feeding strategies of the copepod Acartia clausi on single and mixed diets of toxic and non-toxic strains of the dinoflagellate *Alexandrium minutum. Marine Ecology-Progress Series* **316**:115–125.

Bartell, S. M. 1981. Potential impact of size-selective planktivory on phosphorus release by zooplankton. *Hydrobiologia* **80**:139–145.

Barton, A. D., A. J. Pershing, E. Litchman, N. R. Record, K. F. Edwards, Z. V. Finkel, T. Kiørboe, and B. A. Ward. 2013a. The biogeography of marine plankton traits. *Ecology Letters* **16**:522–534.

Barton, P. S., S. A. Cunningham, D. B. Lindenmayer, and A. D. Manning. 2013b. The role of carrion in maintaining biodiversity and ecological processes in terrestrial ecosystems. *Oecologia* **171**:761–772.

Bauerfeind, S. S. and K. Fischer. 2013. Increased temperature reduces herbivore host-plant quality. *Global Change Biology* **19**:3272–3282.

Baxter, I. and B. P. Dilkes. 2012. Elemental profiles reflect plant adaptations to the environment. *Science* **336**:1661–1663.

Beasley, J. C., Z. H. Olson, and T. L. DeVault. 2012. Carrion cycling in food webs: Comparisons among terrestrial and marine ecosystems. *Oikos* **121**:1021–1026.

Becerra, J. X. 1997. Insects on plants: Macroevolutionary chemical trends in host use. *Science* **276**:253–256.

Becerra, J. X. 2003. Synchronous coadaptation in an ancient case of herbivory. *Proceedings of the National Academy of Sciences* **100**:12804–12807.

Becerra, J. X. and D. L. Venable. 1999. Macroevolution of insect-plant associations: The relevance of host biogeography to host affiliation. *Proceedings of the National Academy of Sciences* **96**:12626–12631.

Bechara, J. A., D. Planas, and S. Paquet. 2007. Indirect effects of brook trout (*Salvelinus fontinalis*) on the structure of epilithic algal communities in an oligotrophic boreal forest stream. *Fundamental and Applied Limnology* **169**:89–99.

Beckerman, A. P., M. Uriarte, and O. J. Schmitz. 1997. Experimental evidence for a behavior-mediated trophic cascade in a terrestrial food chain. *Proceedings of the National Academy of Sciences* **94**:10735–10738.

Bednarska, A. and P. Dawidowicz. 2007. Change in filter-screen morphology and depth selection: Uncoupled responses of *Daphnia* to the presence of filamentous cyanobacteria. *Limnology and Oceanography* **52**:2358–2363.

Behie, S. W., P. M. Zelisko, and M. J. Bidochka. 2012. Endophytic insect-parasitic fungi translocate nitrogen directly from insects to plants. *Science* **336**:1576–1577.

Behmer, S. T., S. J. Simpson, and D. Raubenheimer. 2002. Herbivore foraging in chemically heterogeneous environments: Nutrients and secondary metabolites. *Ecology* **83**:2489–2501.

Bell, W. H., J. M. Lang, and R. Mitchell. 1974. Selective stimulation of marine bacteria by algal extracellular products. *Limnology and Oceanography* **19**:833–839.

Belovsky, G. E. and A. Joern. 1995. The dominance of different regulating factors for rangeland grasshoppers. Pages 359–386 *in* N. Cappuccino and P. W. Price, editors. *Population Dynamics: New Approaches and Synthesis*. Academic Press, New York.

Belovsky, G. E. and J. B. Slade. 2000. Insect herbivory accelerates nutrient cycling and increases plant production. *Proceedings of the National Academy of Sciences* **97**:14412–14417.

Bennett, A. E., J. Alers-Garcia, and J. D. Bever. 2006. Three-way interactions among mutualistic mycorrhizal fungi, plants, and plant enemies: Hypotheses and synthesis. *American Naturalist* **167**:141–152.

Benrey, B. and R. F. Denno. 1997. The slow-growth-high-mortality hypothesis: A test using the cabbage butterfly. *Ecology* **78**:987–999.

Benton, T. G., S. J. Plaistow, A. P. Beckerman, C. T. Lapsley, and S. Littlejohns. 2005. Changes in maternal investment in eggs can affect population dynamics. *Proceedings of the Royal Society of London. Series B: Biological Sciences* **272**:1351–1356.

Berenbaum, M. and J. J. Neal. 1985. Synergism between myristicin and xanthotoxin, a naturally co-occurring plant toxicant. *Journal of Chemical Ecology* **11**:1349–1358.

Berenbaum, M. R. 1995. The chemistry of defense: theory and practice. *Proceedings of the National Academy of Science* **92**:2–8.

Berg, B., M. P. Davey, A. De Marco, B. Emmett, M. Faituri, S. E. Hobbie, M. B. Johansson, C. Liu, C. McClaugherty, L. Norell, F. A. Rutigliano, L. Vesterdal, and A. V. De Santo. 2010. Factors influencing limit values for pine needle litter decomposition: A synthesis for boreal and temperate pine forest systems. *Biogeochemistry* **100**:57–73.

Berg, B. and C. McClaugherty. 2003. *Plant litter: decomposition, humus formation, carbon sequestration*. Springer-Verlag, Berlin, Germany.

Bergman, B., W. W. Zheng, J. Klint, and L. Ran. 2008. On the origin of plants and relations to contemporary cyanobacterial-plant symbioses. *Plant Biotechnology* **25**:213–220.

Bergström, A.-K., C. Faithfull, D. Karlsson, and J. Karlsson. 2013. Nitrogen deposition and warming—Effects on phytoplankton nutrient limitation in subarctic lakes. *Global Change Biology* **19**:2557–2568.

Bernal, J. S., R. F. Luck, and J. G. Morse. 1999. Host influences on sex ratio, longevity, and egg load of two Metaphycus species parasitic on soft scales: Implications for insectary rearing. *Entomologia Experimentalis Et Applicata* **92**:191–204.

Bernards, M. A. 2002. Demystifying suberin. *Canadian Journal of Botany-Revue Canadienne De Botanique* **80**:227–240.

Bernays, E. A. and M. Graham. 1988. On the evolution of host specificity in phytophagous arthropods. *Ecology* **69**:886–892.

Beschta, R. L. and W. J. Ripple. 2013. Are wolves saving Yellowstone's aspen? A landscape-level test of a behaviorally mediated trophic cascade: Comment. *Ecology* **94**:1420–1425.

Bever, J. D. 1994. Feedback between plants and their soil communities in an old field community. *Ecology* **75**:1965–1977.

Bever, J. D. 2002. Negative feedback within a mutualism: Host-specific growth of mycorrhizal fungi reduces plant benefit. *Proceedings of the Royal Society of London Series B: Biological Sciences* **269**:2595–2601.

Bezemer, T. M., R. Wagenaar, N. M. Van Dam, and F. L. Wackers. 2003. Interactions between above- and belowground insect herbivores as mediated by the plant defense system. *Oikos* **101**:555–562.

Biller, S. J., F. Schubotz, S. E. Roggensack, A. W. Thompson, R. E. Summons, and S. W. Chisholm. 2014. Bacterial Vesicles in Marine Ecosystems. *Science* **343**:183–186.

Binkley, D. and C. Giardina. 1998. Why do tree species affect soils? The warp and woof of tree-soil interactions. *Biogeochemistry* **42**:89–106.

Bishop, J. G. 2002. Early primary succession on Mount St. Helens: Impact of insect herbivores on colonizing lupines. *Ecology* **83**:191–202.

Bjornstad, O. N., R. A. Ims, and X. Lambin. 1999. Spatial population dynamics: Analyzing patterns and processes of population synchrony. *Trends in Ecology & Evolution* **14**:427–432.

Blake, S., C. B. Yackulic, F. Cabrera, W. Tapia, J. P. Gibbs, F. Kümmeth, and M. Wikelski. 2013. Vegetation dynamics drive segregation by body size in Galapagos tortoises migrating across altitudinal gradients. *Journal of Animal Ecology* **82**:310–321.

Blunt, J. W., B. R. Copp, W. P. Hu, M. H. G. Munro, P. T. Northcote, and M. R. Prinsep. 2008. Marine natural products. *Natural Product Reports* **25**:35–94.

Boersma, M., C. Becker, A. M. Malzahn, and S. Vernooij. 2009. Food chain effects of nutrient limitation in primary producers. *Marine and Freshwater Research* **60**:983–989.

Boersma, M. and J. J. Elser. 2006. Too much of a good thing: On stoichiometrically balanced diets and maximal growth. *Ecology* **87**:1325–1330.

Boersma, M. and C. Kreutzer. 2002. Life at the edge: Is food quality really of minor importance at low quantities? *Ecology* **83**:2552–2561.

Bolser, R. C. and M. E. Hay. 1996. Are tropical plants better defended? Palatability and defenses of temperate vs tropical seaweeds. *Ecology* **77**:2269–2286.

Bolser, R. C., M. E. Hay, N. Lindquist, W. Fenical, and D. Wilson. 1998. Chemical defenses of freshwater macrophytes against crayfish herbivory. *Journal of Chemical Ecology* **24**:1639–1658.

Borell, E. M., A. Foggo, and R. A. Coleman. 2004. Induced resistance in intertidal macroalgae modifies feeding behaviour of herbivorous snails. *Oecologia* **140**:328–334.

Borer, E. T., E. W. Seabloom, and D. Tilman. 2012. Plant diversity controls arthropod biomass and temporal stability. *Ecology Letters* **15**:1457–1464.

Borlongan, I. G. and S. Satoh. 2001. Dietary phosphorus requirement of juvenile milkfish, *Chanos chanos* (Forsskal). *Aquaculture Research* **32**:26–32.

Bowers, M. D. 2003. Hostplant suitability and defensive chemistry of the Catalpa sphinx, *Ceratomia catalpae*. *Journal of Chemical Ecology* **29**:2359–2367.

Bracken, M. E. S., R. E. Dolecal, and J. D. Long. 2014. Community context mediates the top-down vs. bottom-up effects of grazers on rocky shores. *Ecology* **95**:1458–1463.

Bradley, C. A. and S. Altizer. 2005. Parasites hinder monarch butterfly flight: Implications for disease spread in migratory hosts. *Ecology Letters* **8**:290–300.

Bradley, R. L., B. D. Titus, and C. P. Preston. 2000. Changes to mineral N cycling and microbial communities in black spruce humus after additions of (NH4)(2)SO4 and condensed tannins extracted from *Kalmia angustifolia* and balsam fir. *Soil Biology & Biochemistry* **32**:1227–1240.

Brandle, M., S. Knoll, S. Eber, J. Stadler, and R. Brandl. 2005. Flies on thistles: Support for synchronous speciation? *Biological Journal of the Linnean Society* **84**:775–783.

Breckels, M. N., E. C. Roberts, S. D. Archer, G. Malin, and M. Steinke. 2011. The role of dissolved infochemicals in mediating predator-prey interactions in the heterotrophic dinoflagellate *Oxyrrhis marina*. *Journal of Plankton Research* **33**:629–639.

Bretherton, W. D., J. S. Kominoski, D. G. Fischer, and C. J. LeRoy. 2011. Salmon carcasses alter leaf litter species diversity effects on in-stream decomposition. *Canadian Journal of Fisheries and Aquatic Sciences* **68**:1495–1506.

Brett, M. T. and D. C. Muller-Navarra. 1997. The role of highly unsaturated fatty acids in aquatic food web processes. *Freshwater Biology* **38**:483–499.

Brodbeck, B. V., P. C. Andersen, and R. F. Mizell. 2011. Nutrient mediation of behavioral plasticity and resource allocation in a xylem-feeding leafhopper. *Oecologia* **165**:111–122.

Bronstein, J. L. 2001. The costs of mutualism. *American Zoologist* **41**:825–839.

Brower, L. P. 1958. Bird predation and food plant specificity in closely related procryptic insects. *American Naturalist* **42**:183–187.

Brower, L. P., M. Edmunds, and C. M. Moffitt. 1975. Cardenolide content and palatability of a population of *Danaus chrysippus* butterflies from West Africa. *Journal of Entomology* **49**:183–196.

Brown, J. H., J. F. Gillooly, A. P. Allen, V. M. Savage, and G. B. West. 2004. Toward a metabolic theory of ecology. *Ecology* **85**:1771–1789.

Brown, J. H. and E. J. Heske. 1990. Control of a desert-grassland transition by a keystone rodent guild. *Science* **250**:1705–1707.

Brown, J. S. and B. P. Kotler. 2004. Hazardous duty pay and the foraging cost of predation. *Ecology Letters* **7**:999–1014.

Brown, V. K. and A. C. Gange. 1989a. Differential effects of above- and below-ground insect herbivory during early plant succession. *Oikos* **54**:67–76.

Brown, V. K. and A. C. Gange. 1989b. Herbivory by soil-dwelling insects depresses plant species richness. *Functional Ecology* **3**:667–671.

Brown, V. K. and A. C. Gange. 1992. Secondary plant succession: How is it modified by insect herbivory? *Vegetatio* **101**:3–13.

Bruessow, F., C. Gouhier-Darimont, A. Buchala, J.-P. Metraux, and P. Reymond. 2010. Insect eggs suppress plant defence against chewing herbivores. *Plant Journal* **62**:876–885.

Bryant, J. P., F. S. Chapin, and D. R. Klein. 1983. Carbon nutrient balance of boreal plants in relation to vertebrate herbivory. *Oikos* **40**:357–368.

Bryant, J. P., T. P. Clausen, R. K. Swihart, S. M. Landhausser, M. T. Stevens, C. D. B. Hawkins, S. Carriere, A. P. Kirilenko, A. M. Veitch, R. A. Popko, D. T. Cleland, J. H. Williams, W. J. Jakubas, M. R. Carlson, K. L. Bodony, M. Cebrian, T. F. Paragi,

P. M. Picone, J. E. Moore, E. C. Packee, and T. Malone. 2009. Fire drives transconti-
    nental variation in tree birch defense against browsing by snowshoe hares. *American
    Naturalist* **174**:13–23.

Bryant, J. P., F. D. Provenza, J. Pastor, P. B. Reichardt, T. P. Clausen, and J. T. Dutoit. 1991.
    Interactions between woody-plants and browsing mammals mediated by secondary
    metabolites. *Annual Review of Ecology and Systematics* **22**:431–446.

Bryant, J. P., P. B. Reichardt, T. P. Clausen, and R. A. Werner. 1993. Effects of mineral nutri-
    tion on delayed inducible resistance in Alaska paper birch. *Ecology* **74**:2072–2084.

Buchner, P. 1965. *Endosymbiosis of animals with plant microorganisms*. Wiley Inter-
    science, New York.

Bultman, H., D. Hoekman, J. Dreyer, and C. Gratton. 2014. Terrestrial deposition of
    aquatic insects increases plant quality for insect herbivores and herbivore density.
    Ecological Entomology **39**:419–426.

Bump, J. K., R. O. Peterson, and J. A. Vucetich. 2009a. Wolves modulate soil nutrient het-
    erogeneity and foliar nitrogen by configuring the distribution of ungulate carcasses.
    Ecology **90**:3159–3167.

Bump, J. K., C. R. Webster, J. A. Vucetich, R. O. Peterson, J. M. Shields, and M. D. Pow-
    ers. 2009b. Ungulate carcasses perforate ecological filters and create biogeochem-
    ical hotspots in forest herbaceous layers allowing trees a competitive advantage.
    Ecosystems **12**:996–1007.

Burd, A. B., D. A. Hansell, D. K. Steinberg, T. R. Anderson, J. Arístegui, F. Baltar, S. R.
    Beaupré, K. O. Buesseler, F. DeHairs, G. A. Jackson, D. C. Kadko, R. Koppelmann,
    R. S. Lampitt, T. Nagata, T. Reinthaler, C. Robinson, B. H. Robison, C. Tambu-
    rini, and T. Tanaka. 2010. Assessing the apparent imbalance between geochemical
    and biochemical indicators of meso- and bathypelagic biological activity: What the
    @$#! is wrong with present calculations of carbon budgets? *Deep Sea Research Part
    II: Topical Studies in Oceanography* **57**:1557–1571.

Burdige, D. J. 2007. Preservation of organic matter in marine sediments: Controls, mech-
    anisms, and an imbalance in sediment organic carbon budgets? *Chemical Reviews*
    107:467–485.

Burghardt, K. T. and O. J. Schmitz. 2015. Influence of plant defenses and nutrients on
    trophic control of ecosystems. Pages 203–232 *in* T. C. Hanley and K. J. La Pierre,
    editors. *Trophic ecology: Bottom-up and top-down interactions across aquatic and
    terrestrial ecosystems*. Cambridge University Press, Cambridge, UK.

Burkholder, D. A., M. R. Heithaus, J. W. Fourqurean, A. Wirsing, and L. M. Dill. 2013.
    Patterns of top-down control in a seagrass ecosystem: Could a roving apex pred-
    ator induce a behaviour-mediated trophic cascade? *Journal of Animal Ecology*
    82:1192–1202.

Burns, J. M., S. Hall, and J. L. Ferry. 2009. The adsorption of saxitoxin to clays and sedi-
    ments in fresh and saline waters. *Water Research* **43**:1899–1904.

Campbell, B. C. and S. S. Duffey. 1979. Tomatine and parasitic wasps: Potential incompat-
    ibility of plant antibiosis with biological control. *Science* **205**:700–702.

Capone, D. G. 2014. An iron curtain in the Atlantic Ocean forms a biogeochemical divide.
    Proceedings of the National Academy of Sciences **111**:1231–1232.

Cardinale, B. J., K. L. Matulich, D. U. Hooper, J. E. Byrnes, E. Duffy, L. Gamfeldt, P. Bal-
    vanera, M. I. O'Connor, and A. Gonzalez. 2011. The functional role of producer
    diversity in ecosystems. *American Journal of Botany* **98**:572–592.

Cardinale, B. J., D. S. Srivastava, J. E. Duffy, J. P. Wright, A. L. Downing, M. Sankaran, and C. Jouseau. 2006. Effects of biodiversity on the functioning of trophic groups and ecosystems. *Nature* **443**:989–992.

Carlson, C. A., S. J. Giovannoni, D. A. Hansell, S. J. Goldberg, R. Parsons, and K. Vergin. 2004. Interactions among dissolved organic carbon, microbial processes, and community structure in the mesopelagic zone of the northwestern Sargasso Sea. *Limnology and Oceanography* **49**:1073–1083.

Carpenter, R. C. 1986. Partitioning herbivory and its effects on coral reef algal communities. *Ecological Monographs* **56**:345–363.

Carpenter, S. R., J. F. Kitchell, and J. R. Hodgson. 1985. Cascading trophic interactions and lake productivity. *Bioscience* **35**:634–639.

Carter, D. O., D. Yellowlees, and M. Tibbett. 2008. Temperature affects microbial decomposition of cadavers (*Rattus rattus*) in contrasting soils. *Applied Soil Ecology* **40**:129–137.

Casper, B. B. and J. P. Castelli. 2007. Evaluating plant-soil feedback together with competition in a serpentine grassland. *Ecology Letters* **10**:394–400.

Castagneyrol, B. and H. Jactel. 2012. Unraveling plant–animal diversity relationships: A meta-regression analysis. *Ecology* **93**:2115–2124.

Casteleyn, G., V. A. Chepurnov, F. Leliaert, D. G. Mann, S. S. Bates, N. Lundholm, L. Rhodes, K. Sabbe, and W. Vyverman. 2008. *Pseudo-nitzschia pungens* (Bacillariophyceae): A cosmopolitan diatom species? *Harmful Algae* **7**:241–257.

Castella, G., M. Chapuisat, and P. Christe. 2008. Prophylaxis with resin in wood ants. *Animal Behaviour* **75**:1591–1596.

Castro-Díez, P., O. Godoy, A. Alonso, A. Gallardo, and A. Saldaña. 2013. What explains variation in the impacts of exotic plant invasions on the nitrogen cycle? A meta-analysis. *Ecology Letters* **17**:1–12.

Cavender-Bares, J., D. D. Ackerly, and K. H. Kozak. 2012. Integrating ecology and phylogenetics: The footprint of history in modern-day communities. *Ecology* **93**:S1-S3.

Cease, A. J., J. J. Elser, C. F. Ford, S. Hao, L. Kang, and J. F. Harrison. 2012. Heavy Livestock Grazing Promotes Locust Outbreaks by Lowering Plant Nitrogen Content. *Science* **335**:467–469.

Cebrian, J. 2004. Role of first-order consumers in ecosystem carbon flow. *Ecology Letters* **7**:232–240.

Cebrian, J. and C. M. Duarte. 1995. Plant growth rate dependence of detrital carbon storage in ecosystems. *Science* **268**:1606–1608.

Cebrian, J. and C. M. Duarte. 1998. Patterns in leaf herbivory on seagrasses. *Aquatic Botany* **60**:67–82.

Cebrian, J. and C. M. Duarte. 2001. Detrital stocks and dynamics of the seagrass *Posidonia oceanica* (L.) Delile in the Spanish Mediterranean. *Aquatic Botany* **70**:295–309.

Cebrian, J., J. B. Shurin, E. T. Borer, B. J. Cardinale, J. T. Ngai, M. D. Smith, and W. F. Fagan. 2009. Producer nutritional quality controls ecosystem trophic structure. *PLOS ONE* **4**.

Cembella, A. D. 2003. Chemical ecology of eukaryotic microalgae in marine ecosystems. *Phycologia* **42**:420–447.

Ceulemans, T., C. J. Stevens, L. Duchateau, H. Jacquemyn, D. J. G. Gowing, R. Merckx, H. Wallace, N. van Rooijen, T. Goethem, R. Bobbink, E. Dorland, C. Gaudnik, D. Alard, E. Corcket, S. Muller, N. B. Dise, C. Dupré, M. Diekmann, and O. Honnay.

2014. Soil phosphorus constrains biodiversity across European grasslands. *Global Change Biology* **20**:3814–3822.

Chabot, B. F. and D. J. Hicks. 1982. The ecology of leaf life spans. *Annual Review of Ecology and Systematics* **13**:229–259.

Challis, G. L. and D. A. Hopwood. 2003. Synergy and contingency as driving forces for the evolution of multiple secondary metabolite production by *Streptomyces* species. *Proceedings of the National Academy of Sciences* **100**:14555–14561.

Chapman, S. K., G. S. Newman, S. C. Hart, J. A. Schweitzer, and G. W. Koch. 2013. Leaf litter mixtures alter microbial community development: Mechanisms for non-additive effects in litter decomposition. *PLOS ONE* **8**:e62671.

Chapman, S. K., J. A. Schweitzer, and T. G. Whitham. 2006. Herbivory differentially alters plant litter dynamics of evergreen and deciduous trees. *Oikos* **114**:566–574.

Chavarria Pizarro, L., H. F. McCreery, S. P. Lawson, M. E. Winston, and S. O'Donnell. 2012. Sodium-specific foraging by leafcutter ant workers (*Atta cephalotes*, Hymenoptera: Formicidae). *Ecological Entomology* **37**:435–438.

Chen, R., M. Senbayram, S. Blagodatsky, O. Myachina, K. Dittert, X. Lin, E. Blagodatskaya, and Y. Kuzyakov. 2014. Soil C and N availability determine the priming effect: Microbial N mining and stoichiometric decomposition theories. *Global Change Biology* **20**:2356–2367.

Chen, W., L. Song, L. Peng, N. Wan, X. Zhang, and N. Gan. 2008. Reduction in microcystin concentrations in large and shallow lakes: Water and sediment-interface contributions. *Water Research* **42**:763–773.

Chen, X. F., L. Y. Yang, L. Xiao, A. J. Miao, and B. D. Xi. 2012. Nitrogen removal by denitrification during cyanobacterial bloom in Lake Taihu. *Journal of Freshwater Ecology* **27**:243–258.

Cheplick, G. P. and S. H. Faeth. 2009. *Ecology and Evolution of the Grass-Endophyte Symbiosis*. Oxford University Press, New York.

Chislock, M. F., O. Sarnelle, B. K. Olsen, E. Doster, and A. E. Wilson. 2013. Large effects of consumer offense on ecosystem structure and function. *Ecology* **94**:2375–2380.

Chorus, I. and J. Bartram, editors. 1999. *Toxic Cyanobacteria in Water—A Guide to Their Public Health Consequences, Monitoring, and Management*. E & FN Spon, New York.

Choudhury, D. 1988. Herbivore induced changes in leaf litter resource quality: a neglected aspect of herbivory in ecosystem dynamics. *Oikos* **51**:389–393.

Christenson, L. C., G. M. Lovett, M. J. Mitchell, and P. G. Groffman. 2002. The fate of nitrogen in gypsy moth frass deposited to an oak forest floor. *Oecologia* **131**:444–452.

Christenson, L. M., M. J. Mitchell, P. M. Groffman, and G. M. Lovett. 2010. Winter climate change implications for decomposition in northeastern forests: Comparisons of sugar maple litter with herbivore fecal inputs. *Global Change Biology* **16**:2589–2601.

Christoffersen, K. 1996. Ecological implications of cyanobacterial toxins in aquatic food webs. *Phycologia* **35**:42–50.

Chung, S. H., C. Rosa, E. D. Scully, M. Peiffer, J. F. Tooker, K. Hoover, D. S. Luthe, and G. W. Felton. 2013. Herbivore exploits orally secreted bacteria to suppress plant defenses. *Proceedings of the National Academy of Sciences* **110**:15728–15733.

Cipollini, D., D. Walters, and C. Voelckel. 2014. Costs of resistance in plants: From theory to evidence. Pages 263–307 *Annual Plant Reviews*. John Wiley & Sons, Ltd, New York.

Civitello, D. J., R. M. Penczykowski, J. L. Hite, M. A. Duffy, and S. R. Hall. 2013. Potassium stimulates fungal epidemics in *Daphnia* by increasing host and parasite reproduction. *Ecology* **94**:380–388.

Clancy, K. M. and P. W. Price. 1986. Temporal variation in three-trophic-level interactions among willows, sawflies, and parasites. *Ecology* **67**:1601–1607.

Clancy, K. M. and P. W. Price. 1987. Rapid herbivore growth enhances enemy attack—Sublethal plant defenses remain a paradox. *Ecology* **68**:733–737.

Clark, C. J., J. R. Poulsen, and D. J. Levey. 2012. Vertebrate herbivory impacts seedling recruitment more than niche partitioning or density-dependent mortality. *Ecology* **93**:554–564.

Clark, C. M. and D. Tilman. 2008. Loss of plant species after chronic low-level nitrogen deposition to prairie grasslands. *Nature* **451**:712–715.

Classen, A. T., S. K. Chapman, T. G. Whitham, S. C. Hart, and G. W. Koch. 2007a. Genetic-based plant resistance and susceptibility traits to herbivory influence needle and root litter nutrient dynamics. *Journal of Ecology* **95**:1181–1194.

Classen, A. T., J. DeMarco, S. C. Hart, T. G. Whitham, N. S. Cobb, and G. W. Koch. 2006. Impacts of herbivorous insects on decomposer communities during the early stages of primary succession in a semi-arid woodland. *Soil Biology & Biochemistry* **38**:972–982.

Classen, A. T., S. C. Hart, T. G. Whitman, N. S. Cobb, and G. W. Koch. 2005. Insect infestations linked to shifts in microclimate: Important climate change implications. *Soil Science Society of America Journal* **69**:2049–2057.

Classen, A. T., S. T. Overby, S. C. Hart, G. W. Koch, and T. G. Whitham. 2007b. Season mediates herbivore effects on litter and soil microbial abundance and activity in a semi-arid woodland. *Plant and Soil* **295**:217–227.

Clay, K. 1990. Fungal endophytes of grasses. *Annual Review of Ecology and Systematics* **21**:275–297.

Clay, K. 1991. Fungal endophytes, grasses, and herbivores. Pages 199–223 *in* P. Barbose, V. A. Krischik, and C. G. Jones, editors. *Microbial Mediation of Plant-Herbivore Interactions*. Wiley, New York.

Clay, K. and C. Schardl. 2002. Evolutionary origins and ecological consequences of endophyte symbiosis with grasses. *American Naturalist* **160**:S99–S127.

Clay, N. A., J. Lucas, M. Kaspari, and A. D. Kay. 2013. Manna from heaven: Refuse from an arboreal ant links aboveground and belowground processes in a lowland tropical forest. *Ecosphere* **4**:art141.

Clemmensen, K. E., A. Bahr, O. Ovaskainen, A. Dahlberg, A. Ekblad, H. Wallander, J. Stenlid, R. D. Finlay, D. A. Wardle, and B. D. Lindahl. 2013. Roots and associated fungi drive long-term carbon sequestration in boreal forest. *Science* **339**:1615–1618.

Cleveland, C. C., B. Z. Houlton, W. K. Smith, A. R. Marklein, S. C. Reed, W. Parton, S. J. Del Grosso, and S. W. Running. 2013. Patterns of new versus recycled primary production in the terrestrial biosphere. *Proceedings of the National Academy of Sciences* **110**:12733–12737.

Cobb, R. C., D. A. Orwig, and S. Currie. 2006. Decomposition of green foliage in eastern hemlock forests of southern New England impacted by hemlock woolly adelgid infestations. *Canadian Journal of Forest Research-Revue Canadienne De Recherche Forestiere* **36**:1331–1341.

Cole, J. J. 1982. Interactions between bacteria and algae in aquatic ecosystems. *Annual Review of Ecology and Systematics* **13**:291–314.

Cole, J. J., G. E. Likens, and J. E. Hobbie. 1984. Decomposition of planktonic algae in an oligotrophic lake. *Oikos* **42**:257–266.

Coleman, D. C. 2010. *Big Ecology: The Emergence of Ecosystem Science*. University of California Press, Berkeley.

Coleman, D. C., D. A. Crossley, and P. F. Hendrix. 2004. *Fundamentals of soil ecology*. Elsevier Academic Press, Amsterdam.

Coleman, R. A., S. J. Ramchunder, K. M. Davies, A. J. Moody, and A. Foggo. 2007a. Herbivore-induced infochemicals influence foraging behaviour in two intertidal predators. *Oecologia* **151**:454–463.

Coleman, R. A., S. J. Ramchunder, A. J. Moody, and A. Foggo. 2007b. An enzyme in snail saliva induces herbivore-resistance in a marine alga. *Functional Ecology* **21**:101–106.

Coley, P. D., J. P. Bryant, and F. S. Chapin. 1985. Resource availability and plant antiherbivore defense. *Science* **230**:895–899.

Coley, P. D. and T. A. Kursar. 2014. On tropical forests and their pests. *Science* **343**:35–36.

Colin, S. P. and H. G. Dam. 2003. Effects of the toxic dinoflagellate *Alexandrium fundyense* on the copepod Acartia hudsonica: A test of the mechanisms that reduce ingestion rates. *Marine Ecology-Progress Series* **248**:55–65.

Comins, H. N., M. P. Hassell, and R. M. May. 1992. The spatial dynamics of host-parasitoid systems. *Journal of Animal Ecology* **61**:735–748.

Conn, V. M., A. R. Walker, and C. M. M. Franco. 2008. Endophytic actinobacteria induce defense pathways in *Arabidopsis thaliana*. *Molecular Plant-Microbe Interactions* **21**:208–218.

Cooke, J. and M. R. Leishman. 2011. Is plant ecology more siliceous than we realise? *Trends in Plant Science* **16**:61–68.

Cornelissen, J. H. C., N. Perez-Harguindeguy, S. Diaz, J. P. Grime, B. Marzano, M. Cabido, F. Vendrimi, and B. Cerabolini. 1999. Leaf structure and defence control litter decomposition rate across species and life forms in regional floras on two continents. *New Phytologist* **143**:191–200.

Cornelissen, J. H. C., H. M. Quested, D. Gwynn-Jones, R. S. P. Van Logtestijn, M. A. H. De Beus, A. Kondratchuk, T. V. Callaghan, and R. Aerts. 2004. Leaf digestibility and litter decomposability are related in a wide range of subarctic plant species and types. *Functional Ecology* **18**:779–786.

Cornwell, W. K., J. H. C. Cornelissen, K. Amatangelo, E. Dorrepaal, V. T. Eviner, O. Godoy, S. E. Hobbie, B. Hoorens, H. Kurokawa, N. Perez-Harguindeguy, H. M. Quested, L. S. Santiago, D. A. Wardle, I. J. Wright, R. Aerts, S. D. Allison, P. van Bodegom, V. Brovkin, A. Chatain, T. V. Callaghan, S. Diaz, E. Garnier, D. E. Gurvich, E. Kazakou, J. A. Klein, J. Read, P. B. Reich, N. A. Soudzilovskaia, M. V. Vaieretti, and M. Westoby. 2008. Plant species traits are the predominant control on litter decomposition rates within biomes worldwide. *Ecology Letters* **11**:1065–1071.

Costa, L. G., G. Giordano, and E. M. Faustman. 2010. Domoic acid as a developmental neurotoxin. *Neurotoxicology* **31**:409–423.

Cotrufo, M. F., M. D. Wallenstein, C. M. Boot, K. Denef, and E. Paul. 2013. The Microbial Efficiency-Matrix Stabilization (MEMS) framework integrates plant litter decomposition with soil organic matter stabilization: Do labile plant inputs form stable soil organic matter? *Global Change Biology* **19**:988–995.

Couture, J. J. and R. L. Lindroth. 2014. Atmospheric change alters frass quality of forest canopy herbivores. *Arthropod-Plant Interactions* **8**:33–47.

Covelo, F. and A. Gallardo. 2004. Green and senescent leaf phenolics showed spatial auto-correlation in a *Quercus robur* population in northwestern Spain. *Plant and Soil* **259**:267–276.

Covelo, F., J. Manuel Avila, and A. Gallardo. 2011. Temporal changes in the spatial pattern of leaf traits in a *Quercus robur* population. *Annals of Forest Science* **68**:453–460.

Craft, J. D., V. J. Paul, and E. E. Sotka. 2013. Biogeographic and phylogenetic effects on feeding resistance of generalist herbivores toward plant chemical defenses. *Ecology* **94**:18–24.

Craig, T. P., J. K. Itami, and P. W. Price. 1989. A strong relationship between oviposition preference and larval performance in a shoot-galling sawfly. *Ecology* **70**:1691–1699.

Crawford, K. M., J. M. Land, and J. A. Rudgers. 2010. Fungal endophytes of native grasses decrease insect herbivore preference and performance. *Oecologia* **164**:431–444.

Crawford, K. M. and J. A. Rudgers. 2013. Genetic diversity within a dominant plant out-weighs plant species diversity in structuring an arthropod community. *Ecology* **94**:1025–1035.

Crawley, M. J. 1983. *Herbivory, the Dynamics of Animal-Plant Interactions*. Blackwell Scientific, Oxford, UK.

Crawley, M. J., A. E. Johnston, J. Silvertown, M. Dodd, C. de Mazancourt, M. S. Heard, D. F. Henman, and G. R. Edwards. 2005. Determinants of species richness in the park grass experiment. *American Naturalist* **165**:179–192.

Cressie, N. A. C. 1993. *Statistics for spatial data*. Wiley-Interscience, New York.

Croll, D. A., J. L. Maron, J. A. Estes, E. M. Danner, and G. V. Byrd. 2005. Introduced predators transform subarctic islands from grassland to tundra. *Science* **307**:1959–1961.

Cronin, G. and M. E. Hay. 1996a. Effects of light and nutrient availability on the growth, secondary chemistry, and resistance to herbivory of two brown seaweeds. *Oikos* **77**:93–106.

Cronin, G. and M. E. Hay. 1996b. Induction of seaweed chemical defenses by amphipod grazing. *Ecology* **77**:2287–2301.

Cronin, J. T. and J. D. Reeve. 2005. Host-parasitoid spatial ecology: A plea for a landscape-level synthesis. *Proceedings of the Royal Society B: Biological Sciences* **272**:2225–2235.

Cross, W. F., J. P. Benstead, A. D. Rosemond, and J. B. Wallace. 2003. Consumer-resource stoichiometry in detritus-based streams. *Ecology Letters* **6**:721–732.

Crossley, D. A., C. S. Gist, W. W. Hargrove, L. S. Risley, T. D. Schowalter, and T. R. Seastedt. 1988. Foliage consumption and nutrient dynamics in canopy insects. Pages 193–205 *in* W. T. Swank and D. A. Crossley Jr, editors. *Ecological Studies*. Vol. 66, *Forest Hydrology and Ecology at Coweeta*. Springer-Verlag, New York.

Crutsinger, G. M., M. D. Collins, J. A. Fordyce, Z. Gompert, C. C. Nice, and N. J. Sanders. 2006. Plant genotypic diversity predicts community structure and governs an eco-system process. *Science* **313**:966–968.

Cruz-Rivera, E. and M. E. Hay. 2003. Prey nutritional quality interacts with chemical defenses to affect consumer feeding and fitness. *Ecological Monographs* **73**:483–506.

da Cunha-Santino, M. B. and I. Bianchini. 2002. Humic substance mineralization in a tropical oxbow lake (Sao Paulo, Brazil). *Hydrobiologia* **468**:33–43.

Dahlin, K. M., G. P. Asner, and C. B. Field. 2013. Environmental and community controls on plant canopy chemistry in a Mediterranean-type ecosystem. *Proceedings of the National Academy of Sciences* **110**:6895–6900.

Daines, S. J., J. R. Clark, and T. M. Lenton. 2014. Multiple environmental controls on phytoplankton growth strategies determine adaptive responses of the N : P ratio. *Ecology Letters* **17**:414–425

Dajoz, R. 2000. *Insects and Forests: The Role and Diversity of Insects in the Forest Environment*. Lavoisier, Paris.

Danger, M., J. Cornut, E. Chauvet, P. Chavez, A. Elger, and A. Lecerf. 2013. Benthic algae stimulate leaf litter decomposition in detritus-based headwater streams: A case of aquatic priming effect? *Ecology* **94**:1604–1613.

Danger, M., T. Daufresne, F. Lucas, S. Pissard, and G. Lacroix. 2008. Does Liebig's law of the minimum scale up from species to communities? *Oikos* **117**:1741–1751.

Darwin, C. 1859. *The Origin of Species*. John Murray, London.

Davis, T. S. and R. W. Hofstetter. 2012. Plant secondary chemistry mediates the performance of a nutritional symbiont associated with a tree-killing herbivore. *Ecology* **93**:421–429.

De Deyn, G. B., A. Biere, W. H. van der Putten, R. Wagenaar, and J. N. Klironomos. 2009. Chemical defense, mycorrhizal colonization and growth responses in *Plantago lanceolata* L. *Oecologia* **160**:433–442.

De Deyn, G. B., C. E. Raaijmakers, H. R. Zoomer, M. P. Berg, P. C. de Ruiter, H. A. Verhoef, T. M. Bezemer, and W. H. van der Putten. 2003. Soil invertebrate fauna enhances grassland succession and diversity. *Nature* **422**:711–713.

de Goeij, J. M., D. van Oevelen, M. J. A. Vermeij, R. Osinga, J. J. Middelburg, A. F. P. M. de Goeij, and W. Admiraal. 2013. Surviving in a marine desert: The sponge loop retains resources within coral reefs. *Science* **342**:108–110.

de Leeuw, J. W., G. J. M. Versteegh, and P. F. van Bergen. 2006. Biomacromolecules of algae and plants and their fossil analogues. *Plant Ecology* **182**:209–233.

De Luca, V., V. Salim, S. M. Atsumi, and F. Yu. 2012. Mining the biodiversity of plants: A revolution in the making. *Science* **336**:1658–1661.

de Roode, J. C., L. R. Gold, and S. Altizer. 2007. Virulence determinants in a natural butterfly-parasite system. *Parasitology* **134**:657–668.

de Roode, J. C., T. Lefèvre, and M. D. Hunter. 2013. Self-medication in animals. *Science* **340**:150–151.

de Roode, J. C., A. B. Pedersen, M. D. Hunter, and S. Altizer. 2008. Host plant species affects virulence in monarch butterfly parasites. *Journal of Animal Ecology* **77**:120–126.

de Roode, J. C., R. M. Rarick, A. J. Mongue, N. M. Gerardo, and M. D. Hunter. 2011. Aphids indirectly increase virulence and transmission potential of a monarch butterfly parasite by reducing defensive chemistry of a shared food plant. *Ecology Letters* **14**:453–461.

de Sassi, C., O. T. Lewis, and J. M. Tylianakis. 2012. Plant-mediated and nonadditive effects of two global change drivers on an insect herbivore community. *Ecology* **93**:1892–1901.

Dearing, M. D., W. J. Foley, and S. McLean. 2005. The influence of plant secondary metabolites on the nutritional ecology of herbivorous terrestrial vertebrates. *Annual Review of Ecology Evolution and Systematics* **36**:169–189.

Dearing, M. D., J. S. Forbey, J. D. McLister, and L. Santos. 2008. Ambient temperature influences diet selection and physiology of an herbivorous mammal, *Neotoma albigula*. *Physiological and Biochemical Zoology* **81**:891–897.

Degabriel, J. L., B. D. Moore, W. J. Foley, and C. N. Johnson. 2009. The effects of plant defensive chemistry on nutrient availability predict reproductive success in a mammal. *Ecology* **90**:711–719.

DeMott, W. R. 1999. Foraging strategies and growth inhibition in five daphnids feeding on mixtures of a toxic cyanobacterium and a green alga. *Freshwater Biology* **42**:263–274.

Denno, R. F. and W. F. Fagan. 2003. Might nitrogen limitation promote omnivory among carnivorous arthropods? *Ecology* **84**:2522–2531.

Denno, R. F., C. Gratton, M. A. Peterson, G. A. Langellotto, D. L. Finke, and A. F. Huberty. 2002. Bottom-up forces mediate natural-enemy impact in a phytophagous insect community. *Ecology* **83**:1443–1458.

Denno, R. F. and M. S. McClure. 1983. *Variable Plants and Herbivores in Natural and Managed Systems*. Academic Press, New York.

Denoeud, F., L. Carretero-Paulet, A. Dereeper, G. Droc, R. Guyot, M. Pietrella, C. Zheng, A. Alberti, F. Anthony, G. Aprea, J.-M. Aury, P. Bento, M. Bernard, S. Bocs, C. Campa, A. Cenci, M.-C. Combes, D. Crouzillat, C. Da Silva, L. Daddiego, F. De Bellis, S. Dussert, O. Garsmeur, T. Gayraud, V. Guignon, K. Jahn, V. Jamilloux, T. Joët, K. Labadie, T. Lan, J. Leclercq, M. Lepelley, T. Leroy, L.-T. Li, P. Librado, L. Lopez, A. Muñoz, B. Noel, A. Pallavicini, G. Perrotta, V. Poncet, D. Pot, Priyono, M. Rigoreau, M. Rouard, J. Rozas, C. Tranchant-Dubreuil, R. VanBuren, Q. Zhang, A. C. Andrade, X. Argout, B. Bertrand, A. de Kochko, G. Graziosi, R. J. Henry, Jayarama, R. Ming, C. Nagai, S. Rounsley, D. Sankoff, G. Giuliano, V. A. Albert, P. Wincker, and P. Lashermes. 2014. The coffee genome provides insight into the convergent evolution of caffeine biosynthesis. *Science* **345**:1181–1184.

Dewsbury, B. M. and J. W. Fourqurean. 2010. Artificial reefs concentrate nutrients and alter benthic community structure in an oligotrophic, subtropical estuary. *Bulletin of Marine Science* **86**:813–829.

Ding, S.-Y., Y.-S. Liu, Y. Zeng, M. E. Himmel, J. O. Baker, and E. A. Bayer. 2012. How does plant cell wall nanoscale architecture correlate with enzymatic digestibility? *Science* **338**:1055–1060.

Dinnage, R., M. W. Cadotte, N. M. Haddad, G. M. Crutsinger, and D. Tilman. 2012. Diversity of plant evolutionary lineages promotes arthropod diversity. *Ecology Letters* **15**:1308–1317.

Doane, C. C. 1970. Primary pathogens and their role in the development of an epizootic in the gypsy moth. *Journal of Invertebrate Pathology* **15**:21–33.

Donihue, C. M., L. M. Porensky, J. Foufopoulos, C. Riginos, and R. M. Pringle. 2013. Glade cascades: Indirect legacy effects of pastoralism enhance the abundance and spatial structuring of arboreal fauna. *Ecology* **94**:827–837.

Dorenbosch, M. and E. S. Bakker. 2011. Herbivory in omnivorous fishes: Effect of plant secondary metabolites and prey stoichiometry. *Freshwater Biology* **56**:1783–1797.

Douglas, A. E. 1998. Nutritional interactions in insect–microbial symbioses: Aphids and their symbiotic bacteria *Buchnera. Annual Review of Entomology* **43**:17–37.

Downing, J. A., S. B. Watson, and E. McCauley. 2001. Predicting Cyanobacteria dominance in lakes. *Canadian Journal of Fisheries and Aquatic Sciences* **58**:1905–1908.

Driebe, E. M. and T. G. Whitham. 2000. Cottonwood hybridization affects tannin and nitrogen content of leaf litter and alters decomposition. *Oecologia* **123**:99–107.

Duarte, C. M. and J. Cebrian. 1996. The fate of marine autotrophic production. *Limnology and Oceanography* **41**:1758–1766.

Dudt, J. F. and D. J. Shure. 1994. The influence of light and nutrients on foliar phenolics and insect herbivory. *Ecology* **75**:86–98.

Duffy, J. E. and M. E. Hay. 1994. Herbivore resistance to seaweed chemical defense—The roles of mobility and predation risk. *Ecology* **75**:1304–1319.

Duffy, J. E. and M. E. Hay. 2000. Strong impacts of grazing amphipods on the organization of a benthic community. *Ecological Monographs* **70**:237–263.

Duffy, M. A., J. H. Ochs, R. M. Penczykowski, D. J. Civitello, C. A. Klausmeier, and S. R. Hall. 2012. Ecological context influences epidemic size and parasite-driven evolution. *Science* **335**:1636–1638.

Dugdale, R. C. and F. P. Wilkerson. 1998. Silicate regulation of new production in the equatorial Pacific upwelling. *Nature* **391**:270–273.

Dungait, J. A. J., D. W. Hopkins, A. S. Gregory, and A. P. Whitmore. 2012. Soil organic matter turnover is governed by accessibility not recalcitrance. *Global Change Biology* **18**:1781–1796.

Dunham, A. E. 2008. Above and below ground impacts of terrestrial mammals and birds in a tropical forest. *Oikos* **117**:571–579.

Dybas, H. S. and D. D. Davis. 1962. Population census of 17-year periodical cicadas (Homoptera: Cicadidae, *Magicicada). Ecology* **43**:432-&.

Dyer, L. A., C. D. Dodson, J. O. Stireman, M. A. Tobler, A. M. Smilanich, R. M. Fincher, and D. K. Letourneau. 2003. Synergistic effects of three *Piper* amides on generalist and specialist herbivores. *Journal of Chemical Ecology* **29**:2499–2514.

Dyer, L. A. and D. Letourneau. 2003. Top-down and bottom-up diversity cascades in detrital vs. living food webs. *Ecology Letters* **6**:60–68.

Dyer, L. A. and D. K. Letourneau. 1999. Trophic cascades in a complex terrestrial community. *Proceedings of the National Academy of Sciences* **96**:5072–5076.

Eckhardt, M., M. Haider, S. Dorn, and A. Müller. 2014. Pollen mixing in pollen generalist solitary bees: a possible strategy to complement or mitigate unfavourable pollen properties? *Journal of Animal Ecology* **83**:588–597.

Edelstein-Keshet, L. and M. D. Rausher. 1989. The effects of inducible plant defenses on herbivore populations. 1. Mobile herbivores in continuous time. *American Naturalist* **133**:787–810.

Edmondson, W. T. and A. H. Litt. 1982. *Daphnia* in Lake Washington. *Limnology and Oceanography* **27**:272–293.

Edwards, I. P., D. R. Zak, H. Kellner, S. D. Eisenlord, and K. S. Pregitzer. 2011. Simulated atmospheric N deposition alters fungal community composition and suppresses ligninolytic gene expression in a northern hardwood forest. *PLOS ONE* **6**:e20421.

Edwards, K. F., E. Litchman, and C. A. Klausmeier. 2013. Functional traits explain phytoplankton community structure and seasonal dynamics in a marine ecosystem. *Ecology Letters* **16**:56–63.

Edwards, P. J. and S. D. Wratten. 1983. Wound-induced defenses in plants and their consequences for patterns in insect grazing. *Oecologia* **59**:88–93.

Efting, A. A., D. D. Snow, and S. C. Fritz. 2011. Cyanobacteria and microcystin in the Nebraska (USA) Sand Hills Lakes before and after modern agriculture. *Journal of Paleolimnology* **46**:17–27.

Ehrlich, P. R. and P. H. Raven. 1964. Butterflies and plants: A study in coevolution. *Evolution* **18**:586–608.

Eichenseer, H., M. C. Mathews, J. S. Powell, and G. W. Felton. 2010. Survey of a salivary effector in caterpillars: Glucose oxidase variation and correlation with host range. *Journal of Chemical Ecology* **36**:885–897.

Eisenhauer, N., T. Dobies, S. Cesarz, S. E. Hobbie, R. J. Meyer, K. Worm, and P. B. Reich. 2013. Plant diversity effects on soil food webs are stronger than those of elevated CO2 and N deposition in a long-term grassland experiment. *Proceedings of the National Academy of Sciences* **110**:6889–6894.

Eissfeller, V., C. Langenbruch, A. Jacob, M. Maraun, and S. Scheu. 2013. Tree identity surpasses tree diversity in affecting the community structure of oribatid mites (Oribatida) of deciduous temperate forests. *Soil Biology & Biochemistry* **63**:154–162.

Elderd, B. D., B. J. Rehill, K. J. Haynes, and G. Dwyer. 2013. Induced plant defenses, host–pathogen interactions, and forest insect outbreaks. *Proceedings of the National Academy of Sciences* **110**:14978–14983.

Elser, J. and E. Bennett. 2011. A broken biogeochemical cycle. *Nature* **478**:29–31.

Elser, J. J., K. Acharya, M. Kyle, J. Cotner, W. Makino, T. Markow, T. Watts, S. Hobbie, W. Fagan, J. Schade, J. Hood, and R. W. Sterner. 2003. Growth rate-stoichiometry couplings in diverse biota. *Ecology Letters* **6**:936–943.

Elser, J. J., M. E. S. Bracken, E. E. Cleland, D. S. Gruner, W. S. Harpole, H. Hillebrand, J. T. Ngai, E. W. Seabloom, J. B. Shurin, and J. E. Smith. 2007. Global analysis of nitrogen and phosphorus limitation of primary producers in freshwater, marine and terrestrial ecosystems. *Ecology Letters* **10**:1135–1142.

Elser, J. J., I. Loladze, A. L. Peace, and Y. Kuang. 2012. Lotka re-loaded: Modeling trophic interactions under stoichiometric constraints. *Ecological Modelling* **245**:3–11.

Elser, J. J., A. L. Peace, M. Kyle, M. Wojewodzic, M. L. McCrackin, T. Andersen, and D. O. Hessen. 2010. Atmospheric nitrogen deposition is associated with elevated phosphorus limitation of lake zooplankton. *Ecology Letters* **13**:1256–1261.

Elser, J. J., R. W. Sterner, E. Gorokhova, W. F. Fagan, T. A. Markow, J. B. Cotner, J. F. Harrison, S. E. Hobbie, G. M. Odell, and L. J. Weider. 2000. Biological stoichiometry from genes to ecosystems. *Ecology Letters* **3**:540–550.

Endara, M. J. and P. D. Coley. 2011. The resource availability hypothesis revisited: A meta-analysis. *Functional Ecology* **25**:389–398.

Enge, S., G. M. Nylund, T. Harder, and H. Pavia. 2012. An exotic chemical weapon explains low herbivore damage in an invasive alga. *Ecology* **93**:2736–2745.

Enge, S., G. M. Nylund, and H. Pavia. 2013. Native generalist herbivores promote invasion of a chemically defended seaweed via refuge-mediated apparent competition. *Ecology Letters* **16**:487–492.

Enriquez, S., C. M. Duarte, and K. Sandjensen. 1993. Patterns in decomposition rates among photosynthetic organisms—The importance of detritus C-N-P content. *Oecologia* **94**:457–471.

Erb, M., D. Balmer, E. S. De Lange, G. Von Merey, C. Planchamp, C. A. M. Robert, G. Roder, I. Sobhy, C. Zwahlen, B. Mauch-Mani, and T. C. J. Turlings. 2011. Synergies and trade-offs between insect and pathogen resistance in maize leaves and roots. *Plant Cell and Environment* **34**:1088–1103.

Erb, M., J. Ton, J. Degenhardt, and T. C. J. Turlings. 2008. Interactions between arthropod-induced aboveground and belowground defenses in plants. *Plant Physiology* **146**:867–874.

Erelli, M. C., M. P. Ayres, and G. K. Eaton. 1998. Altitudinal patterns in host suitability for forest insects. *Oecologia* **117**:133–142.

Eshleman, K. N., R. P. Morgan, J. R. Webb, F. A. Deviney, and J. N. Galloway. 1998. Temporal patterns of nitrogen leakage from mid-Appalachian forested watersheds: Role of insect defoliation. *Water Resources Research* **34**:2005–2016.

Eskelinen, A., S. Harrison, and M. Tuomi. 2012. Plant traits mediate consumer and nutrient control on plant community productivity and diversity. *Ecology* **93**:2705–2718.

Esselman, P. C., R. J. Stevenson, F. Lupi, C. M. Riseng, and M. J. Wiley. 2015. Landscape prediction and mapping of game fish biomass, an ecosystem service of Michigan rivers. *North American Journal of Fisheries Management* **35**:302–320.

Estes, J. A., D. F. Doak, A. M. Springer, and T. M. Williams. 2009. Causes and consequences of marine mammal population declines in southwest Alaska: A food-web perspective. *Philosophical Transactions of the Royal Society B: Biological Sciences* **364**:1647–1658.

Estes, J. A. and D. O. Duggins. 1995. Sea otters and kelp forests in Alaska. Generality and variation in a community ecology paradigm. *Ecological Monographs* **65**:75–100.

Estes, J. A., J. Terborgh, J. S. Brashares, M. E. Power, J. Berger, W. J. Bond, S. R. Carpenter, T. E. Essington, R. D. Holt, J. B. C. Jackson, R. J. Marquis, L. Oksanen, T. Oksanen, R. T. Paine, E. K. Pikitch, W. J. Ripple, S. A. Sandin, M. Scheffer, T. W. Schoener, J. B. Shurin, A. R. E. Sinclair, M. E. Soule, R. Virtanen, and D. A. Wardle. 2011. Trophic downgrading of planet Earth. *Science* **333**:301–306.

Etter, R. J. and L. S. Mullineaux. 2001. Deep-sea communities. Pages 367–393 *in* M. D. Bertness, S. D. Gaines, and M. E. Hay, editors. *Marine Community Ecology.* Sinauer Associates, Sunderland, MA.

Eubanks, M. D. and R. F. Denno. 2000. Host plants mediate omnivore-herbivore interactions and influence prey suppression. *Ecology* **81**:936–947.

Faeth, S. H. 1986. Indirect interactions between temporally separated herbivores mediated by the host plant. *Ecology* **67**:479–494.

Faeth, S. H., E. F. Connor, and D. Simberloff. 1981. Early leaf abcission: A neglected source of mortality for folivores. *American Naturalist* **117**:409–415.

Faeth, S. H. and W. F. Fagan. 2002. Fungal endophytes: Common host plant symbionts but uncommon mutualists. *Integrative and Comparative Biology* **42**:360–368.

Fagan, W. F. and J. G. Bishop. 2000. Trophic interactions during primary succession: Herbivores slow a plant reinvasion at Mount St. Helens. *American Naturalist* **155**:238–251.

Fagan, W. F. and R. F. Denno. 2004. Stoichiometry of actual vs. potential predator-prey interactions: Insights into nitrogen limitation for arthropod predators. *Ecology Letters* **7**:876–883.

Fagan, W. F., E. Siemann, C. Mitter, R. F. Denno, A. F. Huberty, H. A. Woods, and J. J. Elser. 2002. Nitrogen in insects: Implications for trophic complexity and species diversification. *American Naturalist* **160**:784–802.

Fanin, N., N. Fromin, B. Buatois, and S. Hättenschwiler. 2013. An experimental test of the hypothesis of non-homeostatic consumer stoichiometry in a plant litter–microbe system. *Ecology Letters* **16**:764–772.

Farkas, T. E. and M. S. Singer. 2013. Can caterpillar density or host-plant quality explain host-plant-related parasitism of a generalist forest caterpillar assemblage? *Oecologia* **173**:971–983.

Farrow, E. P. 1925. *Plant Life on East Anglian Heaths*. Cambridge University Press, Cambridge, UK.

Feeny, P. 1970. Seasonal changes in oak leaf tannins and nutrients as a cause of spring feeding by winter moth caterpillars. *Ecology* **51**:565–581.

Feng, Z. L., J. A. Alfaro-Murillo, D. L. DeAngelis, J. Schmidt, M. Barga, Y. Q. Zheng, M. Tamrin, M. Olson, T. Glaser, K. Kielland, F. S. Chapin, and J. Bryant. 2012. Plant toxins and trophic cascades alter fire regime and succession on a boreal forest landscape. *Ecological Modelling* **244**:79–92.

Feng, Z. L., R. S. Liu, D. L. DeAngelis, J. P. Bryant, K. Kielland, F. Stuart Chapin, and R. Swihart. 2009. Plant toxicity, adaptive herbivory, and plant community dynamics. *Ecosystems* **12**:534–547.

Fenical, W. 2012. Culturing the unculturable: Expanding the diversity of marine microorganisms for drug discovery. *Planta Medica* **78**:1032–1032.

Fenner, N. and C. Freeman. 2013. Carbon preservation in humic lakes: A hierarchical regulatory pathway. *Global Change Biology* **19**:775–784.

Fenoglio, M. S., D. Srivastava, G. Valladares, L. Cagnolo, and A. Salvo. 2012. Forest fragmentation reduces parasitism via species loss at multiple trophic levels. *Ecology* **93**:2407–2420.

Fensham, R. J. and D. Bowman. 1992. Stand structure and the influence of overwood on regeneration in tropical eucalypt forest on Melville Island. *Australian Journal of Botany* **40**:335–352.

Fernandes, I., S. Duarte, F. Cássio, and C. Pascoal. 2013. Effects of riparian plant diversity loss on aquatic microbial decomposers become more pronounced with increasing time. *Microbial Ecology*:1–10.

Field, C. B., M. J. Behrenfeld, J. T. Randerson, and P. Falkowski. 1998. Primary production of the biosphere: Integrating terrestrial and oceanic components. *Science* **281**:237–240.

Findlay, S., M. Carreiro, V. Krischik, and C. G. Jones. 1996. Effects of damage to living plants on leaf litter quality. *Ecological Applications* **6**:269–275.

Fine, P. V. A., I. Mesones, and P. D. Coley. 2004. Herbivores promote habitat specialization by trees in Amazonian forests. *Science* **305**:663–665.

Fine, P. V. A., M. R. Metz, J. Lokvam, I. Mesones, J. M. A. Zuñiga, G. P. A. Lamarre, M. V. s. Pilco, and C. Baraloto. 2013. Insect herbivores, chemical innovation, and the evolution of habitat specialization in Amazonian trees. *Ecology* **94**:1764–1775.

Finke, D. L. and R. F. Denno. 2005. Predator diversity and the functioning of ecosystems: the role of intraguild predation in dampening trophic cascades. *Ecology Letters* **8**:1299–1306.

Fischer, D. G., S. K. Chapman, A. T. Classen, C. A. Gehring, K. C. Grady, J. A. Schweitzer, and T. G. Whitham. 2014. Plant genetic effects on soils under climate change. *Plant and Soil* **379**:1–19.

Florencio, M., C. Díaz-Paniagua, C. Gómez-Rodríguez, and L. Serrano. 2013. Biodiversity patterns in a macroinvertebrate community of a temporary pond network. *Insect Conservation and Diversity* **7**:4–21.

Flynn, K. J. and X. Irigoien. 2009. Aldehyde-induced insidious effects cannot be considered as a diatom defence mechanism against copepods. *Marine Ecology Progress Series* **377**:79–89.

Foley, J. W., G. R. Iason, and C. McArthur. 1999. Role of plant secondary metabolites in the nutritional ecology of mammalian herbivores: How far have we come in 25 years? Pages 130–209 *in* H. G. Jung and G. C. Fahey, editors. *Nutritional Ecology of Herbivores*. American Society of Animal Science, Savoy, IL.

Foley, W. J., A. McIlwee, I. Lawler, L. Aragones, A. P. Woolnough, and N. Berding. 1998. Ecological applications of near infrared reflectance spectroscopy a tool for rapid, cost-effective prediction of the composition of plant and animal tissues and aspects of animal performance. *Oecologia* **116**:293–305.

Fonte, S. J. and T. D. Schowalter. 2004. Decomposition of greenfall vs. senescent foliage in a tropical forest ecosystem in Puerto Rico. *Biotropica* **36**:474–482.

Fonte, S. J. and T. D. Schowalter. 2005. The influence of a neotropical herbivore (*Lamponius portoricensis*) on nutrient cycling and soil processes. *Oecologia* **146**:423–431.

Forbes, W. T. M. 1956. Caterpillars as botanists. *Proceedings of the 10th International Congress of Entomology* **1**:313–317.

Forbey, J. S. and W. J. Foley. 2009. PharmEcology: A pharmacological approach to understanding plant-herbivore interactions: An introduction to the symposium. *Integrative and Comparative Biology* **49**:267–273.

Forbey, J. S. and M. D. Hunter. 2012. The herbivore's prescription: A pharm-ecological perspective on host plant use by vertebrate and invertebrate herbivores. Pages 78–100 *in* G. R. Iason, M. Dicke, and S. E. Hartley, editors. *The Ecology of Plant Secondary Metabolites: From Genes to Global Processes*. Cambridge University Press, Cambridge, UK.

Forister, M. L., L. A. Dyer, M. S. Singer, J. O. Stireman, III, and J. T. Lill. 2012. Revisiting the evolution of ecological specialization, with emphasis on insect-plant interactions. *Ecology* **93**:981–991.

Forkner, R. E. and M. D. Hunter. 2000. What goes up must come down? Nutrient addition and predation pressure on oak herbivores. *Ecology* **81**:1588–1600.

Forkner, R. E., R. J. Marquis, and J. T. Lill. 2004. Feeny revisited: Condensed tannins as anti-herbivore defences in leaf-chewing herbivore communities of *Quercus*. *Ecological Entomology* **29**:174–187.

Fornara, D. A. and J. T. Du Toit. 2008. Browsing-induced effects on leaf litter quality and decomposition in a southern african savanna. *Ecosystems* **11**:238–249.

Foster, M. A., J. C. Schultz, and M. D. Hunter. 1992. Modeling gypsy moth-virus-leaf chemistry interactions: implications of plant quality for pest and pathogen dynamics. *Journal of Animal Ecology* **61**:509–520.

Fox, R. 2013. The decline of moths in Great Britain: A review of possible causes. *Insect Conservation and Diversity* **6**:5–19.

Fraenkel, G. S. 1959. Raison d'être of secondary plant substances. *Science* **129**:1466–1470.

Francino, M. P., editor. 2012. *Horizontal Gene Transfer in Microorganisms*. Caister Academic Press, Norfolk.

Franco, C., P. Michelsen, N. Percy, V. Conn, E. Listiana, S. Moll, R. Loria, and J. Coombs. 2007. Actinobacterial endophytes for improved crop performance. *Australasian Plant Pathology* **36**:524–531.

Frangoulis, C., N. Skliris, G. Lepoint, K. Elkalay, A. Goffart, J. K. Pinnegar, and J. H. Hecq. 2011. Importance of copepod carcasses versus faecal pellets in the upper water column of an oligotrophic area. *Estuarine Coastal and Shelf Science* **92**:456–463.

Frank, D. A. 2008. Evidence for top predator control of a grazing ecosystem. *Oikos* **117**:1718–1724.

Fraser, L. H. and J. P. Grime. 1998. Top-down control and its effect on the biomass and composition of three grasses at high and low soil fertility in outdoor microcosms. *Oecologia* **113**:239–246.

Freeman, C., N. Ostle, and H. Kang. 2001. An enzymic 'latch' on a global carbon store—A shortage of oxygen locks up carbon in peatlands by restraining a single enzyme. *Nature* **409**:149.

Frelich, L. E. 2008. *Forest dynamics and disturbance regimes: Studies from temperate evergreen-deciduous forests*. Cambridge University Press, Cambridge, UK.

Freschet, G. T., R. Aerts, and J. H. C. Cornelissen. 2012. A plant economics spectrum of litter decomposability. *Functional Ecology* **26**:56–65.

Friesen, M. L., S. S. Porter, S. C. Stark, E. J. von Wettberg, J. L. Sachs, and E. Martinez-Romero. 2011. Microbially Mediated Plant Functional Traits. Pages 23–46 *in* D. J. Futuyma, H. B. Shaffer, and D. Simberloff, editors. *Annual Review of Ecology, Evolution, and Systematics*. Vol. 42. Annual Reviews, Palo Alto, CA.

Frisch, D., P. K. Morton, P. R. Chowdhury, B. W. Culver, J. K. Colbourne, L. J. Weider, and P. D. Jeyasingh. 2014. A millennial-scale chronicle of evolutionary responses to cultural eutrophication in *Daphnia*. *Ecology Letters* **17**:360–368.

Fritz, R. S. and E. L. Simms, editors. 1992. *Plant Resistance to Herbivores and Pathogens: Ecology, Evolution, and Genetics*. University of Chicago Press, Chicago.

Frost, C. J. and M. D. Hunter. 2004. Insect canopy herbivory and frass deposition affect soil nutrient dynamics and export in oak mesocosms. *Ecology* **85**:3335–3347.

Frost, C. J. and M. D. Hunter. 2007. Recycling of nitrogen in herbivore feces: plant recovery, herbivore assimilation, soil retention, and leaching losses. *Oecologia* **151**:42–53.

Frost, C. J. and M. D. Hunter. 2008a. Herbivore-induced shifts in carbon and nitrogen allocation in red oak seedlings. *New Phytologist* **178**:835–845.

Frost, C. J. and M. D. Hunter. 2008b. Insect herbivores and their frass affect *Quercus rubra* leaf quality and initial stages of subsequent litter decomposition. *Oikos* **117**:13–22.

Frost, P. C., D. Ebert, and V. H. Smith. 2008. Responses of a bacterial pathogen to phosphorus limitation of its aquatic invertebrate host. *Ecology* **89**:313–318.

Froyd, C. A., E. E. D. Coffey, W. O. van der Knaap, J. F. N. van Leeuwen, A. Tye, and K. J. Willis. 2014. The ecological consequences of megafaunal loss: Giant tortoises and wetland biodiversity. *Ecology Letters* **17**:144–154.

Frye, G. G., J. W. Connelly, D. D. Musil, and J. S. Forbey. 2013. Phytochemistry predicts habitat selection by an avian herbivore at multiple spatial scales. *Ecology* **94**:308–314.

Fryxell, J. M., J. Greever, and A. R. E. Sinclair. 1988. Why are migratory ungulates so abundant? *American Naturalist* **131**:781–798.

Fryxell, J. M. and A. R. E. Sinclair. 1988. Causes and consequences of migration by large herbivores. *Trends in Ecology & Evolution* **3**:237–241.

Fuhrman, J. A. 1999. Marine viruses and their biogeochemical and ecological effects. *Nature* **399**:541–548.

Fujita, Y., H. O. Venterink, P. M. van Bodegom, J. C. Douma, G. W. Heil, N. Holzel, E. Jablonska, W. Kotowski, T. Okruszko, P. Pawlikowski, P. C. de Ruiter, and M. J. Wassen. 2014. Low investment in sexual reproduction threatens plants adapted to phosphorus limitation. *Nature* **505**:547–550.

Fukami, T., D. A. Wardle, P. J. Bellingham, C. P. H. Mulder, D. R. Towns, G. W. Yeates, K. I. Bonner, M. S. Durrett, M. N. Grant-Hoffman, and W. M. Williamson. 2006. Above- and below-ground impacts of introduced predators in seabird-dominated island ecosystems. *Ecology Letters* **9**:1299–1307.

Furey, P. C., R. L. Lowe, M. E. Power, and A. M. Campbell-Craven. 2012. Midges, Clado-phora, and epiphytes: shifting interactions through succession. *Freshwater Science* **31**:93–107.

Gallardo, A. and F. Covelo. 2005. Spatial pattern and scale of leaf N and P concentration in a *Quercus robur* population. *Plant and Soil* **273**:269–277.

Galloway, J. N., F. J. Dentener, D. G. Capone, E. W. Boyer, R. W. Howarth, S. P. Seitzinger, G. P. Asner, C. C. Cleveland, P. A. Green, E. A. Holland, D. M. Karl, A. F. Michaels, J. H. Porter, A. R. Townsend, and C. J. Vorosmarty. 2004. Nitrogen cycles: Past, present, and future. *Biogeochemistry* **70**:153–226.

Gange, A. C. and V. K. Brown. 1989. Effects of root herbivory by an insect on a foliar-feeding species, mediated through changes in the host plant. *Oecologia* **81**:38–42.

Gange, A. C. and V. K. Brown. 2002. Soil food web components affect plant community structure during early succession. *Ecological Research* **17**:217–227.

Gange, A. C., V. K. Brown, and G. S. Sinclair. 1994. Reduction of black vine weevil larval growth by vesicular-arbuscular mycorrhizal infection. *Entomologia Experimentalis Et Applicata* **70**:115–119.

Gange, A. C. and H. M. West. 1994. Interactions between arbuscular mycorrhizal fungi and foliar-feeding insects in *Plantago lanceolata* L. *New Phytologist* **128**:79–87.

Gans, J. H., R. Korson, M. R. Cater, and C. C. Ackerly. 1980. Effects of short-term and long-term theobromine administration to male dogs. *Toxicology and Applied Pharmacology* **53**:481–496.

Gao, X., X. Lu, M. Wu, H. Y. Zhang, R. Q. Pan, J. Tian, S. X. Li, and H. Liao. 2012. Co-inoculation with rhizobia and AMF inhibited soybean red crown rot: From field study to plant defense-related gene expression analysis. *PLOS ONE* **7**.

García-Palacios, P., F. T. Maestre, J. Kattge, and D. H. Wall. 2013. Climate and litter quality differently modulate the effects of soil fauna on litter decomposition across biomes. *Ecology Letters* **16**:1045–1053.

Garrido, E., A. E. Bennett, J. Fornoni, and S. Y. Strauss. 2010. Variation in arbuscular mycorrhizal fungi colonization modifies the expression of tolerance to above-ground defoliation. *Journal of Ecology* **98**:43–49.

Gauld, I. D., K. J. Gaston, and D. H. Janzen. 1992. Plant allelochemicals, tritrophic inter-actions and the anomalous diversity of tropical parasitoids—The nasty host hypoth-esis. *Oikos* **65**:353–357.

Gehring, C. and A. Bennett. 2009. Mycorrhizal fungal-plant-insect interactions: The importance of a community approach. *Environmental Entomology* **38**:93–102.

Geiselhardt, S., T. Otte, and M. Hilker. 2012. Looking for a similar partner: Host plants shape mating preferences of herbivorous insects by altering their contact pheromones. *Ecology Letters* **15**:971–977.

Gentry, G. L. and L. A. Dyer. 2002. On the conditional nature of neotropical caterpillar defenses against their natural enemies. *Ecology* **83**:3108–3119.

Gershenson, A. and W. X. Cheng. 2010. Rhizosphere processes in natural and managed systems: Implications of new research for soil carbon dynamics. *Hortscience* **45**:S35-S35.

Gessner, M. O., C. M. Swan, C. K. Dang, B. G. McKie, R. D. Bardgett, D. H. Wall, and S. Hättenschwiler. 2010. Diversity meets decomposition. *Trends in Ecology & Evolution* **25**:372–380.

Gibson, C. W. D., T. C. Guilford, C. Hambler, and P. H. Sterling. 1983. Transition matrix models and succession after release from grazing on Aldabra Atoll. *Vegetation* **52**:151–159.

Gibson, C. W. D. and J. Hamilton. 1983. Feeding ecology and seasonal movements of giant tortoises on Aldabra Atoll. *Oecologia* **61**:230–240.

Ginsburg, D. W. and V. J. Paul. 2001. Chemical defenses in the sea hare *Aplysia parvula*: Importance of diet and sequestration of algal secondary metabolites. *Marine Ecology-Progress Series* **215**:261–274.

Glenn-Lewin, D. C., R. K. Peet, and T. T. Veblen, editors. 1992. *Plant succession: theory and prediction*. Chapman & Hall, London.

Gliwicz, Z. M. and C. Guisande. 1992. Family planning in *Daphnia*. Resistance to starvation in offspring born to mothers grown at different food levels. *Oecologia* **91**:463–467.

Goecker, M. E., K. L. Heck, and J. F. Valentine. 2005. Effects of nitrogen concentrations in turtlegrass *Thalassia testudinum* on consumption by the bucktooth parrotfish *Sparisoma radians*. *Marine Ecology Progress Series* **286**:239–248.

Goldstein, J. L. and T. Swain. 1965. The inhibition of enzymes by tannins. *Phytochemistry* **4**:185–192.

Gonthier, D. J., G. M. Dominguez, J. D. Witter, A. L. Spongberg, and S. M. Philpott. 2013. Bottom-up effects of soil quality on a coffee arthropod interaction web. *Ecosphere* **4**:art107.

Gough, L., J. C. Moore, G. R. Shaver, R. T. Simpson, and D. R. Johnson. 2012. Above- and belowground responses of arctic tundra ecosystems to altered soil nutrients and mammalian herbivory. *Ecology* **93**:1683–1694.

Gough, L., C. W. Osenberg, K. L. Gross, and S. L. Collins. 2000. Fertilization effects on species density and primary productivity in herbaceous plant communities. *Oikos* **89**:428–439.

Grandmaison, J., G. M. Olah, M. R. Vancalsteren, and V. Furlan. 1993. Characterization and localization of plant phenolics likely invloved in the pathogen resistance expressed by endomycorrhizal roots. *Mycorrhiza* **3**:155–164.

Green, R. A. and J. K. Detling. 2000. Defoliation-induced enhancement of total aboveground nitrogen yield of grasses. *Oikos* **91**:280–284.

Green, T. R. and C. A. Ryan. 1972. Wound-induced proteinase inhibitor in plant leaves: A possible defense mechanism against insects. *Science* **175**:776–777.

Greve, M., A. M. Lykke, C. W. Fagg, J. Bogaert, I. Friis, R. Marchant, A. R. Marshall, J. Ndayishimiye, B. S. Sandel, C. Sandom, M. Schmidt, J. R. Timberlake, J. J.

Wieringa, G. Zizka, and J.-C. Svenning. 2012. Continental-scale variability in browser diversity is a major driver of diversity patterns in acacias across Africa. *Journal of Ecology* **100**:1093–1104.

Grman, E. 2012. Plant species differ in their ability to reduce allocation to non-beneficial arbuscular mycorrhizal fungi. *Ecology* **93**:711–718.

Grossart, H. P. and M. Simon. 1993. Limnetic macroscopic organic aggregates (lake snow)—Occurrence, characteristics, and microbial dynamics in Lake Constance. *Limnology and Oceanography* **38**:532–546.

Grutzmacher, G., G. Wessel, S. Klitzke, and I. Chorus. 2010. Microcystin elimination during sediment contact. *Environmental Science & Technology* **44**:657–662.

Guo, N. C. and P. Xie. 2006. Development of tolerance against toxic *Microcystis aeruginosa* in three cladocerans and the ecological implications. *Environmental Pollution* **143**:513–518.

Gurevitch, J., S. M. Scheiner, and G. A. Fox. 2006. *The Ecology of Plants*. 2nd edition. Sinauer Associates, Sunderland, MA.

Gustafsson, S., K. Rengefors, and L. A. Hansson. 2005. Increased consumer fitness following transfer of toxin tolerance to offspring via maternal effects. *Ecology* **86**:2561–2567.

Haan, N. L., M. R. Hunter, and M. D. Hunter. 2012. Investigating predictors of plant establishment during roadside restoration. *Restoration Ecology* **20**:315–321.

Hackett, J. D., D. M. Anderson, D. L. Erdner, and D. Bhattacharya. 2004. Dinoflagellates: A remarkable evolutionary experiment. *American Journal of Botany* **91**:1523–1534.

Haddad, N. M., G. M. Crutsinger, K. Gross, J. Haarstad, J. M. H. Knops, and D. Tilman. 2009. Plant species loss decreases arthropod diversity and shifts trophic structure. *Ecology Letters* **12**:1029–1039.

Hakanen, P., S. Suikkanen, J. Franzen, H. Franzen, H. Kankaanpaa, and A. Kremp. 2012. Bloom and toxin dynamics of *Alexandrium ostenfeldii* in a shallow embayment at the SW coast of Finland, northern Baltic Sea. *Harmful Algae* **15**:91–99.

Hall, M. C., P. Stiling, B. A. Hungate, B. G. Drake, and M. D. Hunter. 2005a. Effects of elevated CO_2 and herbivore damage on litter quality in a scrub oak ecosystem. *Journal of Chemical Ecology* **31**:2343–2356.

Hall, M. C., P. Stiling, D. C. Moon, B. G. Drake, and M. D. Hunter. 2005b. Effects of elevated CO2 on foliar quality and herbivore damage in a scrub oak ecosystem. *Journal of Chemical Ecology* **31**:267–286.

Hall, M. C., P. Stiling, D. C. Moon, B. G. Drake, and M. D. Hunter. 2006. Elevated CO_2 increases the long-term decomposition rate of *Quercus myrtifolia* leaf litter. *Global Change Biology* **12**:568–577.

Hamilton, E. W. and D. A. Frank. 2001. Can plants stimulate soil microbes and their own nutrient supply? Evidence from a grazing tolerant grass. *Ecology* **82**:2397–2402.

Hamilton, E. W., D. A. Frank, P. M. Hinchey, and T. R. Murray. 2008. Defoliation induces root exudation and triggers positive rhizospheric feedbacks in a temperate grassland. *Soil Biology & Biochemistry* **40**:2865–2873.

Hamilton, J. G., A. R. Zangerl, E. H. DeLucia, and M. R. Berenbaum. 2001. The carbon-nutrient balance hypothesis: Its rise and fall. *Ecology Letters* **4**:86–95.

Handa, I. T., R. Aerts, F. Berendse, M. P. Berg, A. Bruder, O. Butenschoen, E. Chauvet, M. O. Gessner, J. Jabiol, M. Makkonen, B. G. McKie, B. Malmqvist, E. T. H. M.

Peeters, S. Scheu, B. Schmid, J. van Ruijven, V. C. A. Vos, and S. Hättenschwiler. 2014. Consequences of biodiversity loss for litter decomposition across biomes. *Nature* **509**:218–221.

Hannah, L., editor. 2011. *Saving a million species. Extinction risk from climate change.* 2nd edition. Island Press, Washington, D.C.

Hansell, D. A. and C. A. Carlson. 2014. *Biogeochemistry of Marine Dissolved Organic Matter.* 2nd edition. Elsevier, New York.

Hansen, L., G. F. Krog, and M. Sondergaard. 1986. Decomposition of lake phytoplankton. 1. Dynamics of short-term decomposition. *Oikos* **46**:37–44.

Hansson, L.-A., S. Gustafsson, K. Rengefors, and L. Bomark. 2007. Cyanobacterial chemical warfare affects zooplankton community composition. *Freshwater Biology* **52**:1290–1301.

Hansson, L. A. 1996. Algal recruitment from lake sediments in relation to grazing, sinking, and dominance patterns in the phytoplankton community. *Limnology and Oceanography* **41**:1312–1323.

Harpole, W. S., J. T. Ngai, E. E. Cleland, E. W. Seabloom, E. T. Borer, M. E. S. Bracken, J. J. Elser, D. S. Gruner, H. Hillebrand, J. B. Shurin, and J. E. Smith. 2011. Nutrient co-limitation of primary producer communities. *Ecology Letters* **14**:852–862.

Harrison, K. A. and R. D. Bardgett. 2003. How browsing by red deer impacts on litter decomposition in a native regenerating woodland in the Highlands of Scotland. *Biology and Fertility of Soils* **38**:393–399.

Harrison, K. A. and R. D. Bardgett. 2004. Browsing by red deer negatively impacts on soil nitrogen availability in regenerating native forest. *Soil Biology & Biochemistry* **36**:115–126.

Harrison, P. G. 1989. Detrital processing in seagrass systems. A review of factors affecting decay rates, remineralization, and detritivory. *Aquatic Botany* **35**:263–288.

Hart, J. H. and W. E. Hillis. 1974. Inhibition of wood rotting fungi by stilbenes and other polyphenols in *Eucalyptus sideroxylon*. *Phytopathology* **64**:939–948.

Hartley, S. E. and A. C. Gange. 2009. Impacts of plant symbiotic fungi on insect herbivores: Mutualism in a multitrophic context. *Annual Review of Entomology* **54**:323–342.

Hartley, S. E., S. M. Gardner, and R. J. Mitchell. 2003. Indirect effects of grazing and nutrient addition on the hemipteran community of heather moorlands. *Journal of Applied Ecology* **40**:793–803.

Hartzfeld, P. W., R. Forkner, M. D. Hunter, and A. E. Hagerman. 2002. Determination of hydrolyzable tannins (gallotannins and ellagitannins) after reaction with potassium iodate. *Journal of Agricultural and Food Chemistry* **50**:1785–1790.

Harvey, J. A., I. F. Harvey, and D. J. Thompson. 1995. The effect of host nutrition on growth and development of the parasitoid wasp *Venturia canescens*. *Entomologia Experimentalis Et Applicata* **75**:213–220.

Haslam, E. 1989. *Plant polyphenols. Vegetable tannins revisited.* Cambridge University Press, Cambridge, UK.

Hättenschwiler, S., B. Aeschlimann, M. M. Couteaux, J. Roy, and D. Bonal. 2008. High variation in foliage and leaf litter chemistry among 45 tree species of a neotropical rainforest community. *New Phytologist* **179**:165–175.

Hättenschwiler, S., A. V. Tiunov, and S. Scheu. 2005. Biodiversity and litter decomposition in terrestrial ecosystems. *Annual Review of Ecology, Evolution and Systematics* **36**:191–218.

Hättenschwiler, S. and P. M. Vitousek. 2000. The role of polyphenols in terrestrial ecosystem nutrient cycling. *Trends in Ecology and Evolution* **15**:238–243.

Haukioja, E. 1980. On the role of plant defenses in the fluctuation of herbivore populations. *Oikos* **35**:202–213.

Haukioja, E. 1991. Cyclic fluctuations in density: Interactions between a defoliator and its host tree. *Acta Oecologica-International Journal of Ecology* **12**:77–88.

Haukioja, E. and P. Niemela. 1977. Retarded growth of a geometrid larva after mechanical damage to leaves of its host tree. *Annales Zoologici Fennici* **14**:48–52.

Haukioja, E., P. Niemela, and S. Siren. 1985. Foliage phenols and nitrogen in relation to growth, insect damage, and ability to recover after defoliation, in the mountain birch *Betula pubescens* ssp *tortuosa*. *Oecologia* **65**:214–222.

Hawkins, B. A. 1992. Parasitoid-host food webs and donor control. *Oikos* **65**:159–162.

Hawlena, D. and O. J. Schmitz. 2010a. Herbivore physiological response to predation risk and implications for ecosystem nutrient dynamics. *Proceedings of the National Academy of Sciences* **107**:15503–15507.

Hawlena, D. and O. J. Schmitz. 2010b. Physiological stress as a fundamental mechanism linking predation to ecosystem functioning. *American Naturalist* **176**:537–556.

Hawlena, D., M. S. Strickland, M. A. Bradford, and O. J. Schmitz. 2012. Fear of predation slows plant-litter decomposition. *Science* **336**:1434–1438.

Hay, M. E. 1991. Marine-terrestrial contrasts in the ecology of plant chemical defenses against herbivores. *Trends in Ecology & Evolution* **6**:362–365.

Hay, M. E. 1996. Marine chemical ecology: What's known and what's next? *Journal of Experimental Marine Biology and Ecology* **200**:103–134.

Hay, M. E. 2009. Marine chemical ecology: Chemical signals and cues structure marine populations, communities, and ecosystems. *Annual Review of Marine Science* **1**:193–212.

Hay, M. E., J. E. Duffey, and W. Fenical. 1990. Host-plant specialization decreases predation on a marine amphipod: An herbivore in plant's clothing. *Ecology* **71**:733–743.

Hay, M. E., J. E. Duffy, K. Pfister, and W. Fenical. 1987. Chemical defense against different marine herbivores: Are amphipods insect equivalents? *Ecology* **68**:1567–1580.

Hay, M. E. and W. Fenical. 1988. Marine plant-herbivore interactions: The ecology of chemical defense. *Annual Review of Ecology and Systematics* **19**:111–145.

Hay, M. E. and J. Kubanek. 2002. Community and ecosystem level consequences of chemical cues in the plankton. *Journal of Chemical Ecology* **28**:2001–2016.

Heath, K. D. and P. Tiffin. 2007. Context dependence in the coevolution of plant and rhizobial mutualists. *Proceedings of the Royal Society B: Biological Sciences* **274**:1905–1912.

Hebblewhite, M. and E. H. Merrill. 2009. Trade-offs between predation risk and forage differ between migrant strategies in a migratory ungulate. *Ecology* **90**:3445–3454.

Heiskanen, A. S. and K. Kononen. 1994. Sedimentation of vernal and late summer phytoplankton communities in the coastal Baltic Sea. *Archiv Für Hydrobiologie* **131**:175–198.

Helms, S. E., S. J. Connelly, and M. D. Hunter. 2004. Effects of variation among plant species on the interaction between a herbivore and its parasitoid. *Ecological Entomology* **29**:44–51.

Helms, S. E. and M. D. Hunter. 2005. Variation in plant quality and the population dynamics of herbivores: There is nothing average about aphids. *Oecologia* **145**:197–204.

Hemmi, A. and V. Jormalainen. 2002. Nutrient enhancement increases performance of a marine herbivore via quality of its food alga. *Ecology* **83**:1052–1064.

Hendrick, R. L. and K. S. Pregitzer. 1992. The demography of fine roots in a northern hardwood forest. *Ecology* **73**:1094–1104.

Hendrick, R. L. and K. S. Pregitzer. 1993. Patterns of fine root mortality in two sugar maple forests. *Nature* **361**:59–61.

Hendricks, J. J., J. D. Aber, K. J. Nadelhoffer, and R. D. Hallett. 2000. Nitrogen controls on fine root substrate quality in temperate forest ecosystems. *Ecosystems* **3**:57–69.

Herms, D. A. and W. J. Mattson. 1992. The dilemma of plants: To grow or to defend. *Quarterly Review of Biology* **67**:283–335.

Hidding, B., J.-P. Tremblay, and S. D. Côté. 2013. A large herbivore triggers alternative successional trajectories in the boreal forest. *Ecology* **94**:2852–2860.

Hill, R. W. and J. W. H. Dacey. 2006. Metabolism of dimethylsulfoniopropionate (DMSP) by juvenile Atlantic menhaden *Brevoortia tyrannus*. *Marine Ecology Progress Series* **322**:239–248.

Hillebrand, H., D. S. Gruner, E. T. Borer, M. E. S. Bracken, E. E. Cleland, J. J. Elser, W. S. Harpole, J. T. Ngai, E. W. Seabloom, J. B. Shurin, and J. E. Smith. 2007. Consumer versus resource control of producer diversity depends on ecosystem type and producer community structure. *Proceedings of the National Academy of Sciences* **104**:10904–10909.

Hines, J. and M. O. Gessner. 2012. Consumer trophic diversity as a fundamental mechanism linking predation and ecosystem functioning. *Journal of Animal Ecology* **81**:1146–1153.

Hixon, M. A. and W. N. Brostoff. 1983. Damselfish as keystone species in reverse: Intermediate disturbance and diversity of reef algae. *Science* **220**:511–513.

Hobbie, S. E. 1992. Effects of plant species on nutrient cycling. *Trends in Ecology and Evolution* **7**.

Hobbie, S. E., P. B. Reich, J. Oleksyn, M. Ogdahl, R. Zytkowiak, C. Hale, and P. Karolewski. 2006. Tree species effects on decomposition and forest floor dynamics in a common garden. *Ecology* **87**:2288–2297.

Hodgkiss, I. J. and K. C. Ho. 1997. Are changes in N:P ratios in coastal waters the key to increased red tide blooms? *Hydrobiologia* **852**:141–147.

Hol, W. H. G., M. Macel, J. A. van Veen, and E. van der Meijden. 2004. Root damage and aboveground herbivory change concentration and composition of pyrrolizidine alkaloids of *Senecio jacobaea*. *Basic and Applied Ecology* **5**:253–260.

Holland, J. N., W. X. Cheng, and D. A. Crossley. 1996. Herbivore-induced changes in plant carbon allocation: Assessment of below-ground C fluxes using carbon-14. *Oecologia* **107**:87–94.

Holland, J. N., D. L. DeAngelis, and J. L. Bronstein. 2002. Population dynamics and mutualism: Functional responses of benefits and costs. *American Naturalist* **159**:231–244.

Hollinger, D. Y. 1986. Herbivory and the cycling of nitrogen and phosphorus in isolated California oak trees. *Oecologia* **70**:291–297.

Horne, A. and C. R. Goldman. 1994. *Limnology*. 2nd edition. McGraw Hill.

Horsley, S. B., S. L. Stout, and D. S. deCalesta. 2003. White-tailed deer impact on the vegetation dynamics of a northern hardwood forest. *Ecological Applications* **13**:98–118.

Horton, J. L., B. D. Clinton, J. F. Walker, C. M. Beier, and E. T. Nilsen. 2009. Variation in soil and forest floor characteristics along gradients of ericaceous, evergreen shrub cover in the southern Appalachians. *Castanea* **74**:340–352.

Horton, M. W., A. M. Hancock, Y. S. Huang, C. Toomajian, S. Atwell, A. Auton, N. W. Muliyati, A. Platt, F. G. Sperone, B. J. Vilhjalmsson, M. Nordborg, J. O. Borevitz, and J. Bergelson. 2012. Genome-wide patterns of genetic variation in worldwide *Arabidopsis thaliana* accessions from the RegMap panel. *Nature Genetics* **44**:212–216.

Hu, P., W. Zhou, Z. Cheng, M. Fan, L. Wang, and D. Xie. 2013. JAV1 controls jasmonate-regulated plant defense. *Molecular Cell* **50**:504–515.

Huberty, A. F. and R. F. Denno. 2006. Consequences of nitrogen and phosphorus limitation for the performance of two planthoppers with divergent life-history strategies. *Oecologia* **149**:444–455.

Hudson, T. M., B. L. Turner, H. Herz, and J. S. Robinson. 2009. Temporal patterns of nutrient availability around nests of leaf-cutting ants (*Atta colombica*) in secondary moist tropical forest. *Soil Biology & Biochemistry* **41**:1088–1093.

Huffman, M. A. 2003. Animal self-medication and ethno-medicine: Exploration and exploitation of the medicinal properties of plants. *Proceedings of the Nutrition Society* **62**:371–381.

Hughes, A. R., R. J. Best, and J. J. Stachowicz. 2010. Genotypic diversity and grazer identity interactively influence seagrass and grazer biomass. *Marine Ecology Progress Series* **403**:43–51.

Hughes, A. R., B. D. Inouye, M. T. J. Johnson, N. Underwood, and M. Vellend. 2008. Ecological consequences of genetic diversity. *Ecology Letters* **11**:609–623.

Hughes, A. R. and J. J. Stachowicz. 2004. Genetic diversity enhances the resistance of a seagrass ecosystem to disturbance. *Proceedings of the National Academy of Sciences* **101**:8998–9002.

Hughes, B. B., R. Eby, E. Van Dyke, M. T. Tinker, C. I. Marks, K. S. Johnson, and K. Wasson. 2013. Recovery of a top predator mediates negative eutrophic effects on seagrass. *Proceedings of the National Academy of Sciences* **110**:15313–15318.

Hunt-Joshi, T. R., B. Blossey, and R. B. Root. 2004. Root and leaf herbivory on *Lythrum salicaria*: Implications for plant performance and communities. *Ecological Applications* **14**:1574–1589.

Hunter, E. A., J. P. Gibbs, L. J. Cayot, and W. Tapia. 2013. Equivalency of Galápagos giant tortoises used as ecological replacement species to restore ecosystem functions. *Conservation Biology* **27**:701–709.

Hunter, M. D. 1987. Opposing effects of spring defoliation on late season oak caterpillars. *Ecological Entomology* **12**:373–382.

Hunter, M. D. 1990. Differential susceptibility to variable plant phenology and its role in competition between two insect herbivores on oak. *Ecological Entomology* **15**:401–408.

Hunter, M. D. 1992a. Interactions within herbivore communities mediated by the host plant: The keystone herbivore concept. Pages 287–325 *in* M. D. Hunter, T. Ohgushi, and P. W. Price, editors. *Effects of resource distribution on animal-plant interactions*. Academic Press, San Diego, CA.

Hunter, M. D. 1992b. A variable insect-plant interaction: The relationship between tree budburst phenology and population levels of insect herbivores among trees. *Ecological Entomology* **17**:91–95.

Hunter, M. D. 1994. The search for pattern in pest outbreaks. Pages 443–448 *in* S. R. Leather, A. D. Watt, N. J. Mills, and K. F. A. Walters, editors. *Individuals, Populations and Patterns in Ecology*. Intercept, Andover, MA.

Hunter, M. D. 1997. Incorporating variation in plant chemistry into a spatially-explicit ecology of phytophagous insects. Pages 81–96 *in* A. D. Watt, N. E. Stork, and M. D. Hunter, editors. *Forests and Insects*. Chapman and Hall, London.

Hunter, M. D. 1998. Interactions between *Operophtera brumata* and *Tortrix viridana* on oak: New evidence from time-series analysis. *Ecological Entomology* **23**:168–173.

Hunter, M. D. 2000. Mixed signals and cross-talk: Interactions between plants, insect herbivores, and plant pathogens. *Agricultural and Forest Entomology* **2**:155–160.

Hunter, M. D. 2001a. Insect population dynamics meets ecosystem ecology: Effects of herbivory on soil nutrient dynamics. *Agricultural and Forest Entomology* **3**:77–84.

Hunter, M. D. 2001b. Multiple approaches to estimating the relative importance of top-down and bottom-up forces on insect populations: Experiments, life tables, and time-series analysis. *Basic and Applied Ecology* **2**:295–309.

Hunter, M. D. 2001c. Out of sight, out of mind: The impacts of root-feeding insects in natural and managed systems. *Agricultural and Forest Entomology* **3**:3–10.

Hunter, M. D. 2002a. Ecological causes of pest outbreaks. Pages 214–217 *in* D. Pimentel, editor. *Encyclopedia of Pest Management*. Marcel Dekker, Inc, New York.

Hunter, M. D. 2002b. Maternal effects and the population dynamics of insects on plants. *Agricultural and Forest Entomology* **4**:1–9.

Hunter, M. D. 2003. Effects of plant quality on the population ecology of parasitoids. *Agricultural and Forest Entomology* **5**:1–8.

Hunter, M. D. 2008. Root herbivory in forest ecosystems. Pages 68–95 *in* S. N. Johnson and P. J. Murray, editors. *Root Feeders. An Ecosystem Perspective*. CAB Biosciences, Ascot, UK.

Hunter, M. D. 2009. Trophic promiscuity, intraguild predation and the problem of omnivores. *Agricultural and Forest Entomology* **11**:125–131.

Hunter, M. D., S. Adl, C. M. Pringle, and D. C. Coleman. 2003a. Relative effects of macro invertebrates and habitat on the chemistry of litter during decomposition. *Pedobiologia* **47**:101–115.

Hunter, M. D. and R. E. Forkner. 1999. Hurricane damage influences foliar polyphenolics and subsequent herbivory on surviving trees. *Ecology* **80**:2676–2682.

Hunter, M. D., R. E. Forkner, and J. N. McNeil. 2000. Heterogeneity in plant quality and its impact on the population ecology of insect herbivores. Pages 155–179 *in* M. A. Hutchings, E. A. John, and A. J. A. Stewart, editors. *The Ecological Consequences of Environmental Heterogeneity*. Blackwell Scientific, Oxford, UK.

Hunter, M. D., C. R. Linnen, and B. C. Reynolds. 2003b. Effects of endemic densities of canopy herbivores on nutrient dynamics along a gradient in elevation in the southern Appalachians. *Pedobiologia* **47**:231–244.

Hunter, M. D., S. B. Malcolm, and S. E. Hartley. 1996. Population-level variation in plant secondary chemistry and the population biology of herbivores. *Chemoecology* **7**:45–56.

Hunter, M. D. and J. N. McNeil. 1997. Host-plant quality influences diapause and voltinism in a polyphagous insect herbivore. *Ecology* **78**:977–986.

Hunter, M. D., T. Ohgushi, and P. W. Price. 1992. *The Effects of Resource Distribution on Animal-Plant Interactions*. Academic Press, San Diego, CA.

Hunter, M. D. and P. W. Price. 1992a. Natural variability in plants and animals. Pages 1–12 *in* M. D. Hunter, T. Ohgushi, and P. W. Price, editors. *Resource Distribution and Animal-Plant Interactions.* Academic Press, San Diego, CA.

Hunter, M. D. and P. W. Price. 1992b. Playing chutes and ladders: bottom-up and top-down forces in natural communities. *Ecology* **73**:724–732.

Hunter, M. D. and P. W. Price. 1998. Cycles in insect populations: delayed density dependence or exogenous driving variables? *Ecological Entomology* **23**:216–222.

Hunter, M. D. and P. W. Price. 2000. Detecting cycles and delayed density dependence: A reply to Turchin and Berryman. *Ecological Entomology* **25**:122–124.

Hunter, M. D., B. C. Reynolds, M. C. Hall, and C. J. Frost. 2012. Effects of herbivores on ecosystem processes: The role of trait-mediated indirect effects. Pages 339–370 *in* T. Ohgushi, O. J. Schmitz, and R. Holt, editors. *Trait-Mediated Indirect Interactions. Ecological and Evolutionary Perspectives.* Cambridge University Press, Cambridge, UK.

Hunter, M. D. and J. C. Schultz. 1993. Induced plant defenses breached? Phytochemical induction protects an herbivore from disease. *Oecologia* **94**:195–203.

Hunter, M. D. and J. C. Schultz. 1995. Fertilization mitigates chemical induction and herbivore responses within damaged oak trees. *Ecology* **76**:1226–1232.

Hunter, M. D., G. C. Varley, and G. R. Gradwell. 1997. Estimating the relative roles of top-down and bottom-up forces on insect herbivore populations: A classic study revisited. *Proceedings of the National Academy of Sciences* **94**:9176–9181.

Hunter, M. D., A. D. Watt, and M. Docherty. 1991. Outbreaks of the winter moth on Sitka spruce in Scotland are not influenced by nutrient deficiencies of trees, tree budburst, or pupal predation. *Oecologia* **86**:62–69.

Hunter, M. D. and P. G. Willmer. 1989. The potential for interspecific competition between two abundant defoliators on oak: Leaf damage and habitat quality. *Ecological Entomology* **14**:267–277.

Hunter, M. R. and M. D. Hunter. 2008. Designing for conservation of insects in the built environment. *Insect Conservation and Diversity* **1**:189–196.

Hunting, E. R., M. H. Whatley, H. G. van der Geest, C. Mulder, M. H. S. Kraak, A. M. Breure, and W. Admiraal. 2012. Invertebrate footprints on detritus processing, bacterial community structure, and spatiotemporal redox profiles. *Freshwater Science* **31**:724–732.

Hutchinson, G. E. 1948. Circular causal systems in ecology. *Annals of the New York Academy of Sciences* **50**:221–246.

Hutchinson, G. E. 1965. *The Ecological Theater and the Evolutionary Play.* Yale University Press, New Haven, CT.

Ianora, A., A. Miralto, S. A. Poulet, Y. Carotenuto, I. Buttino, G. Romano, R. Casotti, G. Pohnert, T. Wichard, L. Colucci-D'Amato, G. Terrazzano, and V. Smetacek. 2004. Aldehyde suppression of copepod recruitment in blooms of a ubiquitous planktonic diatom. *Nature* **429**:403–407.

Ianora, A., S. A. Poulet, and A. Miralto. 2003. The effects of diatoms on copepod reproduction: a review. *Phycologia* **42**:351–363.

Iason, G. R., M. Dicke, and S. E. Hartley, editors. 2012. *The ecology of plant secondary metabolites. From genes to global processes.* Cambridge University Press, Cambridge, UK.

Ibrahim, M. A., M. Na, J. Oh, R. F. Schinazi, T. R. McBrayer, T. Whitaker, R. J. Doerksen, D. J. Newman, L. G. Zachos, and M. T. Hamann. 2013. Significance of endangered and threatened plant natural products in the control of human disease. *Proceedings of the National Academy of Sciences* **110**:16832–16837.

Iglesias-Rodriguez, M. D., P. R. Halloran, R. E. M. Rickaby, I. R. Hall, E. Colmenero-Hidalgo, J. R. Gittins, D. R. H. Green, T. Tyrrell, S. J. Gibbs, P. von Dassow, E. Rehm, E. V. Armbrust, and K. P. Boessenkool. 2008. Phytoplankton calcification in a high-CO2 world. *Science* **320**:336–340.

Isbell, F., P. B. Reich, D. Tilman, S. E. Hobbie, S. Polasky, and S. Binder. 2013a. Nutrient enrichment, biodiversity loss, and consequent declines in ecosystem productivity. *Proceedings of the National Academy of Sciences* **110**:11911–11916.

Isbell, F., D. Tilman, S. Polasky, S. Binder, and P. Hawthorne. 2013b. Low biodiversity state persists two decades after cessation of nutrient enrichment. *Ecology Letters* **16**:454–460.

Jackrel, S. L. and J. T. Wootton. 2013. Local adaptation of stream communities to intraspecific variation in a terrestrial ecosystem subsidy. *Ecology* **95**:37–43.

Jackson, R. B., H. A. Mooney, and E. D. Schulze. 1997. A global budget for fine root biomass, surface area, and nutrient contents. *Proceedings of the National Academy of Sciences* **94**:7362–7366.

Jamieson, M. A., T. R. Seastedt, and M. D. Bowers. 2012. Nitrogen enrichment differentially affects above- and belowground plant defense. *American Journal of Botany* **99**:1630–1637.

Jani, A. J., S. H. Faeth, and D. Gardner. 2010. Asexual endophytes and associated alkaloids alter arthropod community structure and increase herbivore abundances on a native grass. *Ecology Letters* **13**:106–117.

Janson, E. M., J. O. Stireman, M. S. Singer, and P. Abbott. 2008. Phytophagous insect–microbe mutualisms and adaptive evolutionary diversification. *Evolution* **62**:997–1012.

Janz, N. 2011. Ehrlich and Raven Revisited: Mechanisms Underlying Codiversification of Plants and Enemies. Pages 71–89 *in* D. J. Futuyma, H. B. Shaffer, and D. Simberloff, editors. *Annual Review of Ecology, Evolution, and Systematics*. Vol. 42. Annual Reviews, Palo Alto, CA.

Jeffries, M. J. and J. H. Lawton. 1984. Enemy free space and the structure of ecological communities. *Biological Journal of the Linnean Society* **23**:269–286.

Jepsen, J. U., L. Kapari, S. B. Hagen, T. Schott, O. P. L. Vindstad, A. Nilssen, and R. A. Ims. 2011. Rapid northwards expansion of a forest insect pest attributed to spring phenology matching with sub-arctic birch. *Global Change Biology* **17**:2071–2083.

Joanisse, G. D., R. L. Bradley, C. M. Preston, and A. D. Munson. 2007. Soil enzyme inhibition by condensed litter tannins may drive ecosystem structure and processes: The case of *Kalmia angustifolia*. *New Phytologist* **175**:535–546.

Jochimsen, M. C., R. Kümmerlin, and D. Straile. 2013. Compensatory dynamics and the stability of phytoplankton biomass during four decades of eutrophication and oligotrophication. *Ecology Letters* **16**:81–89.

Joern, A. and S. T. Behmer. 1998. Impact of diet quality on demographic attributes in adult grasshoppers and the nitrogen limitation hypothesis. *Ecological Entomology* **23**:174–184.

Joern, A., T. Provin, and S. T. Behmer. 2012. Not just the usual suspects: Insect herbivore populations and communities are associated with multiple plant nutrients. *Ecology* **93**:1002–1015.

Johnson, N. D., B. Liu, and B. L. Bentley. 1987. The effects of nitrogen fixation, soil nitrate, and defoliation on the growth, alkaloids, and nitrogen levels of *Lupinus succulentus* (Fabaceae). *Oecologia* **74**:425–431.

Johnson, S. N., T. M. Bezemer, and T. H. Jones. 2008. Linking aboveground and belowground herbivory. Pages 153–170 *in* S. N. Johnson and P. J. Murray, editors. *Root Feeders, an Ecosystem Perspective.* CAB Biosciences, Ascot, UK.

Johnson, S. N., C. Mitchell, J. W. McNicol, J. Thompson, and A. J. Karley. 2013. Downstairs drivers—Root herbivores shape communities of above-ground herbivores and natural enemies via changes in plant nutrients. *Journal of Animal Ecology* **82**:1021–1030.

Jormalainen, V. and T. Ramsay. 2009. Resistance of the brown alga *Fucus vesiculosus* to herbivory. *Oikos* **118**:713–722.

Jumars, P. A., D. L. Penry, J. A. Baross, M. J. Perry, and B. W. Frost. 1989. Closing the microbial loop: dissolved carbon pathway to heterotrophic bacteria from incomplete ingestion, digestion and absorption in animals. *Deep Sea Research Part A. Oceanographic Research Papers* **36**:483–495.

Jung, S. C., A. Martinez-Medina, J. A. Lopez-Raez, and M. J. Pozo. 2012. Mycorrhiza-induced resistance and priming of plant defenses. *Journal of Chemical Ecology* **38**:651–664.

Kacsoh, B. Z., Z. R. Lynch, N. T. Mortimer, and T. A. Schlenke. 2013. Fruit flies medicate offspring after seeing parasites. *Science* **339**:947–950.

Kaebernick, M., B. A. Neilan, T. Bürner, and E. Dittmann. 2000. Light and the transcriptional response of the microcystin biosynthesis gene cluster. *Applied Environmental Microbiology* **66**:3387–3392.

Kagata, H. and T. Ohgushi. 2006. Bottom-up trophic cascades and material transfer in terrestrial food webs. *Ecological Research* **21**:26–34.

Kagata, H. and T. Ohgushi. 2011. Ecosystem consequences of selective feeding of an insect herbivore: Palatability-decomposability relationship revisited. *Ecological Entomology* **36**:768–775.

Kagata, H. and T. Ohgushi. 2012. Carbon to nitrogen excretion ratio in lepidopteran larvae: Relative importance of ecological stoichiometry and metabolic scaling. *Oikos* **121**:1869–1877.

Karban, R. and A. A. Agrawal. 2002. Herbivore offense. *Annual Review of Ecology and Systematics* **33**:641–664.

Karban, R. and I. T. Baldwin. 1997. *Induced Responses to Herbivory.* University of Chicago Press, Chicago.

Karban, R., P. Grof-Tisza, J. L. Maron, and M. Holyoak. 2012. The importance of host plant limitation for caterpillars of an arctiid moth (*Platyprepia virginalis*) varies spatially. *Ecology* **93**:2216–2226.

Karban, R., M. Huntzinger, and A. C. McCall. 2004. The specificity of eavesdropping on sagebrush by other plants. *Ecology* **85**:1846–1852.

Kaspari, M., M. N. Garcia, K. E. Harms, M. Santana, S. J. Wright, and J. B. Yavitt. 2008a. Multiple nutrients limit litterfall and decomposition in a tropical forest. *Ecology Letters* **11**:35–43.

Kaspari, M., S. P. Yanoviak, and R. Dudley. 2008b. On the biogeography of salt limitation: A study of ant communities. *Proceedings of the National Academy of Sciences* **105**:17848–17851.

Kaspari, M., S. P. Yanoviak, R. Dudley, M. Yuan, and N. A. Clay. 2009. Sodium shortage as a constraint on the carbon cycle in an inland tropical rainforest. *Proceedings of the National Academy of Sciences* **106**:19405–19409.

Katayama, N., A. O. Silva, O. Kishida, and T. Ohgushi. 2013. Aphids decelerate litter nitrogen mineralisation through changes in litter quality. *Ecological Entomology* **38**:627–630.

Kauffman, M. J., J. F. Brodie, and E. S. Jules. 2010. Are wolves saving Yellowstone's aspen? A landscape-level test of a behaviorally mediated trophic cascade. *Ecology* **91**:2742–2755.

Kaukonen, M., A. L. Ruotsalainen, P. R. Wäli, M. K. Männistö, H. Setälä, K. Saravesi, K. Huusko, and A. Markkola. 2013. Moth herbivory enhances resource turnover in subarctic mountain birch forests? *Ecology* **94**:267–272.

Kay, A. D., J. Mankowski, and S. E. Hobbie. 2008. Long-term burning interacts with herbivory to slow decomposition. *Ecology* **89**:1188–1194.

Kearns, K. D. and M. D. Hunter. 2000. Green algal extracellular products regulate antialgal toxin production in a cyanobacterium. *Environmental Microbiology* **2**:291–297.

Kearns, K. D. and M. D. Hunter. 2001. Toxin-producing *Anabaena flos-aquae* induces settling of *Chlamydomonas reinhardtii*, a competing motile alga. *Microbial Ecology* **42**:80–86.

Kearns, K. D. and M. D. Hunter. 2002. Algal extracellular products suppress *Anabaena flos-aquae* heterocyst spacing. *Microbial Ecology* **43**:174–180.

Keating, S. T., M. D. Hunter, and J. C. Schultz. 1990. Leaf phenolic inhibition of gypsy moth nuclear polyhedrosis virus: Role of polyhedral inclusion body aggregation. *Journal of Chemical Ecology* **16**:1445–1457.

Keating, S. T. and W. G. Yendol. 1987. Influence of selected host plants on gypsy moth (Lepidoptera, Lymantriidae) larval mortality caused by a baculovirus. *Environmental Entomology* **16**:459–462.

Keating, S. T., W. G. Yendol, and J. C. Schultz. 1988. Relationship between susceptibility of gypsy moth larvae (Lepidoptera: Lymantriidae) to a baculovirus and host plant foliage constituents. *Environmental Entomology* **17**:952–958.

Kelaher, B. P., M. J. Bishop, J. Potts, P. Scanes, and G. Skilbeck. 2013. Detrital diversity influences estuarine ecosystem performance. *Global Change Biology* **19**:1909–1918.

Kellmann, R., A. Stuken, R. J. S. Orr, H. M. Svendsen, and K. S. Jakobsen. 2010. Biosynthesis and molecular genetics of polyketides in marine dinoflagellates. *Marine Drugs* **8**:1011–1048.

Kerbes, R. H., P. M. Kotanen, and R. L. Jeffries. 1990. Destruction of wetland habitats by lesser snow geese. A keystone species on the west coast of Hudson Bay. *Journal of Applied Ecology* **27**:242–258.

Kersch-Becker, M. F. and T. M. Lewinsohn. 2012. Bottom-up multitrophic effects in resprouting plants. *Ecology* **93**:9–16.

Kersten, B., A. Ghirardo, J.-P. Schnitzler, B. Kanawati, P. Schmitt-Kopplin, M. Fladung, and H. Schroeder. 2013. Integrated transcriptomics and metabolomics decipher differences in the resistance of pedunculate oak to the herbivore *Tortrix viridana* L. *BMC Genomics* **14**:737.

Kessler, A. and I. T. Baldwin. 2001. Defensive function of herbivore-induced plant volatile emissions in nature. *Science* **291**:2141–2144.

Kessler, A., R. Halitschke, and I. T. Baldwin. 2004. Silencing the jasmonate cascade: Induced plant defenses and insect populations. *Science* **305**:665–668.

Kielland, K. and J. P. Bryant. 1998. Moose herbivory in taiga: Effects on biogeochemistry and vegetation dynamics in primary succession. *Oikos* **82**:377–383.

Kielland, K., J. P. Bryant, and R. W. Ruess. 1997. Moose herbivory and carbon turnover of early successional stands in interior Alaska. *Oikos* **80**:25–30.

Kiers, E. T. and R. F. Denison. 2008. Sanctions, cooperation, and the stability of plant-rhizosphere mutualisms. *Annual Review of Ecology Evolution and Systematics* **39**:215–236.

Killingbeck, K. T. 1996. Nutrients in senesced leaves: Keys to the search for potential resorption and resorption proficiency. *Ecology* **77**:1716–1727.

Kingdon, J., B. Agwanda, M. Kinnaird, T. O'Brien, C. Holland, T. Gheysens, M. Boulet-Audet, and F. Vollrath. 2012. A poisonous surprise under the coat of the African crested rat. *Proceedings of the Royal Society B: Biological Sciences* **279**:675–680.

Kinney, K. K., R. L. Lindroth, S. M. Jung, and E. V. Nordheim. 1997. Effects of CO2 and NO3 availability on deciduous trees: Phytochemistry and insect performance. *Ecology* **78**:215–230.

Kiorboe, T. 1993. Turbulence, phytoplankton cell size, and the structure of pelagic food webs. *Advances in Marine Biology* **29**:1–72.

Kirk, K. L. and J. J. Gilbert. 1992. Variation in herbivore response to chemical defenses: Zooplankton foraging on toxic cyanobacteria. *Ecology* **73**:2208–2217.

Kitchell, J. F., R. V. O'Neill, D. Webb, G. W. Gallepp, S. M. Bartell, J. F. Koonce, and B. S. Ausmus. 1979. Consumer regulation of nutrient cycling. *Bioscience* **29**:28–34.

Kjøller, A. H. and S. Struwe. 2002. Fungal communities, succession, enzymes, and decomposition. Pages 267–284 *in* R. G. Burns and R. P. Dick, editors. *Enzymes in the Environment: Activity, Ecology and Applications.* Marcel Dekker, New York.

Klaassen, M. and B. A. Nolet. 2008. Stoichiometry of endothermy: Shifting the quest from nitrogen to carbon. *Ecology Letters* **11**:785–792.

Klaper, R., K. Ritland, T. A. Mousseau, and M. D. Hunter. 2001. Heritability of phenolics in *Quercus laevis* inferred using molecular markers. *Journal of Heredity* **92**:421–426.

Klaper, R. D. and M. D. Hunter. 1998. Genetic versus environmental effects on the phenolic chemistry of turkey oak, *Quercus laevis*. Pages 262–268 *in* K. C. Steiner, editor. *Diversity and adaptation in oak species.* Pennsylvania State University Press, University Park.

Klausmeier, C. A., E. Litchman, T. Daufresne, and S. A. Levin. 2004. Optimal nitrogen-to-phosphorus stoichiometry of phytoplankton. *Nature* **429**:171–174.

Klemola, N., T. Andersson, K. Ruohomaki, and T. Klemola. 2010. Experimental test of parasitism hypothesis for population cycles of a forest lepidopteran. *Ecology* **91**:2506–2513.

Klironomos, J. N. and M. M. Hart. 2001. Food-web dynamics—Animal nitrogen swap for plant carbon. *Nature* **410**:651–652.

Kneip, C., P. Lockhart, C. VoSZ, and U.-G. Maier. 2007. Nitrogen fixation in eukaryotes—New models for symbiosis. *BMC Evolutionary Biology* **7**:55.

Knoepp, J. D., J. M. Vose, B. D. Clinton, and M. D. Hunter. 2011. Hemlock infestation and mortality: Impacts on nutrient pools and cycling in Appalachian forests. *Soil Science Society of America Journal* **75**:1935–1945.

Knops, J. M. H., M. E. Ritchie, and D. Tilman. 2000. Selective herbivory on a nitrogen fixing legume (*Lathyrus venosus*) influences productivity and ecosystem nitrogen pools in an oak savanna. *Ecoscience* **7**:166–174.

Knops, J. M. H., D. Tilman, N. M. Haddad, S. Naeem, C. E. Mitchell, J. Haarstad, M. E. Ritchie, K. M. Howe, P. B. Reich, E. Siemann, and J. Groth. 1999. Effects of plant species richness on invasion dynamics, disease outbreaks, insect abundances and diversity. *Ecology Letters* **2**:286–293.

Kodner, R. B., R. E. Summons, and A. H. Knoll. 2009. Phylogenetic investigation of the aliphatic, non-hydrolyzable biopolymer algaenan, with a focus on green algae. *Organic Geochemistry* **40**:854–862.

Kohl, K. D. and M. D. Dearing. 2012. Experience matters: Prior exposure to plant toxins enhances diversity of gut microbes in herbivores. *Ecology Letters* **15**:1008–1015.

Kohl, K. D., R. B. Weiss, J. Cox, C. Dale, and M. Denise Dearing. 2014. Gut microbes of mammalian herbivores facilitate intake of plant toxins. *Ecology Letters* **17**:1238–1246.

Kohout, P., Z. Sykorova, M. Ctvrtlikova, J. Rydlova, J. Suda, M. Vohnik, and R. Sudova. 2012. Surprising spectra of root-associated fungi in submerged aquatic plants. *Fems Microbiology Ecology* **80**:216–235.

Kokaly, R. F., G. P. Asner, S. V. Ollinger, M. E. Martin, and C. A. Wessman. 2009. Characterizing canopy biochemistry from imaging spectroscopy and its application to ecosystem studies. *Remote Sensing of Environment* **113**:S78-S91.

Kominoski, J. S., C. M. Pringle, B. A. Ball, M. A. Bradford, D. C. Coleman, D. B. Hall, and M. D. Hunter. 2007. Nonadditive effects of leaf litter species diversity on breakdown dynamics in a detritus-based stream. *Ecology* **88**:1167–1176.

Koricheva, J. 2002. Meta-analysis of sources of variation in fitness costs of plant antiherbivore defenses. *Ecology* **83**:176–190.

Koricheva, J., A. C. Gange, and T. Jones. 2009. Effects of mycorrhizal fungi on insect herbivores: a meta-analysis. *Ecology* **90**:2088–2097.

Koricheva, J., S. Larsson, E. Haukioja, and M. Keinanen. 1998. Regulation of woody plant secondary metabolism by resource availability: Hypothesis testing by means of meta-analysis. *Oikos* **83**:212–226.

Kramer, M. G., J. Sanderman, O. A. Chadwick, J. Chorover, and P. M. Vitousek. 2012. Long-term carbon storage through retention of dissolved aromatic acids by reactive particles in soil. *Global Change Biology* **18**:2594–2605.

Krock, B., U. Tillmann, T. J. Alpermann, D. Voss, O. Zielinski, and A. D. Cembella. 2013. Phycotoxin composition and distribution in plankton fractions from the German Bight and western Danish coast. *Journal of Plankton Research* **35**:1093–1108.

Kubanek, J., M. E. Hay, P. J. Brown, N. Lindquist, and W. Fenical. 2001. Lignoid chemical defenses in the freshwater macrophyte *Saururus cernuus*. *Chemoecology* **11**:1–8.

Kubanek, J., S. E. Lester, W. Fenical, and M. E. Hay. 2004. Ambiguous role of phlorotannins as chemical defenses in the brown alga *Fucus vesiculosus*. *Marine Ecology Progress Series* **277**:79–93.

Kurmayer, R. 2011. The toxic cyanobacterium *Nostoc* sp. strain 152 produces highest amounts of microcystin and nostophycin under stress conditions. *Journal of Phycology* **47**:200–207.

Kurokawa, H., D. A. Peltzer, and D. A. Wardle. 2010. Plant traits, leaf palatability and litter decomposability for co-occurring woody species differing in invasion status and nitrogen fixation ability. *Functional Ecology* **24**:513–523.

Kuzyakov, Y. 2010. Priming effects: Interactions between living and dead organic matter. *Soil Biology & Biochemistry* **42**:1363–1371.

Lampert, E. C. and M. D. Bowers. 2010. Host plant influences on iridoid glycoside sequestration of generalist and specialist caterpillars. *Journal of Chemical Ecology* **36**:1101–1104.

Lampert, E. C., L. A. Dyer, and M. D. Bowers. 2010. Caterpillar chemical defense and parasitoid success: *Cotesia congregata* parasitism of *Ceratomia catalpae*. *Journal of Chemical Ecology* **36**:992–998.

Lampert, E. C., A. R. Zangerl, M. R. Berenbaum, and P. J. Ode. 2011. Generalist and specialist host-parasitoid associations respond differently to wild parsnip (*Pastinaca sativa*) defensive chemistry. *Ecological Entomology* **36**:52–61.

Lampert, W. 1978. Release of dissolved organic carbon by grazing zooplankton. *Limnology and Oceanography* **23**:831–834.

Lampert, W. 1987. Laboratory studies on zooplankton-cyanobacteria interactions. *New Zealand Journal of Marine and Freshwater Research* **21**:483–490.

Lasley-Rasher, R. S., D. B. Rasher, Z. H. Marion, R. B. Taylor, and M. E. Hay. 2011. Predation constrains host choice for a marine mesograzer. *Marine Ecology-Progress Series* **434**:91–99.

Laspoumaderes, C., B. Modenutti, M. S. Souza, M. Bastidas Navarro, F. Cuassolo, and E. Balseiro. 2013. Glacier melting and stoichiometric implications for lake community structure: zooplankton species distributions across a natural light gradient. *Global Change Biology* **19**:316–326.

Leandro, L. F., G. J. Teegarden, P. B. Roth, Z. H. Wang, and G. J. Doucette. 2010. The copepod *Calanus finmarchicus*: A potential vector for trophic transfer of the marine algal biotoxin, domoic acid. *Journal of Experimental Marine Biology and Ecology* **382**:88–95.

Lefèvre, T., A. Chiang, M. Kelavkar, H. Li, J. Li, C. L. F. de Castillejo, L. Oliver, Y. Potini, M. D. Hunter, and J. C. de Roode. 2012. Behavioural resistance against a protozoan parasite in the monarch butterfly. *Journal of Animal Ecology* **81**:70–79.

Lefèvre, T., L. Oliver, M. D. Hunter, and J. C. de Roode. 2010. Evidence for transgenerational medication in nature. *Ecology Letters* **13**:1485–1493.

Lehahn, Y., I. Koren, D. Schatz, M. Frada, U. Sheyn, E. Boss, S. Efrati, Y. Rudich, M. Trainic, S. Sharoni, C. Laber, Giacomo R. DiTullio, Marco J. L. Coolen, Ana M. Martins, Benjamin A. S. Van Mooy, Kay D. Bidle, and A. Vardi. 2014. Decoupling physical from biological processes to assess the impact of viruses on a mesoscale algal bloom. *Current Biology* **24**:2041–2046.

Leibold, M. A. 1989. Resource edibility and the effects of predators and productivity on the outcome of trophic interactions. *American Naturalist* **134**:922–949.

Leimu, R. and J. Koricheva. 2006. A meta-analysis of tradeoffs between plant tolerance and resistance to herbivores: combining the evidence from ecological and agricultural studies. *Oikos* **112**:1–9.

Lemaire, V., S. Brusciotti, I. van Gremberghe, W. Vyverman, J. Vanoverbeke, and L. De Meester. 2012. Genotype genotype interactions between the toxic cyanobacterium *Microcystis* and its grazer, the waterflea *Daphnia*. *Evolutionary Applications* **5**:168–182.

Leroux, S. J., D. Hawlena, and O. J. Schmitz. 2012. Predation risk, stoichiometric plasticity and ecosystem elemental cycling. *Proceedings of the Royal Society B: Biological Sciences* **279**:4183–4191.

LeRoy, C. J., S. C. Wooley, and R. L. Lindroth. 2012. Genotype and soil nutrient environment influence aspen litter chemistry and in-stream decomposition. *Freshwater Science* **31**:1244–1253.

Lewitus, A. J., R. A. Horner, D. A. Caron, E. Garcia-Mendoza, B. M. Hickey, M. Hunter, D. D. Huppert, R. M. Kudela, G. W. Langlois, J. L. Largier, E. J. Lessard, R. RaLonde, J. E. J. Rensel, P. G. Strutton, V. L. Trainer, and J. F. Tweddle. 2012. Harmful algal blooms along the North American West Coast region: History, trends, causes, and impacts. *Harmful Algae* **19**:133–159.

Lewontin, R. 2002. *The Triple Helix: Gene, Organism, and Environment*. Harvard University Press, Cambridge, MA.

Li, Y., M. Dicke, J. Harvey, and R. Gols. 2014. Intra-specific variation in wild *Brassica oleracea* for aphid-induced plant responses and consequences for caterpillar–parasitoid interactions. *Oecologia* **174**:853–862.

Liechti, R. and E. E. Farmer. 2002. The jasmonate pathway. *Science* **296**:1649–1650.

Liere, H. and A. Larsen. 2010. Cascading trait-mediation: disruption of a trait-mediated mutualism by parasite-induced behavioral modification. *Oikos* **119**:1394–1400.

Liess, A., S. Drakare, and M. Kahlert. 2009. Atmospheric nitrogen-deposition may intensify phosphorus limitation of shallow epilithic periphyton in unproductive lakes. *Freshwater Biology* **54**:1759–1773.

Lilburn, T. C., K. S. Kim, N. E. Ostrom, K. R. Byzek, J. R. Leadbetter, and J. A. Breznak. 2001. Nitrogen fixation by symbiotic and free-living spirochetes. *Science* **292**:2495–2498.

Lindegren, M., T. Blenckner, and N. C. Stenseth. 2012. Nutrient reduction and climate change cause a potential shift from pelagic to benthic pathways in a eutrophic marine ecosystem. *Global Change Biology* **18**:3491–3503.

Lindeman, R. 1942. The trophic-dynamic aspect of ecology. *Ecology* **23**:399–418.

Lindo, Z., M.-C. Nilsson, and M. J. Gundale. 2013. Bryophyte-cyanobacteria associations as regulators of the northern latitude carbon balance in response to global change. *Global Change Biology* **19**:2022–2035.

Lindquist, N., P. H. Barber, and J. B. Weisz. 2005. Episymbiotic microbes as food and defence for marine isopods: Unique symbioses in a hostile environment. *Proceedings of the Royal Society B: Biological Sciences* **272**:1209–1216.

Lindström, K. and S. A. Mousavi. 2010. *Rhizobium and Other N-fixing Symbioses*. eLS. John Wiley & Sons Ltd, Chichester. http://www.els.net [doi:10.1002/9780470015902.a0021157].

Lisonbee, L. D., J. J. Villalba, F. D. Provenza, and J. O. Hall. 2009. Tannins and self-medication: Implications for sustainable parasite control in herbivores. *Behavioural Processes* **82**:184–189.

Liu, X., Y. Zhang, W. Han, A. Tang, J. Shen, Z. Cui, P. Vitousek, J. W. Erisman, K. Goulding, P. Christie, A. Fangmeier, and F. Zhang. 2013a. Enhanced nitrogen deposition over China. *Nature* **494**:459–462.

Liu, Y., M. J. Aryee, L. Padyukov, M. D. Fallin, E. Hesselberg, A. Runarsson, L. Reinius, N. Acevedo, M. Taub, M. Ronninger, K. Shchetynsky, A. Scheynius, J. Kere, L. Alfredsson, L. Klareskog, T. J. Ekstrom, and A. P. Feinberg. 2013b. Epigenome-wide association data implicate DNA methylation as an intermediary of genetic risk in rheumatoid arthritis. *Nature Biotechnology* **31**:142–147.

Lloyd, K. G., L. Schreiber, D. G. Petersen, K. U. Kjeldsen, M. A. Lever, A. D. Steen, R. Stepanauskas, M. Richter, S. Kleindienst, S. Lenk, A. Schramm, and B. B. Jorgensen.

2013. Predominant archaea in marine sediments degrade detrital proteins. *Nature* **496**:215–218.

Loladze, I. and J. J. Elser. 2011. The origins of the Redfield nitrogen-to-phosphorus ratio are in a homoeostatic protein-to-rRNA ratio. *Ecology Letters* **14**:244–250.

Longhurst, A. R., A. W. Bedo, W. G. Harrison, E. J. H. Head, and D. D. Sameoto. 1990. Vertical flux of respiratory carbon by oceanic diel migrant biota. *Deep-Sea Research Part A—Oceanographic Research Papers* **37**:685–694.

Longhurst, A. R. and W. G. Harrison. 1988. Vertical nitrogen flux from the oceanic photic zone by diel migrant zooplankton and nekton. *Deep-Sea Research Part A—Oceanographic Research Papers* **35**:881–889.

Loranger, J., S. T. Meyer, B. Shipley, J. Kattge, H. Loranger, C. Roscher, and W. W. Weisser. 2012. Predicting invertebrate herbivory from plant traits: Evidence from 51 grassland species in experimental monocultures. *Ecology* **93**:2674–2682.

Loreau, M. 2010. *From Populations to Ecosystems: Theoretical Foundations for a New Ecological Synthesis* (MPB-46). Princeton University Press, Princeton, NJ.

Lou, Y. G. and I. T. Baldwin. 2004. Nitrogen supply influences herbivore-induced direct and indirect defenses and transcriptional responses to *Nicotiana attenuata*. *Plant Physiology* **135**:496–506.

Lovett, G. M. and A. E. Ruesink. 1995. Carbon and nitrogen mineralization from decomposing gypsy moth frass. *Oecologia* **104**:133–138.

Low, P. A., C. McArthur, K. Fisher, and D. F. Hochuli. 2014. Elevated volatile concentrations in high-nutrient plants: Do insect herbivores pay a high price for good food? *Ecological Entomology* **39**:480–491.

Lubchenco, J. 1983. *Littorina* and *Fucus*: Effects of herbivores, substratum heterogeneity, and plant escapes during succession. *Ecology* **64**:1116–1123.

Lubchenco, J. and S. D. Gaines. 1981. A unified approach to marine plant-herbivore interactions. I. Populations and communities. *Annual Review of Ecology and Systematics* **12**:405–437.

Luck, G. W., S. Lavorel, S. McIntyre, and K. Lumb. 2012. Improving the application of vertebrate trait-based frameworks to the study of ecosystem services. *Journal of Animal Ecology* **81**:1065–1076.

Lyons, D. A., C. Arvanitidis, A. J. Blight, E. Chatzinikolaou, T. Guy-Haim, J. Kotta, H. Orav-Kotta, A. M. Queirós, G. Rilov, P. J. Somerfield, and T. P. Crowe. 2014. Macroalgal blooms alter community structure and primary productivity in marine ecosystems. *Global Change Biology* **20**:2712–2724.

Madritch, M. D., J. R. Donaldson, and R. L. Lindroth. 2007. Canopy herbivory can mediate the influence of plant genotype on soil processes through frass deposition. *Soil Biology & Biochemistry* **39**:1192–1201.

Madritch, M. D., S. L. Greene, and R. L. Lindroth. 2009. Genetic mosaics of ecosystem functioning across aspen-dominated landscapes. *Oecologia* **160**:119–127.

Madritch, M. D. and M. D. Hunter. 2002. Phenotypic diversity influences ecosystem functioning in an oak sandhills community. *Ecology* **83**:2084–2090.

Madritch, M. D. and M. D. Hunter. 2003. Intraspecific litter diversity and nitrogen deposition affect nutrient dynamics and soil respiration. *Oecologia* **136**:124–128.

Madritch, M. D. and M. D. Hunter. 2004. Phenotypic diversity and litter chemistry affect nutrient dynamics during litter decomposition in a two species mix. *Oikos* **105**:125–131.

Madritch, M. D. and M. D. Hunter. 2005. Phenotypic variation in oak litter influences short- and long-term nutrient cycling through litter chemistry. *Soil Biology & Biochemistry* **37**:319–327.

Madritch, M. D., C. C. Kingdon, A. Singh, K. E. Mock, R. L. Lindroth, and P. A. Townsend. 2014. Imaging spectroscopy links aspen genotype with below-ground processes at landscape scales. *Philosophical Transactions of the Royal Society B: Biological Sciences* **369**.

Mahadevan, A. 2014. Eddy effects on biogeochemistry. *Nature* **506**:168–169.

Majdi, N., A. Boiché, W. Traunspurger, and A. Lecerf. 2014. Predator effects on a detritus-based food web are primarily mediated by non-trophic interactions. *Journal of Animal Ecology* **83**:953–962.

Makkonen, M., M. P. Berg, I. T. Handa, S. Hättenschwiler, J. van Ruijven, P. M. van Bodegom, and R. Aerts. 2012. Highly consistent effects of plant litter identity and functional traits on decomposition across a latitudinal gradient. *Ecology Letters* **15**:1033–1041.

Malzahn, A. M., N. Aberle, C. Clemmesen, and M. Boersma. 2007. Nutrient limitation of primary producers affects planktivorous fish condition. *Limnology and Oceanography* **52**:2062–2071.

Mangan, S. A., S. A. Schnitzer, E. A. Herre, K. M. L. Mack, M. C. Valencia, E. I. Sanchez, and J. D. Bever. 2010. Negative plant-soil feedback predicts tree-species relative abundance in a tropical forest. *Nature* **466**:752–755.

Mann, P. J., W. V. Sobczak, M. M. LaRue, E. Bulygina, A. Davydova, J. E. Vonk, J. Schade, S. Davydov, N. Zimov, R. M. Holmes, and R. G. M. Spencer. 2014. Evidence for key enzymatic controls on metabolism of arctic river organic matter. *Global Change Biology* **20**:1089–1100.

Mao, W., M. A. Schuler, and M. R. Berenbaum. 2013. Honey constituents up-regulate detoxification and immunity genes in the western honey bee *Apis mellifera*. *Proceedings of the National Academy of Sciences* **110**:8842–8846.

Marañón, E., P. Cermeño, D. C. López-Sandoval, T. Rodríguez-Ramos, C. Sobrino, M. Huete-Ortega, J. M. Blanco, and J. Rodríguez. 2013. Unimodal size scaling of phytoplankton growth and the size dependence of nutrient uptake and use. *Ecology Letters* **16**:371–379.

Marcarelli, A. M., C. V. Baxter, M. M. Mineau, and R. O. Hall. 2011. Quantity and quality: unifying food web and ecosystem perspectives on the role of resource subsidies in freshwaters. *Ecology* **92**:1215–1225.

Marczak, L. B., K. Więski, R. F. Denno, and S. C. Pennings. 2013. Importance of local vs. geographic variation in salt marsh plant quality for arthropod herbivore communities. *Journal of Ecology* **101**:1169–1182.

Maron, J. L. and R. L. Jeffries. 1999. Bush lupine mortality, altered resource availability, and alternative vegetation states. *Ecology* **80**:443–454.

Marquis, R. J. and C. J. Whelan. 1994. Insecivorous birds increase growth of white oak through consumption of leaf-chewing insects. *Ecology* **75**:2007–2014.

Marsh, K. J., B. D. Moore, I. R. Wallis, and W. J. Foley. 2014. Feeding rates of a mammalian browser confirm the predictions of a 'foodscape' model of its habitat. *Oecologia* **174**:873–882.

Marsh, K. J., I. R. Wallis, and W. J. Foley. 2005. Detoxification rates constrain feeding in common brushtail possums (*Trichosurus vulpecula*). *Ecology* **86**:2946–2954.

Martin, P., S. T. Dyhrman, M. W. Lomas, N. J. Poulton, and B. A. S. Van Mooy. 2014. Accumulation and enhanced cycling of polyphosphate by Sargasso Sea plankton in response to low phosphorus. *Proceedings of the National Academy of Sciences* **111**:8089–8094.

Martiny, A. C., C. T. A. Pham, F. W. Primeau, J. A. Vrugt, J. K. Moore, S. A. Levin, and M. W. Lomas. 2013. Strong latitudinal patterns in the elemental ratios of marine plankton and organic matter. *Nature Geoscience* **6**:279–283.

Mason, N. W. H., D. A. Peltzer, S. J. Richardson, P. J. Bellingham, and R. B. Allen. 2010. Stand development moderates effects of ungulate exclusion on foliar traits in the forests of New Zealand. *Journal of Ecology* **98**:1422–1433.

Masters, G. J. and V. K. Brown. 1992. Plant-mediated interactions between two spatially separated insects. *Functional Ecology* **6**:175–179.

Masters, G. J., V. K. Brown, and A. C. Gange. 1993. Plant mediated interactions between above- and below-ground insect herbivores. *Oikos* **66**:148–151.

Mata, L., H. Gaspar, and R. Santos. 2012. Carbon/nutrient balance in relation to biomass production and halogenated compound content in the red alga *Asparagopsis taxiformis* (Bonnemaisoniaceae). *Journal of Phycology* **48**:248–253.

Mattson, W. J. and R. A. Haack. 1987. The role of drought in outbreaks of plant-eating insects. *Bioscience* **37**:110–118.

Mattson, W. J., Jr. 1980. Herbivory in relation to plant nitrogen content. *Annual Review of Ecology and Systematics* **11**:119–161.

May, R. M. 1974. Biological populations with nonoverlapping generations: stable points, stable cycles, and chaos. *Science* **186**:645–647.

Mayhew, P. J. and H. C. J. Godfray. 1997. Mixed sex allocation strategies in a parasitoid wasp. *Oecologia* **110**:218–221.

McArt, S. H., D. E. Spalinger, W. B. Collins, E. R. Schoen, T. Stevenson, and M. Bucho. 2009. Summer dietary nitrogen availability as a potential bottom-up constraint on moose in south-central Alaska. *Ecology* **90**:1400–1411.

McCann, K. S. 2011. *Food Webs* (MPB-50). Princeton University Press, Princeton, NJ.

McCormick, A. C., S. B. Unsicker, and J. Gershenzon. 2012. The specificity of herbivore-induced plant volatiles in attracting herbivore enemies. *Trends in Plant Science* **17**:303–310.

McDonald, K. E. and J. T. Lehman. 2013. Dynamics of *Aphanizomenon* and *Microcystis* (cyanobacteria) during experimental manipulation of an urban impoundment. *Lake and Reservoir Management* **29**:103–115.

McGillicuddy, D. J., D. W. Townsend, R. He, B. A. Keafer, J. L. Kleindinst, Y. Li, J. P. Manning, D. G. Mountain, M. A. Thomas, and D. M. Anderson. 2011. Suppression of the 2010 Alexandrium fundyense bloom by changes in physical, biological, and chemical properties of the Gulf of Maine. *Limnology and Oceanography* **56**:2411–2426.

McLister, J. D., J. S. Sorensen, and M. D. Dearing. 2004. Effects of consumption of juniper (*Juniperus monosperma*) on cost of thermoregulation in the woodrats *Neotoma albigula* and *Neotoma stephensi* at different acclimation temperatures. *Physiological and Biochemical Zoology* **77**:305–312.

McNaughton, S. J., R. W. Ruess, and S. W. Seagle. 1988. Large mammals and process dynamics in African ecosystems: Herbivorous mammals affect primary productivity and regulate recycling balances. *Bioscience* **38**:794–800.

McSherry, M. E. and M. E. Ritchie. 2013. Effects of grazing on grassland soil carbon: A global review. *Global Change Biology* **19**:1347–1357.

Mech, L. D., M. E. Nelson, and R. E. McRoberts. 1991. Effects of maternal and grandmaternal nutrition on deer mass and vulnerability to wolf predation. *Journal of Mammalogy* **72**:146–151.

Meentemeyer, V. 1978. Macroclimate and lignin control of litter decomposition rates. *Ecology* **59**:465–472.

Melillo, J. M., J. D. Aber, and J. F. Muratore. 1982. Nitrogen and lignin control of hardwood leaf litter decomposition dynamics. *Ecology* **59**:465–472.

Merton, L. F. H., D. M. Bourn, and R. J. Hnatiuk. 1976. Giant tortoise and vegetation interactions on Aldabra Atoll. I. Inland. *Biological Conservation* **9**:293–304.

Mertz, W. 1981. The essential trace elements. *Science* **213**:1332–1338.

Messerli, F. H. 2012. Chocolate consumption, cognitive function, and Nobel laureates. *New England Journal of Medicine* **370**:1454–1457.

Metcalfe, D. B., G. P. Asner, R. E. Martin, J. E. Silva Espejo, W. H. Huasco, F. F. Farfán Amézquita, L. Carranza-Jimenez, D. F. Galiano Cabrera, L. D. Baca, F. Sinca, L. P. Huaraca Quispe, I. A. Taype, L. E. Mora, A. R. Dávila, M. M. Solórzano, B. L. Puma Vilca, J. M. Laupa Román, P. C. Guerra Bustios, N. S. Revilla, R. Tupayachi, C. A. J. Girardin, C. E. Doughty, and Y. Malhi. 2013. Herbivory makes major contributions to ecosystem carbon and nutrient cycling in tropical forests. *Ecology Letters* **17**:324–332.

Michalak, A. M., E. J. Anderson, D. Beletsky, S. Boland, N. S. Bosch, T. B. Bridgeman, J. D. Chaffin, K. Cho, R. Confesor, I. Daloğlu, J. V. DePinto, M. A. Evans, G. L. Fahnenstiel, L. He, J. C. Ho, L. Jenkins, T. H. Johengen, K. C. Kuo, E. LaPorte, X. Liu, M. R. McWilliams, M. R. Moore, D. J. Posselt, R. P. Richards, D. Scavia, A. L. Steiner, E. Verhamme, D. M. Wright, and M. A. Zagorski. 2013. Record-setting algal bloom in Lake Erie caused by agricultural and meteorological trends consistent with expected future conditions. *Proceedings of the National Academy of Sciences* **110**:6448–6452.

Middleton, A. D., M. J. Kauffman, D. E. McWhirter, J. G. Cook, R. C. Cook, A. A. Nelson, M. D. Jimenez, and R. W. Klaver. 2013. Animal migration amid shifting patterns of phenology and predation: Lessons from a Yellowstone elk herd. *Ecology* **94**:1245–1256.

Mikola, J., K. Ilmarinen, M. Nieminen, and M. Vestberg. 2005. Long-term soil feedback on plant N allocation in defoliated grassland miniecosystems. *Soil Biology & Biochemistry* **37**:899–904.

Milan, N. F., B. Z. Kacsoh, and T. A. Schlenke. 2012. Alcohol consumption as self-medication against blood-borne parasites in the fruit fly. *Current Biology* **22**:488–493.

Miller, M. J., M. M. Critchley, J. Hutson, and H. J. Fallowfield. 2001. The adsorption of cyanobacterial hepatotoxins from water onto soil during batch experiments. *Water Research* **35**:1461–1468.

Mills, M. M. and K. R. Arrigo. 2010. Magnitude of oceanic nitrogen fixation influenced by the nutrient uptake ratio of phytoplankton. *Nature Geoscience* **3**:412–416.

Mineau, M. M., C. V. Baxter, A. M. Marcarelli, and G. W. Minshall. 2012. An invasive riparian tree reduces stream ecosystem efficiency via a recalcitrant organic matter subsidy. *Ecology* **93**:1501–1508.

Miralto, A., L. Guglielmo, G. Zagami, I. Buttino, A. Granata, and A. Ianora. 2003. Inhibition of population growth in the copepods *Acartia clausi* and *Calanus helgolandicus* during diatom blooms. *Marine Ecology Progress Series* **254**:253–268.

Moller, E. F. 2007. Production of dissolved organic carbon by sloppy feeding in the copepods *Acartia tonsa*, *Centropages typicus*, and *Temora longicornis*. *Limnology and Oceanography* **52**:79–84.

Moon, D. C., J. Barnouti, and B. Younginger. 2013. Context-dependent effects of mycorrhizae on herbivore density and parasitism in a tritrophic coastal study system. *Ecological Entomology* **38**:31–39.

Mooney, K. A., R. Halitschke, A. Kessler, and A. A. Agrawal. 2010. Evolutionary trade-offs in plants mediate the strength of trophic cascades. *Science* **327**:1642–1644.

Moore, B. D., I. R. Lawler, I. R. Wallis, C. M. Beale, and W. J. Foley. 2010. Palatability mapping: a koala's eye view of spatial variation in habitat quality. *Ecology* **91**:3165–3176.

Moore, L. R. 2013. More mixotrophy in the marine microbial mix. *Proceedings of the National Academy of Sciences* **110**:8323–8324.

Mooshammer, M., W. Wanek, J. Schnecker, B. Wild, S. Leitner, F. Hofhansl, A. Blochl, I. Hammerle, A. H. Frank, L. Fuchslueger, K. M. Keiblinger, S. Zechmeister-Boltenstern, and A. Richter. 2012. Stoichiometric controls of nitrogen and phosphorus cycling in decomposing beech leaf litter. *Ecology* **93**:770–782.

Moran, Y., D. Fredman, P. Szczesny, M. Grynberg, and U. Technau. 2012. Recurrent horizontal transfer of bacterial toxin genes to eukaryotes. *Molecular Biology and Evolution* **29**:2223–2230.

Moreira, X., K. A. Mooney, S. Rasmann, W. K. Petry, A. Carrillo-Gavilán, R. Zas, and L. Sampedro. 2014. Trade-offs between constitutive and induced defences drive geographical and climatic clines in pine chemical defences. *Ecology Letters* **17**:537–546.

Moreira, X., K. A. Mooney, R. Zas, and L. Sampedro. 2012. Bottom-up effects of host-plant species diversity and top-down effects of ants interactively increase plant performance. *Proceedings of the Royal Society B: Biological Sciences* **279**:4464–4472.

Mousseau, T. A. and C. W. Fox, editors. 1998. *Maternal Effects as Adaptations*. Oxford University Press, New York.

Mulkey, S. S., A. P. Smith, and T. P. Young. 1984. Predation by elephants on *Senecio keniodedron* (Compositae) in the alpine zone of Mount Kenya. *Biotropica* **16**:246–248.

Muller, R. N., P. J. Kalisz, and J. O. Luken. 1989. Fine root production of astringent phenolics. *Oecologia* **79**:563–565.

Muñoz-Marín, M. d. C., I. Luque, M. V. Zubkov, P. G. Hill, J. Diez, and J. M. García-Fernández. 2013. *Prochlorococcus* can use the Pro1404 transporter to take up glucose at nanomolar concentrations in the Atlantic Ocean. *Proceedings of the National Academy of Sciences* **110**:8597–8602.

Mur, L. A. J., E. Prats, S. Pierre, M. A. Hall, and K. H. Hebelstrup. 2013. Integrating nitric oxide into salicylic acid and jasmonic acid/ethylene plant defense pathways. *Frontiers in Plant Science* Article 215:doi:10.3389/fpls.2013.00215.

Murdoch, W. W., F. C. Evans, and C. H. Peterson. 1972. Diversity and pattern in plants and insects. *Ecology* **53**:819–829.

Murray, B. D., C. R. Webster, and J. K. Bump. 2013. Broadening the ecological context of ungulate–ecosystem interactions: The importance of space, seasonality, and nitrogen. *Ecology* **94**:1317–1326.

Murray, S., C. L. C. Ip, R. Moore, Y. Nagahama, and Y. Fukuyo. 2009. Are prorocentroid dinoflagellates monophyletic? A study of 25 species based on nuclear and mitochondrial genes. *Protist* **160**:245–264.

Murray, S. A., T. K. Mihali, and B. A. Neilan. 2011. Extraordinary conservation, gene loss, and positive selection in the evolution of an ancient neurotoxin. *Molecular Biology and Evolution* **28**:1173–1182.

Murray, T. R., D. A. Frank, and C. A. Gehring. 2010. Ungulate and topographic control of arbuscular mycorrhizal fungal spore community composition in a temperate grassland. *Ecology* **91**:815–827.

Musser, R. O., S. M. Hum-Musser, H. Eichenseer, M. Peiffer, G. Ervin, J. B. Murphy, and G. W. Felton. 2002. Herbivory: Caterpillar saliva beats plant defences— A new weapon emerges in the evolutionary arms race between plants and herbivores. *Nature* **416**:599–600.

Myer, J. L. 1990. A blackwater perspective on river ecosystems. *Bioscience* 40: 643–651.

Myer, J. L. and R. T. Edwards. 1990. Ecosystem metabolism and turnover of organic carbon along a blackwater river continuum. *Ecology* **71**:668–677.

Nabity, P. D., M. J. Haus, M. R. Berenbaum, and E. H. DeLucia. 2013. Leaf-galling phylloxera on grapes reprograms host metabolism and morphology. *Proceedings of the National Academy of Sciences* **110**:16663–16668.

Nelson, C. E. and E. K. Wear. 2014. Microbial diversity and the lability of dissolved organic carbon. *Proceedings of the National Academy of Sciences* **111**:7166–7167.

Nelson, W. A., O. N. Bjørnstad, and T. Yamanaka. 2013. Recurrent insect outbreaks caused by temperature-driven changes in system stability. *Science* **341**:796–799.

Ness, J. H. 2003. *Catalpa bignonioides* alters extrafloral nectar production after herbivory and attracts ant bodyguards. *Oecologia* **134**:210–218.

Nevitt, G. A. 2008. Sensory ecology on the high seas: The odor world of the procellariiform seabirds. *Journal of Experimental Biology* **211**:1706–1713.

Newman, R. M., W. C. Kerfoot, and Z. Hanscom. 1996. Watercress allelochemical defends high-nitrogen foliage against consumption: Effects on freshwater invertebrate herbivores. *Ecology* **77**:2312–2323.

Ngai, J. T. and D. S. Srivastava. 2006. Predators accelerate nutrient cycling in a bromeliad ecosystem. *Science* **314**:963–963.

Nicholson, A. J. and V. A. Bailey. 1935. The balance of animal populations. Part 1. *Proceedings of the Zoological Society of London* **105**:551–598.

Northup, R. R., Z. Yu, R. A. Dahlgren, and K. A. Vogt. 1995. Polyphenol control of nitrogen release from pine litter. *Nature* **377**:227–229.

Novotny, V., Y. Basset, S. E. Miller, G. D. Weiblen, B. Bremer, L. Cizek, and P. Drozd. 2002. Low host specificity of herbivorous insects in a tropical forest. *Nature* **416**:841–844.

Novotny, V., S. E. Miller, L. Baje, S. Balagawi, Y. Basset, L. Cizek, K. J. Craft, F. Dem, R. A. I. Drew, J. Hulcr, J. Leps, O. T. Lewis, R. Pokon, A. J. A. Stewart, G. Allan Samuelson, and G. D. Weiblen. 2010. Guild-specific patterns of species richness and host specialization in plant–herbivore food webs from a tropical forest. *Journal of Animal Ecology* **79**:1193–1203.

Nyman, T. 2010. To speciate, or not to speciate? Resource heterogeneity, the subjectivity of similarity, and the macroevolutionary consequences of niche-width shifts in plant-feeding insects. *Biological Reviews* **85**:393–411.

Nyman, T., H. P. Linder, C. Peña, T. Malm, and N. Wahlberg. 2012. Climate-driven diversity dynamics in plants and plant-feeding insects. *Ecology Letters* **15**:889–898.

Ode, P. J. 2006. Plant chemistry and natural enemy fitness: Effects on herbivore and natural enemy interactions. *Annual Review of Entomology* **51**:163–185.

Ode, P. J., M. R. Berenbaum, A. R. Zangerl, and C. W. Hardy. 2004. Host plant, host plant chemistry and the polyembryonic parasitoid *Copidosoma sosares*: Indirect effects in a tritrophic interaction. *Oikos* **104**:388–400.

Odum, E. P. 1953. *Fundamentals of Ecology*. W. B. Saunders Company, Philadelphia.

Odum, H. T. 1957. Trophic structure and productivity of Silver Springs, Florida. *Ecological Monographs* **27**:55–112.

Ohgushi, T., O. J. Schmitz, and R. Holt, editors. 2012. *Interaction Richness and Complexity: Ecological and Evolutionary Aspects of Trait-Mediated Indirect Effects*. Cambridge University Press, Cambridge, UK.

Oksanen, L., S. D. Fretwell, J. Arruda, and P. Niemela. 1981. Exploitation ecosystems in gradients of primary productivity. *American Naturalist* **118**:240–261.

Okullo, P. and S. R. Moe. 2012. Large herbivores maintain termite-caused differences in herbaceous species diversity patterns. *Ecology* **93**:2095–2103.

Olff, H., M. E. Ritchie, and H. H. T. Prins. 2002. Global environmental controls of diversity in large herbivores. *Nature* **415**:901–904.

Olofsson, J. and L. Oksanen. 2002. Role of litter decomposition for the increased primary production in areas heavily grazed by reindeer: A litterbag experiment. *Oikos* **96**:507–515.

Olsgard, F. 1993. Do toxic algal blooms affect subtidal soft-bottom communities? *Marine Ecology Progress Series* **102**:269–286.

Omand, M. M., E. A. D'Asaro, C. M. Lee, M. J. Perry, N. Briggs, I. Cetinic, and A. Mahadevan. 2015. Eddy-driven subduction exports particulate organic carbon from the spring bloom. *Science* **348**:222–225.

Opitz, S. E. W., S. R. Jensen, and C. Muller. 2010. Sequestration of glucosinolates and iridoid glucosides in sawfly species of the genus *Athalia* and their role in defense against ants. *Journal of Chemical Ecology* **36**:148–157.

Opitz, S. E. W. and C. Muller. 2009. Plant chemistry and insect sequestration. *Chemoecology* **19**:117–154.

Orians, C., A. Thorn, and S. Gómez. 2011. Herbivore-induced resource sequestration in plants: why bother? *Oecologia* **167**:1–9.

Orlofske, S. A., R. C. Jadin, D. L. Preston, and P. T. J. Johnson. 2012. Parasite transmission in complex communities: Predators and alternative hosts alter pathogenic infections in amphibians. *Ecology* **93**:1247–1253.

Ortlund, E. A., J. T. Bridgham, M. R. Redinbo, and J. W. Thornton. 2007. Crystal structure of an ancient protein: Evolution by conformational epistasis. *Science* **317**:1544–1548.

Orwig, D. A., R. C. Cobb, A. W. D'Amato, M. L. Kizlinski, and D. R. Foster. 2008. Multi-year ecosystem response to hemlock woolly adelgid infestation in southern New England forests. *Canadian Journal of Forest Research-Revue Canadienne De Recherche Forestiere* **38**:834–843.

Osono, T. 2007. Ecology of ligninolytic fungi associated with leaf litter decomposition. *Ecological Research* **B22B**:955–974.

Ostrofsky, M. L. 1997. Relationship between chemical characteristics of autumn-shed leaves and aquatic processing rates. *Journal of the North American Benthological Society* **16**:750–759.

Paine, R. T. 1966. Food web complexity and species diversity. *American Naturalist* **100**:65–75.

Paine, R. T. 1980. Food webs: Linkage, interaction strength and community infrastructure—The Third Tansley Lecture. *Journal of Animal Ecology* **49**:667–685.

Painter, L. E., R. L. Beschta, E. J. Larsen, and W. J. Ripple. 2014. Recovering aspen follow changing elk dynamics in Yellowstone: Evidence of a trophic cascade? *Ecology* **96**:252–263.

Palumbo, M. J., F. E. Putz, and S. T. Talcott. 2007. Nitrogen fertilizer and gender effects on the secondary metabolism of yaupon, a caffeine-containing North American holly. *Oecologia* **151**:1–9.

Pan, Y. L., D. V. S. Rao, K. H. Mann, W. K. W. Li, and W. G. Harrison. 1996. Effects of silicate limitation on production of domoic acid, a neurotoxin, by the diatom *Pseudonitzschia* multiseries. 2. Continuous culture studies. *Marine Ecology Progress Series* **131**:235–243.

Paragi, T. F., W. N. Johnson, D. D. Katnik, and A. J. Magoun. 1997. Selection of post-fire seres by lynx and snowshoe hares in the Alaskan taiga. *Northwestern Naturalist* **78**.

Pardee, G. L. and S. M. Philpott. 2011. Cascading indirect effects in a coffee agroecosystem: Effects of parasitic phorid flies on ants and the coffee berry borer in a high-shade and low-shade habitat. *Environmental Entomology* **40**:581–588.

Paré, M. C. and A. Bedard-Haughn. 2013. Soil organic matter quality influences mineralization and ghg emissions in cryosols: A field-based study of sub- to high arctic. *Global Change Biology* **19**:1126–1140.

Parker, J. D., D. E. Burkepile, D. O. Collins, J. Kubanek, and M. E. Hay. 2007. Stream mosses as chemically-defended refugia for freshwater macroinvertebrates. *Oikos* **116**:302–312.

Parker, J. D., D. O. Collins, J. Kubanek, M. C. Sullards, D. Bostwick, and M. E. Hay. 2006. Chemical defenses promote persistence of the aquatic plant *Micranthemum umbrosum*. *Journal of Chemical Ecology* **32**:815–833.

Parmenter, R. R. and J. A. MacMahon. 2009. Carrion decomposition and nutrient cycling in a semiarid shrub-steppe ecosystem. *Ecological Monographs* **79**:637–661.

Parsons, M. L., Q. Dortch, and G. J. Doucette. 2013. An assessment of *Pseudo-nitzschia* population dynamics and domoic acid production in coastal Louisiana. *Harmful Algae* **30**:65–77.

Partensky, F., W. R. Hess, and D. Vaulot. 1999. *Prochlorococcus*, a marine photosynthetic prokaryote of global significance. *Microbiology and Molecular Biology Reviews* **63**:106–127.

Pastor, J. and Y. Cohen. 1997. Herbivores, the functional diversity of plant species, and the cycling of nutrients in ecosystems. *Theoretical Population Biology* **51**:165–179.

Pastor, J., B. Dewey, R. Moen, D. J. Mladenoff, M. White, and Y. Cohen. 1998. Spatial patterns in the moose-forest-soil ecosystem on Isle Royale, Michigan, USA. *Ecological Applications* **8**:411–424.

Pastor, J., B. Dewey, R. J. Naiman, P. F. McInnes, and Y. Cohen. 1993. Moose browsing and soil fertility in the boreal forests of Isle-Royale National Park. *Ecology* **74**:467–480.

Pastor, J. and R. J. Naiman. 1992. Selective foraging and ecosystem processes in boreal forests. *American Naturalist* **139**:690–705.

Paterson, E. and A. Sim. 2013. Soil-specific response functions of organic matter mineralization to the availability of labile carbon. *Global Change Biology* **19**:1562–1571.

Pavia, H. and G. B. Toth. 2000. Inducible chemical resistance to herbivory in the brown seaweed *Ascophyllum nodosum*. *Ecology* **81**:3212–3225.

Peacor, S. D. and E. E. Werner. 2001. The contribution of trait-mediated indirect effects to the net effects of a predator. *Proceedings of the National Academy of Sciences* **98**:3904–3908.

Pearse, I. S., R. C. Cobb, and R. Karban. 2013. The phenology–substrate-match hypothesis explains decomposition rates of evergreen and deciduous oak leaves. *Journal of Ecology* **102**:28–35.

Pedler, B. E., L. I. Aluwihare, and F. Azam. 2014. Single bacterial strain capable of significant contribution to carbon cycling in the surface ocean. *Proceedings of the National Academy of Sciences* **111**:7202–7207.

Peduzzi, P., M. Gruber, M. Gruber, and M. Schagerl. 2014. The virus's tooth: cyanophages affect an African flamingo population in a bottom-up cascade. *The ISME Journal* **8**:1346–1351

Peiffer, J. A., A. Spor, O. Koren, Z. Jin, S. G. Tringe, J. L. Dangl, E. S. Buckler, and R. E. Ley. 2013. Diversity and heritability of the maize rhizosphere microbiome under field conditions. *Proceedings of the National Academy of Sciences* **110**:6548–6553.

Peipp, H., W. Maier, J. Schmidt, V. Wray, and D. Strack. 1997. Arbuscular mycorrhizal fungus-induced changes in the accumulation of secondary compounds in barley roots. *Phytochemistry* **44**:581–587.

Pellissier, L., C. Ndiribe, A. Dubuis, J.-N. Pradervand, N. Salamin, A. Guisan, and S. Rasmann. 2013. Turnover of plant lineages shapes herbivore phylogenetic beta diversity along ecological gradients. *Ecology Letters* **16**:600–608.

Penaflor, M. F. G. V., M. Erb, C. A. M. Robert, L. A. Miranda, A. G. Werneburg, F. C. Alda Dossi, T. C. J. Turlings, and J. Mauricio Simoes Bento. 2011. Oviposition by a moth suppresses constitutive and herbivore-induced plant volatiles in maize. *Planta* **234**:207–215.

Pennings, S. C., S. Nastisch, and V. J. Paul. 2001. Vulnerability of sea hares to fish predators: Importance of diet and fish species. *Coral Reefs* **20**:320–324.

Pennings, S. C. and V. J. Paul. 1993. Sequestration of dietary secondary metabolites by three species of sea hares—location, specificity and dynamics. *Marine Biology* **117**:535–546.

Penuelas, J., J. Sardans, A. Rivas-Ubach, and I. A. Janssens. 2012. The human-induced imbalance between C, N and P in Earth's life system. *Global Change Biology* **18**:3–6.

Percy, D. M., R. D. M. Page, and Q. C. B. Cronk. 2004. Plant-insect interactions: Double-dating associated insect and plant lineages reveals asynchronous radiations. *Systematic Biology* **53**:120–127.

Perini, V. and M. E. S. Bracken. 2014. Nitrogen availability limits phosphorus uptake in an intertidal macroalga. *Oecologia* **175**:667–676.

Perini, V. C. 2013. The role of seasonality, seaweed traits, and seaweed-herbivore interactions in nutrient cycling in the southern Gulf of Maine. Master's Thesis. Northwestern University, Evanston, IL.

Perkins, M. C., H. A. Woods, J. F. Harrison, and J. J. Elser. 2004. Dietary phosphorus affects the growth of larval *Manduca sexta*. *Archives of Insect Biochemistry and Physiology* **55**:153–168.

Persson, A., B. C. Smith, J. H. Alix, C. Senft-Batoh, and G. H. Wikfors. 2012. Toxin content differs between life stages of *Alexandrium fundyense* (Dinophyceae). *Harmful Algae* **19**:101–107.

Petit, R. J., E. Pineau, B. Demesure, R. Bacilier, A. Ducousso, and A. Kremer. 1997. Chloroplast DNA footprints of postglacial recolonization by oaks. *Proceedings of the National Academy of Science* **94**:9996–10001.

Philpott, S. M., I. Perfecto, and J. Vandermeer. 2008. Effects of predatory ants on lower trophic levels across a gradient of coffee management complexity. *Journal of Animal Ecology* **77**:505–511.

Phoenix, G. K., B. A. Emmett, A. J. Britton, S. J. M. Caporn, N. B. Dise, R. Helliwell, L. Jones, J. R. Leake, I. D. Leith, L. J. Sheppard, A. Sowerby, M. G. Pilkington, E. C. Rowe, M. R. Ashmorek, and S. A. Power. 2012. Impacts of atmospheric nitrogen deposition: Responses of multiple plant and soil parameters across contrasting ecosystems in long-term field experiments. *Global Change Biology* **18**:1197–1215.

Pickett, S. T. A. and P. S. White, editors. 1987. *The Ecology of Natural Disturbance and Patch Dynamics*. Academic Press, San Diego, CA.

Pietikainen, A., J. Mikola, M. Vestberg, and H. Setala. 2009. Defoliation effects on *Plantago lanceolata* resource allocation and soil decomposers in relation to AM symbiosis and fertilization. *Soil Biology & Biochemistry* **41**:2328–2335.

Pineiro, G., J. M. Paruelo, M. Oesterheld, and E. G. Jobbagy. 2010. Pathways of grazing effects on soil organic carbon and nitrogen. *Rangeland Ecology & Management* **63**:109–119.

Piovia-Scott, J., D. A. Spiller, G. Takimoto, L. H. Yang, A. N. Wright, and T. W. Schoener. 2013. The effect of chronic seaweed subsidies on herbivory: Plant-mediated fertilization pathway overshadows lizard-mediated predator pathways. *Oecologia* **172**:1129–1135.

Poelman, E. H. and M. Dicke. 2014. Plant-mediated interactions among insects within a community ecological perspective. *Annual Plant Reviews* **47**:309–337.

Pohnert, G., M. Steinke, and R. Tollrian. 2007. Chemical cues, defence metabolites and the shaping of pelagic interspecific interactions. *Trends in Ecology & Evolution* **22**:198–204.

Pokon, R., V. Novotny, and G. A. Samuelson. 2005. Host specialization and species richness of root-feeding chrysomelid larvae (Chrysomelidae, Coleoptera) in a New Guinea rain forest. *Journal of Tropical Ecology* **21**:595–604.

Polis, G. A., W. B. Anderson, and R. D. Holt. 1997a. Toward and integration of landscape and food web ecology: the dynamics of spatially subsidized food webs. *Annual Review of Ecological Systematics* **28**:289–316.

Polis, G. A., S. D. Hurd, C. T. Jackson, and F. Sanches-Piñero. 1997b. El Niño effects on the dynamics and control of an island ecosystem in the Gulf of California. *Ecology* **78**:1884–1897.

Polis, G. A., C. A. Myers, and R. D. Holt. 1989. The ecology and evolution of intraguild predation: potential competitors that eat each other. *Annual Review of Ecology and Systematics* **20**:297–330.

Poore, A. G. B., A. H. Campbell, R. A. Coleman, G. J. Edgar, V. Jormalainen, P. L. Reynolds, E. E. Sotka, J. J. Stachowicz, R. B. Taylor, M. A. Vanderklift, and J. E. Duffy. 2012. Global patterns in the impact of marine herbivores on benthic primary producers. *Ecology Letters* **15**:912–922.

Poore, A. G. B., N. A. Hill, and E. E. Sotka. 2008. Phylogenetic and geographic variation in host breadth and composition by herbivorous amphipods in the family ampithoidae. *Evolution* **62**:21–38.

Porter, K. G. 1977. The plant-animal interface in freshwater ecosystems. *American Scientist* **65**:159–170.

Povey, S., S. C. Cotter, S. J. Simpson, and K. Wilson. 2013. Dynamics of macronutrient self-medication and illness-induced anorexia in virally-infected insects. *Journal of Animal Ecology* **83**:245–255.

Powell, G. V. N., J. W. Fourqrean, W. J. Kenworthy, and J. C. Zieman. 1991. Bird colonies cause seagrass enrichment in a subtropical estuary. Observation and experimental evidence. *Estuarine Coastal and Shelf Science* **32**:567–579.

Power, M. E., A. J. Stewart, and W. J. Matthews. 1988. Grazer control of algae in an Ozark mountain stream: Effects of short-term exclusion. *Ecology* **69**:1894–1989.

Pregitzer, C. C., J. K. Bailey, and J. A. Schweitzer. 2013. Genetic by environment interactions affect plant-soil linkages. *Ecology and Evolution* **7**:2322–2333.

Pregitzer, K. S., J. L. DeForest, A. J. Burton, M. F. Allen, R. W. Ruess, and R. L. Hendrick. 2002. Fine root architecture of nine North American trees. *Ecological Monographs* **72**:293–309.

Prescott, C. E. 2010. Litter decomposition: What controls it and how can we alter it to sequester more carbon in forest soils? *Biogeochemistry* **101**:133–149.

Price, P. W. 1991. The plant vigor hypothesis and herbivore attack. *Oikos* **62**:244–251.

Price, P. W., W. G. Abrahamson, M. D. Hunter, and G. Melika. 2004. Using gall wasps on oaks to test broad ecological concepts. *Conservation Biology* **18**:1405–1416.

Price, P. W., C. E. Bouton, P. Gross, B. A. McPheron, J. N. Thompson, and A. E. Weis. 1980. Interactions among three tropic levels: Influence of plants on interactions between insect herbivores and natural enemies. *Annual Review of Ecology and Systematics* **11**:41–65.

Price, P. W. and K. M. Clancy. 1986. Multiple effects of precipitation on *Salix lasiolepis* and populations of the stem-galling sawfly, *Euura lasiolepis*. *Ecological Research* **1**:1–14.

Price, P. W. and M. D. Hunter. 2005. Long-term population dynamics of a sawfly show strong bottom-up effects. *Journal of Animal Ecology* **74**:917–925.

Price, P. W. and M. D. Hunter. 2015. Population dynamics of an insect herbivore over 32 years are driven by precipitation and host-plant effects: Testing model predictions. *Environmental Entomology* **44**:463–473.

Prol, M. J., C. Guisande, A. Barreiro, B. Miguez, P. de la Iglesia, A. Villar, A. Gago-Martinez, and M. P. Combarro. 2009. Evaluation of the production of paralytic shellfish poisoning toxins by extracellular bacteria isolated from the toxic dinoflagellate *Atexandrium minutum*. *Canadian Journal of Microbiology* **55**:943–954.

Prowe, A. E. F., M. Pahlow, S. Dutkiewicz, M. Follows, and A. Oschlies. 2012. Top-down control of marine phytoplankton diversity in a global ecosystem model. *Progress in Oceanography* **101**:1–13.

Puglisi, M. P. and V. J. Paul. 1997. Intraspecific variation in the red alga *Portieria hornemannii*: Monoterpene concentrations are not influenced by nitrogen or phosphorus enrichment. *Marine Biology* **128**:161–170.

Pulliam, H. R. 1988. Sources, sinks, and population regulation. *American Naturalist* **132**:652–661.

Raine, R., G. McDermott, J. Silke, K. Lyons, G. Nolan, and C. Cusack. 2010. A simple short range model for the prediction of harmful algal events in the bays of southwestern Ireland. *Journal of Marine Systems* **83**:150–157.

Rand, T. A. and S. A. Louda. 2006. Spillover of agriculturally subsidized predators as a potential threat to native insect herbivores in fragmented landscapes. *Conservation Biology* **20**:1720–1729.

Rand, T. A., J. M. Tylianakis, and T. Tscharntke. 2006. Spillover edge effects: The dispersal of agriculturally subsidized insect natural enemies into adjacent natural habitats. *Ecology Letters* **9**:603–614.

Rantala, A., D. P. Fewer, M. Hisbergues, L. Rouhiainen, J. Vaitomaa, T. Borner, and K. Sivonen. 2004. Phylogenetic evidence for the early evolution of microcystin synthesis. *Proceedings of the National Academy of Sciences* **101**:568–573.

Rapala, J., K. Lahti, K. Sivonen, and S. I. Niemela. 1994. Biodegradability and adsorption on lake sediments of cyanobacterial hepatotoxins and anatoxin-a. *Letters in Applied Microbiology* **19**:423–428.

Rasher, D. B., A. S. Hoey, and M. E. Hay. 2013. Consumer diversity interacts with prey defenses to drive ecosystem function. *Ecology* **94**:1347–1358.

Rasmann, S. and A. A. Agrawal. 2011. Evolution of specialization: A phylogenetic study of host range in the red milkweed beetle (*Tetraopes tetraophthalmus*). *American Naturalist* **177**:728–737.

Rasmann, S., A. A. Agrawal, S. C. Cook, and A. C. Erwin. 2009. Cardenolides, induced responses, and interactions between above- and belowground herbivores of milkweed (*Asclepias* spp.). *Ecology* **90**:2393–2404.

Rasmann, S., T. L. Bauerle, K. Poveda, and R. Vannette. 2011. Predicting root defence against herbivores during succession. *Functional Ecology* **25**:368–379.

Rasmann, S., T. G. Kollner, J. Degenhardt, I. Hiltpold, S. Toepfer, U. Kuhlmann, J. Gershenzon, and T. C. J. Turlings. 2005. Recruitment of entomopathogenic nematodes by insect-damaged maize roots. *Nature* **434**:732–737.

Raubenheimer, D. and S. A. Jones. 2006. Nutritional imbalance in an extreme generalist omnivore: Tolerance and recovery through complementary food selection. *Animal Behaviour* **71**:1253–1262.

Raubenheimer, D. and S. J. Simpson. 2004. Organismal stoichiometry: Quantifying non-independence among food components. *Ecology* **85**:1203–1216.

Raubenheimer, D. and S. J. Simpson. 2009. Nutritional PharmEcology: Doses, nutrients, toxins, and medicines. *Integrative and Comparative Biology* **49**:329–337.

Redfern, M. and M. D. Hunter. 2005. Time tells: long-term patterns in the population dynamics of the yew gall midge, *Taxomyia taxi* (Cecidomyiidae), over 35 years. *Ecological Entomology* **30**:86–95.

Redfield, A. C. 1934. On the proportions of organic derivations in sea water and their relation to the composition of plankton. Pages 177–192 *in* R. J. Daniel, editor. *James Johnstone Memorial Volume*. University Press of Liverpool, Liverpool.

Redfield, A. C. 1958. The biological control of chemical factors in the environment. *American Scientist* **46**:205–221.

Relyea, R. A. 2002. The many faces of predation: How induction, selection, and thinning combine to alter prey phenotypes. *Ecology* **83**:1953–1964.

Reusch, T. B. H., A. Ehlers, A. Hammerli, and B. Worm. 2005. Ecosystem recovery after climatic extremes enhanced by genotypic diversity. *Proceedings of the National Academy of Sciences* **102**:2826–2831.

Reynolds, B. C., D. A. Crossley, and M. D. Hunter. 2003. Response of soil invertebrates to forest canopy inputs along a productivity gradient. *Pedobiologia* **47**:127–139.

Reynolds, B. C. and M. D. Hunter. 2001. Responses of soil respiration, soil nutrients, and litter decomposition to inputs from canopy herbivores. *Soil Biology & Biochemistry* **33**:1641–1652.

Reynolds, B. C., M. D. Hunter, and D. A. Crossley, Jr. 2000. Effects of canopy herbivory on nutrient cycling in a northern hardwood forest in western North Carolina. *Selbyana* **21**:74–78.

Reynolds, C. S. 1984. Phytoplankton periodicity: the interactions of form, function and environment variability. *Freshwater Biology* **14**:111–142.

Reynolds, J. J. H., X. Lambin, F. P. Massey, S. Reidinger, J. A. Sherratt, M. J. Smith, A. White, and S. E. Hartley. 2012. Delayed induced silica defences in grasses and their potential for destabilising herbivore population dynamics. *Oecologia* **170**:445–456.

Richards, L. A., L. A. Dyer, M. L. Forister, A. M. Smilanich, C. D. Dodson, M. D. Leonard, and C. S. Jeffrey. 2015. Phytochemical diversity drives tropical plant-insect community diversity. *Proceedings of the National Academy of Sciences* **112**:10973–10978.

Richards, L. A., L. A. Dyer, A. M. Smilanich, and C. D. Dodson. 2010. Synergistic effects of amides from two *Piper* species on generalist and specialist herbivores. *Journal of Chemical Ecology* **36**:1105–1113.

Riedell, W. E., E. A. Beckendorf, and M. A. Catangui. 2013. Relationships between soybean shoot nitrogen components and soybean aphid populations. *Arthropod-Plant Interactions* **7**:667–676.

Rietsma, C. S., I. Valiela, and R. Buchsbaum. 1988. Detrital chemistry, growth, and food choice in the salt marsh snail (*Melampus bidentatus*). *Ecology* **69**:261–266.

Riolo, M. A., P. Rohani, and M. D. Hunter. 2015. Local variation in plant quality influences large-scale population dynamics. *Oikos* **124**:1160–1170.

Ripple, W. J. and R. L. Beschta. 2006. Linking a cougar decline, trophic cascade, and catastrophic regime shift in Zion National Park. *Biological Conservation* **133**:397–408.

Ripple, W. J., R. L. Beschta, J. K. Fortin, and C. T. Robbins. 2014a. Trophic cascades from wolves to grizzly bears in Yellowstone. *Journal of Animal Ecology* **83**:223–233.

Ripple, W. J., J. A. Estes, R. L. Beschta, C. C. Wilmers, E. G. Ritchie, M. Hebblewhite, J. Berger, B. Elmhagen, M. Letnic, M. P. Nelson, O. J. Schmitz, D. W. Smith, A. D. Wallach, and A. J. Wirsing. 2014b. Status and ecological effects of the world's largest carnivores. *Science* **343**:1241484.

Risley, L. S. 1986. The influence of herbivores on seasonal leaf-fall: Premature leaf abscission and petiole clipping. *Journal of Agricultural Entomology* **3**:152–162.

Risley, L. S. and D. A. Crossley. 1988. Herbivore-caused greenfall in the southern Appalachians. *Ecology* **69**:1118–1127.

Risley, L. S. and D. A. Crossley. 1992. Contribution of herbivore-caused greenfall to litterfall: N flux in several southern Appalachian forested watersheds. *American Midland Naturalist* **129**:67–74.

Ritchie, M. E., D. Tilman, and J. M. H. Knops. 1998. Herbivore effects on plant and nitrogen dynamics in oak savanna. *Ecology* **79**:165–177.

Ritland, D. B. and L. P. Brower. 1993. A reassessment of the mimicry relationship among viceroys, queens, and monarchs in Florida. *Natural History Museum of Los Angeles County Science Series* **38**:129–139.

Robertson, G. P., D. C. Coleman, C. S. Bledsoe, and P. Sollins, editors. 1999. *Standard Soil Methods for Long-Term Ecological Research*. Oxford University Press, Oxford, UK.

Rodrigues, J. L. M., V. H. Pellizari, R. Mueller, K. Baek, E. d. C. Jesus, F. S. Paula, B. Mirza, G. S. Hamaoui, S. M. Tsai, B. Feigl, J. M. Tiedje, B. J. M. Bohannan, and K. Nüsslein. 2012. Conversion of the Amazon rainforest to agriculture results in biotic homogenization of soil bacterial communities. *Proceedings of the National Academy of Sciences* **110**:988–993.

Rodriguez, A., J. Duran, F. Covelo, J. Maria Fernandez-Palacios, and A. Gallardo. 2011. Spatial pattern and variability in soil N and P availability under the influence of two dominant species in a pine forest. *Plant and Soil* **345**:211–221.

Rohani, P., T. J. Lewis, D. Grunbaum, and G. D. Ruxton. 1997. Spatial self-organization in ecology: Pretty patterns or robust reality? *Trends in Ecology & Evolution* **12**:70–74.

Romano, G., G. L. Russo, I. Buttino, A. Ianora, and À. Miralto. 2003. A marine diatom-derived aldehyde induces apoptosis in copepod and sea urchin embryos. *Journal of Experimental Biology* **206**:3487–3494.

Rosenheim, J. A. 2007. Intraguild predation: New theoretical and empirical perspectives. *Ecology* **88**:2679–2680.

Rosenthal, G. A. 1977. The biological effects and mode of action of L-canavanine, a structural analogue of L-arginine. *Quarterly Review of Biology* **52**:155–178.

Rosenthal, G. A. and M. R. Berenbaum. 1991. *Herbivores. Their Interactions with Secondary Plant Metabolites*. Vol. 1, 2nd edition. Academic Press, San Diego, CA.

Rosenthal, J. P. and P. M. Kotanen. 1994. Terrestrial plant tolerance to herbivory. *Trends in Ecology and Evolution* **9**:145–148.

Rosenwasser, S., M. A. Mausz, D. Schatz, U. Sheyn, S. Malitsky, A. Aharoni, E. Weinstock, O. Tzfadia, S. Ben-Dor, E. Feldmesser, G. Pohnert, and A. Vardi. 2014. Rewiring host lipid metabolism by large viruses determines the fate of *Emiliania huxleyi*, a bloom-forming alga in the ocean. *The Plant Cell* **26**:2689–2707.

Rosenzweig, M. L. 1971. Paradox of enrichment: Destabilization of exploitation ecosystems in ecological time. *Science* **171**:385–387.

Ross, B. A., J. R. Bray, and W. H. Marshall. 1970. Effects of long-term deer exclusion on a *Pinus resinosa* forest in north-central Minnesota. *Ecology* **51**:1088–1093.

Rossiter, M. C. 1980. Anomalous oviposition behavior by the gypsy moth (Lepidoptera: Lymantriidae). *Journal of the New York Entomological Society* **88**:69–70.

Rossiter, M. C. 1987a. Genetic and phenotypic variation in diet breadth in a generalist herbivore. *Evolutionary Ecology* **1**:272–282.

Rossiter, M. C. 1987b. Use of a secondary host, pitch pine, by non-outbreak populations of the gypsy moth. *Ecology* **68**:857–868.

Rossiter, M. C. 1991. Maternal effects generate variation in life-history—Consequences of egg weight plasticity in the gypsy moth. *Functional Ecology* **5**:386–393.

Rossiter, M. C. 1994. Maternal effects hypothesis of herbivore outbreak. *Bioscience* **44**:752–763.

Rossiter, M. C. 1996. Incidence and consequences of inherited environmental effects. *Annual Review of Ecology and Systematics* **27**:451–476.

Rossiter, M. C., D. L. Cox-Foster, and M. A. Briggs. 1993. Initiation of maternal effects in *Lymantria dispar*—Genetic and ecological components of egg provisioning. *Journal of Evolutionary Biology* **6**:577–589.

Rossiter, M. C., J. C. Schultz, and I. T. Baldwin. 1988. Relationships among defoliation, red oak phenolics, and gypsy moth growth and reproduction. *Ecology* **69**:267–277.

Rotem, K., A. A. Agrawal, and L. Kott. 2003. Parental effects in *Pieris rapae* in response to variation in food quality: Adaptive plasticity across generations? *Ecological Entomology* **28**:211–218.

Rothschild, M. 1964. An extension of Dr. Lincoln Brower's theory on bird predation and food specificity, together with some observations on bird memory in relation to aposematic colour patterns. *The Entomologist* **97**:73–78.

Rouse, G. W., S. K. Goffredi, and R. C. Vrijenhoek. 2004. Osedax: Bone-eating marine worms with dwarf males. *Science* **305**:668–671.

Royama, T. 1992. *Analytical Population Dynamics*. Springer, New York.

Rudgers, J. A., J. Holah, S. P. Orr, and K. Clay. 2007. Forest succession suppressed by an introduced plant-fungal symbiosis. *Ecology* **88**:18–25.

Rudolf, V. H. W. and N. L. Rasmussen. 2013. Ontogenetic functional diversity: Size structure of a keystone predator drives functioning of a complex ecosystem. *Ecology* **94**:1046–1056.

Rzanny, M. and W. Voigt. 2012. Complexity of multitrophic interactions in a grassland ecosystem depends on plant species diversity. *Journal of Animal Ecology* **81**:614–627.

Saage, A., O. Vadstein, and U. Sommer. 2009. Feeding behaviour of adult *Centropages hamatus* (Copepoda, Calanoida): Functional response and selective feeding experiments. *Journal of Sea Research* **62**:16–21.

Saari, S., J. Sundell, O. Huitu, M. Helander, E. Ketoja, H. Ylonen, and K. Saikkonen. 2010. Fungal-mediated multitrophic interactions—Do grass endophytes in diet protect voles from predators? *PLOS ONE* **5**:e9845.

Saba, G. K., D. K. Steinberg, and D. A. Bronk. 2011. The relative importance of sloppy feeding, excretion, and fecal pellet leaching in the release of dissolved carbon and nitrogen by *Acartia tonsa* copepods. *Journal of Experimental Marine Biology and Ecology* **404**:47–56.

Saikkonen, K., S. Saari, and M. Helander. 2010. Defensive mutualism between plants and endophytic fungi? *Fungal Diversity* **41**:101–113.

Sakayaroj, J., S. Preedanon, O. Supaphon, E. B. G. Jones, and S. Phongpaichit. 2010. Phylogenetic diversity of endophyte assemblages associated with the tropical seagrass *Enhalus acoroides* in Thailand. *Fungal Diversity* **42**:27–45.

Salmore, A. K. and M. D. Hunter. 2001a. Elevational trends in defense chemistry, vegetation, and reproduction in *Sanguinaria canadensis*. *Journal of Chemical Ecology* **27**:1713–1727.

Salmore, A. K. and M. D. Hunter. 2001b. Environmental and genotypic influences on isoquinoline alkaloid content in *Sanguinaria canadensis*. *Journal of Chemical Ecology* **27**:1729–1747.

Sarfraz, M., L. M. Dosdall, and B. A. Keddie. 2009. Host plant nutritional quality affects the performance of the parasitoid *Diadegma insulare*. *Biological Control* **51**:34–41.

Sariyildiz, T., E. Akkuzu, M. Kucuk, A. Duman, and Y. Aksu. 2008. Effects of *Ips typographus* (L.) damage on litter quality and decomposition rates of oriental spruce *Picea orientalis* (L.) Link in Hatila Valley National Park, Turkey. *European Journal of Forest Research* **127**:429–440.

Sasso, S., G. Pohnert, M. Lohr, M. Mittag, and C. Hertweck. 2012. Microalgae in the postgenomic era: A blooming reservoir for new natural products. *Fems Microbiology Reviews* **36**:761–785.

Sato, T., T. Egusa, K. Fukushima, T. Oda, N. Ohte, N. Tokuchi, K. Watanabe, M. Kanaiwa, I. Murakami, and K. D. Lafferty. 2012. Nematomorph parasites indirectly alter the food web and ecosystem function of streams through behavioural manipulation of their cricket hosts. *Ecology Letters* **15**:786–793.

Sauchyn, L. K., J. S. Lauzon-Guay, and R. E. Scheibling. 2011. Sea urchin fecal production and accumulation in a rocky subtidal ecosystem. *Aquatic Biology* **13**:215–223.

Sauchyn, L. K. and R. E. Scheibling. 2009. Degradation of sea urchin feces in a rocky subtidal ecosystem: Implications for nutrient cycling and energy flow. *Aquatic Biology* **6**:99–108.

Schadler, M., T. Rottstock, and R. Brandl. 2005. Food web properties in aquatic microcosms with litter mixtures are predictable from component species. *Archiv Für Hydrobiologie* **163**:211–223.

Schaffers, A. P., I. P. Raemakers, K. V. Sykora, and C. J. F. Ter Braak. 2008. Arthropod assemblages are best predicted by plant species composition. *Ecology* **89**:782–794.

Schallhart, N., M. J. Tusch, C. Wallinger, K. Staudacher, and M. Traugott. 2012. Effects of plant identity and diversity on the dietary choice of a soil-living insect herbivore. *Ecology* **93**:2650–2657.

Schatz, A. and S. E. Waksman. 1944. Effect of streptomycin and other antibiotic substances upon *Mycobacterium tuberculosis* and related organisms. *Proceedings of the Society for Experimental Biology and Medicine* **57**:244–248.

Scheller, H. V. and P. Ulvskov. 2010. Hemicelluloses. *Annual Review of Plant Biology* **61**:263–289.

Scherber, C., N. Eisenhauer, W. W. Weisser, B. Schmid, W. Voigt, M. Fischer, E. D. Schulze, C. Roscher, A. Weigelt, E. Allan, H. Bessler, M. Bonkowski, N. Buchmann, F. Buscot, L. W. Clement, A. Ebeling, C. Engels, S. Halle, I. Kertscher, A. M. Klein, R. Koller, S. Konig, E. Kowalski, V. Kummer, A. Kuu, M. Lange, D. Lauterbach, C. Middelhoff, V. D. Migunova, A. Milcu, R. Muller, S. Partsch, J. S. Petermann, C. Renker, T. Rottstock, A. Sabais, S. Scheu, J. Schumacher, V. M. Temperton, and T. Tscharntke. 2010. Bottom-up effects of plant diversity on multitrophic interactions in a biodiversity experiment. *Nature* **468**:553–556.

Schimel, J. P. and S. Hättenschwiler. 2007. Nitrogen transfer between decomposing leaves of different N status. *Soil Biology & Biochemistry* **39**:1428–1436.

Schindler, D. E., S. R. Carpenter, J. J. Cole, J. F. Kitchell, and M. L. Pace. 1997. Influence of food web structure on carbon exchange between lakes and the atmosphere. *Science* **277**:248–251.

Schindler, D. E., J. F. Kitchell, X. He, S. R. Carpenter, J. R. Hodgson, and K. L. Cottingham. 1993. Food web structure and phosphorus cycling in lakes. *Transactions of the American Fisheries Society* **122**:756–772.

Schindler, D. W. 1974. Eutrophication and recovery in experimental lakes—Implications for lake management. *Science* **184**:897–899.

Schlesinger, W. H. and E. S. Bernhardt. 2013. *Biogeochemistry: An Analysis of Global Change*. 3rd edition. Academic Press, San Diego, CA.

Schlosser, C., J. K. Klar, B. D. Wake, J. T. Snow, D. J. Honey, E. M. S. Woodward, M. C. Lohan, E. P. Achterberg, and C. M. Moore. 2014. Seasonal ITCZ migration dynamically controls the location of the (sub)tropical Atlantic biogeochemical divide. *Proceedings of the National Academy of Sciences* **111**:1438–1442.

Schmelz, E. A., R. J. Grebenok, D. W. Galbraith, and W. S. Bowers. 1998. Damage-induced accumulation of phytoecdysteroids in spinach: A rapid root response involving the octadecanoic acid pathway. *Journal of Chemical Ecology* **24**:339–360.

Schmitz, O. J. 1998. Direct and indirect effects of predation and predation risk in old-field interaction webs. *American Naturalist* **151**:327–342.

Schmitz, O. J. 2003. Top predator control of plant biodiversity and productivity in an old-field ecosystem. *Ecology Letters* **6**:156–163.

Schmitz, O. J. 2006. Predators have large effects on ecosystem properties by changing plant diversity, not plant biomass. *Ecology* **87**:1432–1437.

Schmitz, O. J. 2008a. Effects of predator hunting mode on grassland ecosystem function. *Science* **319**:952–954.

Schmitz, O. J. 2008b. Herbivory from individuals to ecosystems. *Annual Review of Ecology Evolution and Systematics* **39**:133–152.

Schmitz, O. J., D. Hawlena, and G. C. Trussell. 2010. Predator control of ecosystem nutrient dynamics. *Ecology Letters* **13**:1199–1209.

Schmitz, O. J., V. Krivan, and O. Ovadia. 2004. Trophic cascades: The primacy of trait-mediated indirect interactions. *Ecology Letters* **7**:153–163.

Scholey, A., S. French, P. Morris, D. Kennedy, A. Milne, and C. Haskell. 2010. Consumption of cocoa flavanols results in acute improvements in mood and cognitive performance during sustained mental effort. *Journal of Psychopharmacology* **24**:1505–1514.

Schönknecht, G., W.-H. Chen, C. M. Ternes, G. G. Barbier, R. P. Shrestha, M. Stanke, A. Bräutigam, B. J. Baker, J. F. Banfield, R. M. Garavito, K. Carr, C. Wilkerson, S. A. Rensing, D. Gagneul, N. E. Dickenson, C. Oesterhelt, M. J. Lercher, and A. P. M. Weber. 2013. Gene transfer from bacteria and archaea facilitated evolution of an extremophilic eukaryote. *Science* **339**:1207–1210.

Schultz, J. C., H. M. Appel, A. P. Ferrieri, and T. M. Arnold. 2013. Flexible resource allocation during plant defense responses. *Frontiers in Plant Science* **4** Article 324:doi:10.3389/fpls.2013.00324.

Schultz, J. C. and I. T. Baldwin. 1982. Oak leaf quality declines in response to defoliation by gypsy moth larvae. *Science* **217**:149–151.

Schwarzenberger, A., C. J. Kuster, and E. Elert. 2012. Molecular mechanisms of tolerance to cyanobacterial protease inhibitors revealed by clonal differences in *Daphnia magna*. *Molecular Ecology* **21**:4898–4911.

Schweitzer, J. A., J. K. Bailey, D. G. Fischer, C. J. LeRoy, T. G. Whitham, and S. C. Hart. 2012a. Functional and heritable consequences of plant genotype on community composition and ecosystem processes. Pages 371–390 *in* T. Ohgushi, O. J. Schmitz, and R. D. Holt, editors. *Trait-Mediated Indirect Interactions. Ecological and Evolutionary Perspectives*. Cambridge University Press, Cambridge, UK.

Schweitzer, J. A., J. K. Bailey, S. C. Hart, and T. G. Whitham. 2005a. Nonadditive effects of mixing cottonwood genotypes on litter decomposition and nutrient dynamics. *Ecology* **86**:2834–2840.

Schweitzer, J. A., J. K. Bailey, S. C. Hart, G. M. Wimp, S. K. Chapman, and T. G. Whitham. 2005b. The interaction of plant genotype and herbivory decelerate leaf litter decomposition and alter nutrient dynamics. *Oikos* **110**:133–145.

Schweitzer, J. A., J. K. Bailey, B. J. Rehill, G. D. Martinsen, S. C. Hart, R. L. Lindroth, P. Keim, and T. G. Whitham. 2004. Genetically based trait in a dominant tree affects ecosystem processes. *Ecology Letters* **7**:127–134.

Schweitzer, J. A., M. D. Madritch, J. K. Bailey, C. J. LeRoy, D. G. Fischer, B. J. Rehill, R. L. Lindroth, A. E. Hagerman, S. C. Wooley, S. C. Hart, and T. G. Whitham. 2008. From genes to ecosystems: The genetic basis of condensed tannins and their role in nutrient regulation in a *Populus* model system. *Ecosystems* **11**:1005–1020.

Schweitzer, J. A., M. D. Madritch, E. Felker-Quinn, and J. K. Bailey. 2012b. From genes to ecosystems: how plant genetics links above and below-ground processes. Pages 82–98 *in* D. H. Wall, editor. *Soil Ecology and Ecosystem Services.* Oxford University Press, Oxford, UK.

Seastedt, T. R., D. A. J. Crossley, and W. W. Hargrove. 1983. The effects of low-level consumption by canopy arthropods on the growth and nutrient dynamics of black locust and red maple trees in the southern Appalachians. *Ecology* **64**:1040–1048.

Seipke, R. F., M. Kaltenpoth, and M. I. Hutchings. 2012. *Streptomyces* as symbionts: An emerging and widespread theme? *Fems Microbiology Reviews* **36**:862–876.

Sekula-Wood, E., A. Schnetzer, C. R. Benitez-Nelson, C. Anderson, W. M. Berelson, M. A. Brzezinski, J. M. Burns, D. A. Caron, I. Cetinic, J. L. Ferry, E. Fitzpatrick, B. H. Jones, P. E. Miller, S. L. Morton, R. A. Schaffner, D. A. Siegel, and R. Thunell. 2009. Rapid downward transport of the neurotoxin domoic acid in coastal waters. *Nature Geoscience* **2**:272–275.

Selås, V., S. Kobro, and G. A. Sonerud. 2013. Population fluctuations of moths and small rodents in relation to plant reproduction indices in southern Norway. *Ecosphere* **4** Article 123:doi:110.1890/es1813–00228.00221.

Semmartin, M. and C. M. Ghersa. 2006. Intraspecific changes in plant morphology, associated with grazing, and effects on litter quality, carbon and nutrient dynamics during decomposition. *Austral Ecology* **31**:99–105.

Seymour, J. R., T. Ahmed, and R. Stocker. 2009. Bacterial chemotaxis towards the extracellular products of the toxic phytoplankton *Heterosigma akashiwo. Journal of Plankton Research* **31**:1557–1561.

Seymour, J. R., R. Simo, T. Ahmed, and R. Stocker. 2010. Chemoattraction to dimethylsulfoniopropionate throughout the marine microbial food web. *Science* **329**:342–345.

Shi, W. B., S. X. Xie, X. Y. Chen, S. Sun, X. Zhou, L. T. Liu, P. Gao, N. C. Kyrpides, E. G. No, and J. S. Yuan. 2013. Comparative genomic analysis of the endosymbionts of herbivorous insects reveals eco-environmental adaptations: Biotechnology applications. *PLOS Genetics* **9**:e1003131.

Shurin, J. B., D. S. Gruner, and H. Hillebrand. 2006. All wet or dried up? Real differences between aquatic and terrestrial food webs. *Proceedings of the Royal Society B: Biological Sciences* **273**:1–9.

Sieg, R. D., K. L. Poulson-Ellestad, and J. Kubanek. 2011. Chemical ecology of the marine plankton. *Natural Product Reports* **28**:388–399.

Siemann, E. 1998. Experimental tests of effects of plant productivity and diversity on grassland arthropod diversity. *Ecology* **79**:2057–2070.

Sime, K. 2002. Chemical defence of *Battus philenor* larvae against attack by the parasitoid *Trogus pennator. Ecological Entomology* **27**:337–345.

Simone-Finstrom, M. and M. Spivak. 2010. Propolis and bee health: The natural history and significance of resin use by honey bees. *Apidologie* **41**:295–311.

Simpson, S. J. and D. Raubenheimer. 1993. A multi-level analysis of feeding behaviour: The geometry of nutritional decisions. *Philosophical Transactions of the Royal Society of London Series B* **342**:381–402.

Simpson, S. J. and D. Raubenheimer. 2001. The geometric analysis of nutrient-allelochemical interactions: A case study using locusts. *Ecology* **82**:422–439.

Sims, P. L. and J. S. Singh. 1978. Structure and function of ten western North American grasslands. 3. Net primary production, turnover and efficiencies of energy capture and water-use. *Journal of Ecology* **66**:573–597.

Sinclair, A. R. E., K. L. Metzger, J. M. Fryxell, C. Packer, A. E. Byrom, M. E. Craft, K. Hampson, T. Lembo, S. M. Durant, G. J. Forrester, J. Bukombe, J. McHetto, J. Dempewolf, R. Hilborn, S. Cleaveland, A. Nkwabi, A. Mosser, and S. A. R. Mduma. 2013. Asynchronous food-web pathways could buffer the response of Serengeti predators to El Niño Southern Oscillation. *Ecology* **94**:1123–1130.

Singer, M. S., Y. Carriere, C. Theuring, and T. Hartmann. 2004a. Disentangling food quality from resistance against parasitoids: diet choice by a generalist caterpillar. *American Naturalist* **164**:423–429.

Singer, M. S., T. E. Farkas, C. M. Skorik, and K. A. Mooney. 2012. Tritrophic interactions at a community level: Effects of host plant species quality on bird predation of caterpillars. *American Naturalist* **179**:363–374.

Singer, M. S., K. C. Mace, and E. A. Bernays. 2009. Self-medication as adaptive plasticity: Increased ingestion of plant toxins by parasitized caterpillars. *PLOS ONE* **4**:e4796.

Singer, M. S., D. Rodrigues, J. O. Stireman, and Y. Carriere. 2004b. Roles of food quality and enemy-free space in host use by a generalist insect herbivore. *Ecology* **85**:2747–2753.

Sinsabaugh, R. L. 2010. Phenol oxidase, peroxidase and organic matter dynamics of soil. *Soil Biology & Biochemistry* **42**:391–404.

Sinsabaugh, R. L., B. H. Hill, and J. J. Follstad Shah. 2009. Ecoenzymatic stoichiometry of microbial organic nutrient acquisition in soil and sediment. *Nature* **462**:795–798.

Sinsabaugh, R. L., C. L. Lauber, M. N. Weintraub, B. Ahmed, S. D. Allison, C. Crenshaw, A. R. Contosta, D. Cusack, S. Frey, M. E. Gallo, T. B. Gartner, S. E. Hobbie, K. Holland, B. L. Keeler, J. S. Powers, M. Stursova, C. Takacs-Vesbach, M. P. Waldrop, M. D. Wallenstein, D. R. Zak, and L. H. Zeglin. 2008. Stoichiometry of soil enzyme activity at global scale. *Ecology Letters* **11**:1252–1264.

Smayda, T. J. 1997. Harmful algal blooms: Their ecophysiology and general relevance to phytoplankton blooms in the sea. *Limnology and Oceanography* **42**:1137–1153.

Smetacek, V., C. Klaas, V. H. Strass, P. Assmy, M. Montresor, B. Cisewski, N. Savoye, A. Webb, F. d/'Ovidio, J. M. Arrieta, U. Bathmann, R. Bellerby, G. M. Berg, P. Croot, S. Gonzalez, J. Henjes, G. J. Herndl, L. J. Hoffmann, H. Leach, M. Losch, M. M. Mills, C. Neill, I. Peeken, R. Rottgers, O. Sachs, E. Sauter, M. M. Schmidt, J. Schwarz, A. Terbruggen, and D. Wolf-Gladrow. 2012. Deep carbon export from a Southern Ocean iron-fertilized diatom bloom. *Nature* **487**:313–319.

Smilanich, A. M., L. A. Dyer, J. Q. Chambers, and M. D. Bowers. 2009a. Immunological cost of chemical defence and the evolution of herbivore diet breadth. *Ecology Letters* **12**:612–621.

Smilanich, A. M., L. A. Dyer, and G. L. Gentry. 2009b. The insect immune response and other putative defenses as effective predictors of parasitism. *Ecology* **90**:1434–1440.

Smilanich, A. M., J. Vargas, L. A. Dyer, and M. D. Bowers. 2011. Effects of ingested secondary metabolites on the immune response of a polyphagous caterpillar *Grammia incorrupta*. *Journal of Chemical Ecology* **37**:239–245.

Smith, C. R. and A. R. Baco. 2003. Ecology of whale falls at the deep-sea floor. *Oceanography and Marine Biology* **41**:311–354.

Smith, S. E. and D. J. Read. 2008. *Mycorrhizal Symbiosis*. 3rd edition. Academic Press, San Diego, CA.

Smith, S. V. and J. T. Hollibaugh. 1993. Coastal metabolism and the oceanic organic carbon balance. *Reviews of Geophysics* **31**:75–89.

Sohlenius, B. 1980. Abundance, biomass and contribution to energy flow by soil nematodes in terrestrial ecosystems. *Oikos* **34**:186–194.

Sommer, U. 1989. The role of competition for resources in phytoplankton succession. Pages 57–106 *in* U. Sommer, editor. *Plankton Ecology: Succession in Plankton Communities*. Springer, New York.

Song, M.-H., F.-H. Yu, H. Ouyang, G.-M. Cao, X.-L. Xu, and J. H. C. Cornelissen. 2012. Different inter-annual responses to availability and form of nitrogen explain species coexistence in an alpine meadow community after release from grazing. *Global Change Biology* **18**:3100–3111.

Sopanen, S., P. Uronen, P. Kuuppo, C. Svensen, A. Ruhl, T. Tamminen, E. Graneli, and C. Legrand. 2009. Transfer of nodularin to the copepod *Eurytemora affinis* through the microbial food web. *Aquatic Microbial Ecology* **55**:115–130.

Sorensen, J. S., M. M. Skopec, and M. D. Dearing. 2006. Application of pharmacological approaches to plant-mammal interactions. *Journal of Chemical Ecology* **32**:1229–1246.

Sotka, E. E., J. Forbey, M. Horn, A. G. B. Poore, D. Raubenheimer, and K. E. Whalen. 2009. The emerging role of pharmacology in understanding consumer-prey interactions in marine and freshwater systems. *Integrative and Comparative Biology* **49**:291–313.

Speight, M. R., M. D. Hunter, and A. D. Watt. 2008. *The Ecology of Insects: Concepts and Applications*. 2nd edition. Wiley-Blackwell, Oxford, UK.

Sperfeld, E., D. Martin-Creuzburg, and A. Wacker. 2012. Multiple resource limitation theory applied to herbivorous consumers: Liebig's minimum rule vs. interactive co-limitation. *Ecology Letters* **15**:142–150.

Spiller, D. A., J. Piovia-Scott, A. N. Wright, L. H. Yang, G. Takimoto, T. W. Schoener, and T. Iwata. 2010. Marine subsidies have multiple effects on coastal food webs. *Ecology* **91**:1424–1434.

Sprent, J. I. and R. Parsons. 2000. Nitrogen fixation in legume and non-legume trees. *Field Crops Research* **65**:183–196.

Srivastava, D. S., B. J. Cardinale, A. L. Downing, J. E. Duffy, C. Jouseau, M. Sankaran, and J. P. Wright. 2009. Diversity has stronger top-down than bottom-up effects on decomposition. *Ecology* **90**:1073–1083.

Stachowicz, J. J. and M. E. Hay. 1999. Reducing predation through chemically mediated camouflage: Indirect effects of plant defenses on herbivores. *Ecology* **80**:495–509.

Stadler, B., T. Muller, and D. Orwig. 2006. The ecology of energy and nutrient fluxes in hemlock forests invaded by hemlock woolly adelgid. *Ecology* **87**:1792–1804.

Stamp, N. 2003. Out of the quagmire of plant defense hypotheses. *Quarterly Review of Biology* **78**:23–55.

Stark, S., R. Julkunen-Tiitto, and J. Kumpula. 2007. Ecological role of reindeer summer browsing in the mountain birch (*Betula pubescens* ssp *czerepanovii*) forests: Effects on plant defense, litter decomposition, and soil nutrient cycling. *Oecologia* **151**:486–498.

Steinauer, E. M. and S. L. Collins. 1995. Effects of urine deposition on small-scale patch structure in prairie vegetation. *Ecology* **76**:1195–1205.

Stelzer, R. S. and G. A. Lamberti. 2002. Ecological stoichiometry in running waters: Periphyton chemical composition and snail growth. *Ecology* **83**:1039–1051.

Steppuhn, A. and I. T. Baldwin. 2007. Resistance management in a native plant: nicotine prevents herbivores from compensating for plant protease inhibitors. *Ecology Letters* **10**:499–511.

Sternberg, E. D., J. C. de Roode, and M. D. Hunter. 2015. Trans-generational parasite protection associated with paternal diet. *Journal of Animal Ecology* **84**:310–321.

Sternberg, E. D., T. Lefèvre, J. Li, C. L. F. de Castillejo, H. Li, M. D. Hunter, and J. C. de Roode. 2012. Food plant derived disease tolerance and resistance in a natural butterly-plant-parasite interaction. *Evolution* **66**:3367–3376.

Sterner, R. W. 1986. Herbivores direct and indirect effects on algal populations. *Science* **231**:605–607.

Sterner, R. W. and J. J. Elser. 2002. *Ecological Stoichiometry: The Biology of Elements from Molecules to the Biosphere*. Princeton University Press, Princeton, NJ.

Stevens, C. J., N. B. Dise, J. O. Mountford, and D. J. Gowing. 2004. Impact of nitrogen deposition on the species richness of grasslands. *Science* **303**:1876–1879.

Stiling, P., D. C. Moon, M. D. Hunter, J. Colson, A. M. Rossi, G. J. Hymus, and B. G. Drake. 2003. Elevated CO_2 lowers relative and absolute herbivore density across all species of a scrub-oak forest. *Oecologia* **134**:82–87.

Stiling, P. and A. M. Rossi. 1997. Experimental manipulations of top-down and bottom-up factors in a tri-trophic system. *Ecology* **78**:1602–1606.

Stiling, P., D. Simberloff, and B. V. Brodbeck. 1991. Variation in rates of leaf abscission between plants may affect the distribution patterns of sessile insects. *Oecologia* **88**:367–370.

Stocks, B. J., J. A. Mason, J. B. Todd, E. M. Bosch, B. M. Wotton, B. D. Amiro, M. D. Flannigan, K. G. Hirsch, K. A. Logan, D. L. Martell, and W. R. Skinner. 2002. Large forest fires in Canada, 1959–1997. *Journal of Geophysical Research-Atmospheres* **108** Article 8149:doi:10.1029/2001JD000484.

Stoddart, D. R. and J. F. Peake. 1979. Historical records of Indian Ocean giant tortoise populations. *Philosophical Transactions of the Royal Society of London Series B: Biological Sciences* **B286**:147–162.

Stoepler, T. M., J. T. Lill, and S. M. Murphy. 2011. Cascading effects of host size and host plant species on parasitoid resource allocation. *Ecological Entomology* **36**:724–735.

Stoler, A. B. and R. A. Relyea. 2013. Leaf litter quality induces morphological and developmental changes in larval amphibians. *Ecology* **94**:1594–1603.

Stone, G. N. and J. M. Cook. 1998. The structure of cynipid oak galls: Patterns in the evolution of an extended phenotype. *Proceedings of the Royal Society of London Series B: Biological Sciences* **265**:979–988.

Strengbom, J., J. Witzell, A. Nordin, and L. Ericson. 2005. Do multitrophic interactions override N fertilization effects on *Operophtera* larvae? *Oecologia* **143**:241–250.

Strickland, M. S., D. Hawlena, A. Reese, M. A. Bradford, and O. J. Schmitz. 2013. Trophic cascade alters ecosystem carbon exchange. *Proceedings of the National Academy of Sciences* **110**:11035–11038.

Strickland, M. S., C. Lauber, N. Fierer, and M. A. Bradford. 2009a. Testing the functional significance of microbial community composition. *Ecology* **90**:441–451.

Strickland, M. S., E. Osburn, C. Lauber, N. Fierer, and M. A. Bradford. 2009b. Litter quality is in the eye of the beholder: Initial decomposition rates as a function of inoculum characteristics. *Functional Ecology* **23**:627–636.

Striebel, M., G. Singer, H. Stibor, and T. Andersen. 2012. "Trophic overyielding": Phytoplankton diversity promotes zooplankton productivity. *Ecology* **93**:2719–2727.

Strong, D. R. 1992. Are trophic cascadees all wet? Differentiation and donor-control in speciose ecosystems. *Ecology* **73**:747–754.

Strong, D. R., J. L. Maron, P. G. Connors, A. Whipple, S. Harrison, and R. L. Jeffries. 1995. High mortality, fluctuation in numbers, and heavy subterranean insect herbivory in bush lupine, *Lupinus arboreus*. *Oecologia* **104**:85–92.

Strong, D. R., A. V. Whipple, A. C. Child, and B. Dennis. 1999. Model selection for a subterranean trophic cascade: Root-feeding caterpillars and entomopathogenic nematodes. *Ecology* **80**:2750–2761.

Suárez-Rodríguez, M., I. López-Rull, and C. M. Garcia. 2013. Incorporation of cigarette butts into nests reduces nest ectoparasite load in urban birds: new ingredients for an old recipe? *Biology Letters* **9** doi:10.1098/rsbl.2012.0931.

Suchanek, T. H., S. L. Williams, J. C. Ogden, D. K. Hubbard, and I. P. Gill. 1985. Utilization of shallow-water seagrass detritus by Carribbean deep-sea macrofauna: δ13C evidence. *Deep Sea Research Part A. Oceanographic Research Papers* **32**:201–214.

Suding, K. N., S. L. Collins, L. Gough, C. Clark, E. E. Cleland, K. L. Gross, D. G. Milchunas, and S. Pennings. 2005. Functional- and abundance-based mechanisms explain diversity loss due to N fertilization. *Proceedings of the National Academy of Sciences* **102**:4387–4392.

Suding, K. N., K. L. Gross, and G. R. Houseman. 2004. Alternative states and positive feedbacks in restoration ecology. *Trends in Ecology & Evolution* **19**:46–53.

Suttle, C. A. 2007. Marine viruses—major players in the global ecosystem. *Nature Reviews Microbiology* **5**:801–812.

Suzuki, N. and K. Kato. 1953. Studies on suspended materials. Marine snow in the sea. I. Sources of marine snow. *Bulletin of the Faculty of Fisheries of Hokkaido University* **4**:132–135.

Swan, B. K., B. Tupper, A. Sczyrba, F. M. Lauro, M. Martinez-Garcia, J. M. González, H. Luo, J. J. Wright, Z. C. Landry, N. W. Hanson, B. P. Thompson, N. J. Poulton, P. Schwientek, S. G. Acinas, S. J. Giovannoni, M. A. Moran, S. J. Hallam, R. Cavicchioli, T. Woyke, and R. Stepanauskas. 2013. Prevalent genome streamlining and latitudinal divergence of planktonic bacteria in the surface ocean. *Proceedings of the National Academy of Sciences* **110**:11463–11468.

Swan, C. M. and M. A. Palmer. 2006. Composition of speciose leaf litter alters stream detritivore growth, feeding activity and leaf breakdown. *Oecologia* **147**:469–478.

Talbot, J. M. and K. K. Treseder. 2012. Interactions among lignin, cellulose, and nitrogen drive litter chemistry-decay relationships. *Ecology* **93**:345–354.

Talbot, J. M., D. J. Yelle, J. Nowick, and K. K. Treseder. 2012. Litter decay rates are determined by lignin chemistry. *Biogeochemistry* **108**:279–295.

Tanentzap, A. J., K. J. Kirby, and E. Goldberg. 2012. Slow responses of ecosystems to reductions in deer (Cervidae) populations and strategies for achieving recovery. *Forest Ecology and Management* **264**:159–166.

Tang, K. W. 2001. Defecation of dimethylsulfoniopropionate (DMSP) by the copepod *Acartia tonsa* as functions of ambient food concentration and body DMSP content. *Journal of Plankton Research* **23**:549–553.

Tao, L., A. Berns, and M. D. Hunter. 2014. Why does a good thing become too much? Interactions between foliar nutrients and toxins determine performance of an insect herbivore. *Functional Ecology* **28**:190–196.

Tao, L. and M. D. Hunter. 2011. Effects of insect herbivores on the nitrogen economy of plants. Pages 257–279 *in* J. C. Polacco and C. D. Todd, editors. *Ecological Aspects of Nitrogen Metabolism in Plants*. Wiley-Blackwell., Oxford, UK.

Tao, L. and M. D. Hunter. 2012a. Allocation of resources away from sites of herbivory under simultaneous attack by aboveground and belowground herbivores in the common milkweed, *Asclepias syriaca*. *Arthropod-Plant Interactions* **7**:217–224.

Tao, L. and M. D. Hunter. 2012b. Does anthropogenic nitrogen deposition induce phosphorus limitation in herbivorous insects? *Global Change Biology* **18**:1843–1853.

Tao, L. and M. D. Hunter. 2015. Effects of soil nutrients on the sequestration of plant defence chemicals by the specialist insect herbivore, *Danaus plexippus*. *Ecological Entomology* **40**:123–132.

Targett, N. M. and T. M. Arnold. 2001. Effects of secondary metabolites on digestion in marine herbivores. Pages 391–412 *in* C. D. Amsler, editor. *Marine chemical ecology*. CRC Press, Boca Raton, FL.

Taylor, J. R. and R. Stocker. 2012. Trade-offs of chemotactic foraging in turbulent water. *Science* **338**:675–679.

Taylor, R. B., N. Lindquist, J. Kubanek, and M. E. Hay. 2003. Intraspecific variation in palatability and defensive chemistry of brown seaweeds: Effects on herbivore fitness. *Oecologia* **136**:412–423.

Taylor, R. B. and P. D. Steinberg. 2005. Host use by Australasian seaweed mesograzers in relation to feeding preferences of larger grazers. *Ecology* **86**:2955–2967.

Teder, T. and T. Tammaru. 2002. Cascading effects of variation in plant vigour on the relative performance of insect herbivores and their parasitoids. *Ecological Entomology* **27**:94–104.

Teeling, H., B. M. Fuchs, D. Becher, C. Klockow, A. Gardebrecht, C. M. Bennke, M. Kassabgy, S. X. Huang, A. J. Mann, J. Waldmann, M. Weber, A. Klindworth, A. Otto, J. Lange, J. Bernhardt, C. Reinsch, M. Hecker, J. Peplies, F. D. Bockelmann, U. Callies, G. Gerdts, A. Wichels, K. H. Wiltshire, F. O. Glockner, T. Schweder, and R. Amann. 2012. Substrate-controlled succession of marine bacterioplankton populations induced by a phytoplankton bloom. *Science* **336**:608–611.

Teichman, K. J., S. E. Nielsen, and J. Roland. 2013. Trophic cascades: Linking ungulates to shrub-dependent birds and butterflies. *Journal of Animal Ecology* **82**:1288–1299.

Tessier, A. J. and P. Woodruff. 2002. Cryptic trophic cascade along a gradient of lake size. *Ecology* **83**:1263–1270.

Thaler, J. 1999. Jasmonate-inducible plant defences cause increased parasitism of herbivores. *Nature* **399**:686–688.

Thamer, S., M. Schadler, D. Bonte, and D. J. Ballhorn. 2011. Dual benefit from a belowground symbiosis: Nitrogen fixing rhizobia promote growth and defense against a specialist herbivore in a cyanogenic plant. *Plant and Soil* **341**:209–219.

Thomas, J. D. 1997. The role of dissolved organic matter, particularly free amino acids and humic substances, in freshwater ecosystems. *Freshwater Biology* **38**:1–36.

Thompson, A. W., R. A. Foster, A. Krupke, B. J. Carter, N. Musat, D. Vaulot, M. M. M. Kuypers, and J. P. Zehr. 2012. Unicellular cyanobacterium symbiotic with a single-celled eukaryotic alga. *Science* **337**:1546–1550.

Thompson, J. N. 1994. *The Coevolutionary Process*. University of Chicago Press, Chicago.

Thor, P., H. G. Dam, and D. R. Rogers. 2003. Fate of organic carbon released from decomposing copepod fecal pellets in relation to bacterial production and ectoenzymatic activity. *Aquatic Microbial Ecology* **33**:279–288.

Throop, H. L. and M. T. Lerdau. 2004. Effects of nitrogen deposition on insect herbivory: Implications for community and ecosystem processes. *Ecosystems* **7**:109–133.

Tikkanen, O. P. and R. Julkunen-Tiitto. 2003. Phenological variation as protection against defoliating insects: The case of *Quercus robur* and *Operophtera brumata*. *Oecologia* **136**:244–251.

Tillmanns, A. R., S. K. Burton, and F. R. Pick. 2011. *Daphnia* pre-exposed to toxic microcystis exhibit feeding selectivity. *International Review of Hydrobiology* **96**:20–28.

Tilman, D. 1977. Resource competition between planktonic algae: An experimental and theoretical approach. *Ecology* **58**:338–348.

Tilman, D. 1987. The importance of the mechanisms of interspecific competition. *American Naturalist* **129**:769–774.

Tilman, D., J. Knops, D. Wedin, P. Reich, M. Ritchie, and E. Siemann. 1997. The influence of functional diversity and composition on ecosystem processes. *Science* **277**:1300–1302.

Tilman, D., D. Wedin, and J. Knops. 1996. Productivity and sustainability influenced by biodiversity in grassland ecosystems. *Nature* **379**:718–720.

Tinta, T., T. Kogovsek, A. Malej, and V. Turk. 2012. Jellyfish modulate bacterial dynamic and community structure. *PLOS ONE* **7**:e39274.:

Tobin, P. C. and A. E. Hajek. 2012. Release, establishment, and initial spread of the fungal pathogen *Entomophaga maimaiga* in island populations of *Lymantria dispar*. *Biological Control* **63**:31–39.

Tomitani, A., A. H. Knoll, C. M. Cavanaugh, and T. Ohno. 2006. The evolutionary diversification of cyanobacteria: Molecular-phylogenetic and paleontological perspectives. *Proceedings of the National Academy of Sciences* **103**:5442–5447.

Toth, G. B., O. Langhamer, and H. Pavia. 2005. Inducible and constitutive defenses of valuable seaweed tissues: Consequences for herbivore fitness. *Ecology* **86**:612–618.

Toth, G. B. and H. Pavia. 2000. Water-borne cues induce chemical defense in a marine alga (*Ascophyllum nodosum*). *Proceedings of the National Academy of Sciences* **97**:14418–14420.

Toth, G. B. and H. Pavia. 2007. Induced herbivore resistance in seaweeds: A meta-analysis. *Journal of Ecology* **95**:425–434.

Toupoint, N., L. Gilmore-Solomon, F. Bourque, B. Myrand, F. Pernet, F. Olivier, and R. Tremblay. 2012. Match/mismatch between the *Mytilus edulis* larval supply and seston quality: Effect on recruitment. *Ecology* **93**:1922–1934.

Towne, E. G. 2000. Prairie vegetation and soil nutrient responses to ungulate carcasses. *Oecologia* **122**:232–239.

Townsend, P. A., K. N. Eshleman, and C. Welcker. 2004. Remote sensing of gypsy moth defoliation to assess variations in stream nitrogen concentrations. *Ecological Applications* **14**:504–516.

Treseder, K. K. and P. M. Vitousek. 2001. Potential ecosystem-level effects of genetic variation among populations of *Metrosideros polymorpha* from a soil fertility gradient in Hawaii. *Oecologia* **126**:266–275.

Triebwasser, D. J., N. Tharayil, C. M. Preston, and P. D. Gerard. 2012. The susceptibility of soil enzymes to inhibition by leaf litter tannins is dependent on the tannin chemistry, enzyme class and vegetation history. *New Phytologist* **196**:1122–1132.

Trussell, G. R. and O. J. Schmitz. 2012. Species functional traits, trophic control, and the ecosystem comsequences of adaptive foraging in the middle of food chains. Pages 324–338 *in* T. Ohgushi, O. J. Schmitz, and R. D. Holt, editors. *Trait-mediated*

Indirect Interactions: Ecological and Evolutionary Perspectives. Cambridge University Press, Cambridge, UK.

Tuchman, N. C., R. G. Wetzel, S. T. Rier, K. A. Wahtera, and J. A. Teeri. 2002. Elevated atmospheric CO2 lowers leaf litter nutritional quality for stream ecosystem food webs. *Global Change Biology* **8**:163–170.

Tukey, H. B. and J. V. Morgan. 1963. Injury to foliage and its effects upon the leaching of nutrients from above-ground plant parts. *Physiologia Planta* **16**:557–564.

Tully, K. L., T. E. Wood, A. M. Schwantes, and D. Lawrence. 2013. Soil nutrient availability and reproductive effort drive patterns in nutrient resorption in *Pentaclethra macroloba. Ecology* **94**:930–940.

Turchin, P. 1990. Rarity of density dependence or population regulation with lags? *Nature* **344**:660–663.

Turchin, P., S. N. Wood, S. P. Ellner, B. E. Kendall, W. W. Murdoch, A. Fischlin, J. Casas, E. McCauley, and C. J. Briggs. 2003. Dynamical effects of plant quality and parasitism on population cycles of larch budmoth. *Ecology* **84**:1207–1214.

Turlings, T. C. J., I. Hiltpold, and S. Rasmann. 2012. The importance of root-produced volatiles as foraging cues for entomopathogenic nematodes. *Plant and Soil* **358**:47–56.

Turlings, T. C. J., J. H. Tumlinson, and W. J. Lewis. 1990. Exploitation of herbivore-induced plant odors by host-seeking parasitic wasps. *Science* **250**:1251–1253.

Turner, J. T. 2002. Zooplankton fecal pellets, marine snow and sinking phytoplankton blooms. *Aquatic Microbial Ecology* **27**:57–102.

Turner, M. G. 1989. Landscape ecology: The effect of pattern on process. *Annual Review of Ecology and Systematics* **20**:171–197.

Tylianakis, J. M., R. K. Didham, J. Bascompte, and D. A. Wardle. 2008. Global change and species interactions in terrestrial ecosystems. *Ecology Letters* **11**:1351–1363.

Underwood, N. 1999. The influence of plant and herbivore characteristics on the interaction between induced resistance and herbivore population dynamics. *American Naturalist* **153**:282–294.

Underwood, N. 2000. Density dependence in induced plant resistance to herbivore damage: Threshold, strength and genetic variation. *Oikos* **89**:295–300.

Underwood, N. 2004. Variance and skew of the distribution of plant quality influence herbivore population dynamics. *Ecology* **85**:686–693.

Underwood, N. 2009. Effect of genetic variance in plant quality on the population dynamics of a herbivorous insect. *Journal of Animal Ecology* **78**:839–847.

Underwood, N. 2010. Density dependence in insect performance within individual plants: Induced resistance to *Spodoptera exigua* in tomato. *Oikos* **119**:1993–1999.

Underwood, N., W. Morris, K. Gross, and J. R. Lockwood. 2000. Induced resistance to Mexican bean beetles in soybean: Variation among genotypes and lack of correlation with constitutive resistance. *Oecologia* **122**:83–89.

Underwood, N. and M. Rausher. 2002. Comparing the consequences of induced and constitutive plant resistance for herbivore population dynamics. *American Naturalist* **160**:20–30.

Underwood, N. and M. D. Rausher. 2000. The effects of host-plant genotype on herbivore population dynamics. *Ecology* **81**:1565–1576.

Underwood, N. C. 1998. The timing of induced resistance and induced susceptibility in the soybean Mexican bean beetle system. *Oecologia* **114**:376–381.

Urban, N. R., M. T. Auer, S. A. Green, X. Lu, D. S. Apul, K. D. Powell, and L. Bub. 2005. Carbon cycling in Lake Superior. *Journal of Geophysical Research-Oceans* **110** doi:10.1029/2003jc002230.

Uriarte, M. 2000. Interactions between goldenrod (*Solidago altissima* L.) and its insect herbivore (*Trirhabda virgata*) over the course of succession. *Oecologia* **122**:521–528.

Uselman, S. M., K. A. Snyder, and R. R. Blank. 2011. Insect biological control accelerates leaf litter decomposition and alters short-term nutrient dynamics in a *Tamarix*-invaded riparian ecosystem. *Oikos* **120**:409–417.

Utsumi, S., Y. Ando, and T. Miki. 2010. Linkages among trait-mediated indirect effects: A new framework for the indirect interaction web. *Population Ecology* **52**:485–497.

Vaccaro, L. E., B. L. Bedford, and C. A. Johnston. 2009. Litter accumulaton promotes dominance of invasive species of cattails (*Typha* Spp.) in Lake Ontario wetlands. *Wetlands* **29**:1036–1048.

Valdez Barillas, J. R., M. W. Paschke, M. H. Ralphs, and R. D. Child. 2007. White locoweed toxicity is facilitated by a fungal endophyte and nitrogen-fixing bacteria. *Ecology* **88**:1850–1856.

Valladares, G. and J. H. Lawton. 1991. Host-plant selection in the holly leaf-miner: Does mother know best? *Journal of Animal Ecology* **60**:227–240.

Van Alstyne, K. L. and K. N. Pelletreau. 2000. Effects of nutrient enrichment on growth and phlorotannin production in *Fucus gardneri* embryos. *Marine Ecology Progress Series* **206**:33–43.

Van Alstyne, K. L., G. V. Wolfe, T. L. Freidenburg, A. Neill, and C. Hicken. 2001. Activated defense systems in marine macroalgae: Evidence for an ecological role for DMSP cleavage. *Marine Ecology Progress Series* **213**:53–65.

van Dam, N. M. 2009. Belowground herbivory and plant defenses. *Annual Review of Ecology Evolution and Systematics* **40**:373–391.

van Dam, N. M., L. Witjes, and A. Svatos. 2004. Interactions between aboveground and belowground induction of glucosinolates in two wild *Brassica* species. *New Phytologist* **161**:801–810.

Van de Waal, D. B., V. H. Smith, S. A. J. Declerck, E. C. M. Stam, and J. J. Elser. 2014. Stoichiometric regulation of phytoplankton toxins. *Ecology Letters* **17**:736–742.

van der Maarel, E. and J. Franklin, editors. 2013. *Vegetation Ecology*. Wiley-Blackwell, Oxford, UK .

Van der Meijden, E., M. Wijn, and H. J. Verkaar. 1988. Defence and regrowth, alternative plant strategies in the struggle against herbivores. *Oikos* **51**:355–363.

Van Der Putten, W. H. 2003. Plant defense belowground and spatiotemporal processes in natural vegetation. *Ecology* **84**:2269–2280.

Van Dolah, F. M. 2000. Marine algal toxins: Origins, health effects, and their increased occurrence. *Environmental Health Perspectives* **108**:133–141.

van Kempen, M. M. L., A. J. P. Smolders, G. M. Bogemann, L. L. M. Lamers, E. J. W. Visser, and J. G. M. Roelofs. 2013. Responses of the *Azolla filiculoides* Stras.-*Anabaena azollae* Lam. association to elevated sodium chloride concentrations: Amino acids as indicators for salt stress and tipping point. *Aquatic Botany* **106**:20–28.

van Nouhuys, S., J. H. Reudler, A. Biere, and J. A. Harvey. 2012. Performance of secondary parasitoids on chemically defended and undefended hosts. *Basic and Applied Ecology* **13**:241–249.

Vance-Chalcraft, H. D., J. A. Rosenheim, J. R. Vonesh, C. W. Osenberg, and A. Sih. 2007. The influence of intraguild predation on prey suppression and prey release: A meta-analysis. *Ecology* **88**:2689–2696.

Vandermeer, J. and I. Perfecto. 2007. The agricultural matrix and a future paradigm for conservation. *Conservation Biology* **21**:274–277.

Vannette, R. L. and M. D. Hunter. 2011a. Genetic variation in expression of defense phenotype may mediate evolutionary adaptation of *Asclepias syriaca* to elevated CO2. *Global Change Biology* **17**:1277–1288.

Vannette, R. L. and M. D. Hunter. 2011b. Plant defence theory re-examined: Nonlinear expectations based on the costs and benefits of resource mutualisms. *Journal of Ecology* **99**:66–76.

Vannette, R. L. and M. D. Hunter. 2013. Mycorrhizal abundance affects the expression of plant resistance traits and herbivore performance. *Journal of Ecology* **101**:1019–1029.

Vannette, R. L. and M. D. Hunter. 2014. Genetic variation in plant below-ground response to elevated CO2 and two herbivore species. *Plant and Soil* **38**:303–314.

Vannette, R. L. and S. Rasmann. 2012. Arbuscular mycorrhizal fungi mediate below-ground plant-herbivore interactions: A phylogenetic study. *Functional Ecology* **26**:1033–1042.

Vannette, R. L., S. Rasmann, and M. D. Hunter. 2013. Arbuscular mycorrhizal fungi alter above- and below-ground chemical defense expression differentially among *Asclepias* species. *Frontiers in Plant-Microbe Interaction* **4** Article 361:doi:10.3389/fpls.2013.00361.

Vanni, M. J. 2002. Nutrient cycling by animals in freshwater ecosystems. *Annual Review of Ecology and Systematics* **33**:341–370.

Vanni, M. J., G. Boros, and P. B. McIntyre. 2013. When are fish sources vs. sinks of nutrients in lake ecosystems? *Ecology* **94**:2195–2206.

Varley, G. C., G. R. Gradwell, and M. P. Hassell. 1973. *Insect Population Ecology: An Analytical Approach*. Blackwell Scientific Publications Oxford, UK.

Villalba, J. J., F. D. Provenza, J. O. Hall, and L. D. Lisonbee. 2010. Selection of tannins by sheep in response to a gastro-intestinal nematode infection. *Journal of Animal Science* **88**:2189–2198.

Vincent, D., G. Slawyk, S. L'Helguen, G. Sarthou, M. Gallinari, L. Seuront, B. Sautour, and O. Ragueneau. 2007. Net and gross incorporation of nitrogen by marine copepods fed on (15)N-labelled diatoms: Methodology and trophic studies. *Journal of Experimental Marine Biology and Ecology* **352**:295–305.

Vitousek, P. M., J. D. Aber, R. W. Howarth, G. E. Likens, P. A. Matson, D. W. Schindler, W. H. Schlesinger, and D. Tilman. 1997. Human alteration of the global nitrogen cycle: Sources and consequences. *Ecological Applications* **7**:737–750.

Vitousek, P. M., S. Porder, B. Z. Houlton, and O. A. Chadwick. 2010. Terrestrial phosphorus limitation: Mechanisms, implications, and nitrogen-phosphorus interactions. *Ecological Applications* **20**:5–15.

von der Heyden, B. P., A. N. Roychoudhury, T. N. Mtshali, T. Tyliszczak, and S. C. B. Myneni. 2012. Chemically and geographically distinct solid-phase iron pools in the Southern Ocean. *Science* **338**:1199–1201.

Wada, Y., K. Iwasaki, and Y. Yusa. 2013. Changes in algal community structure via density- and trait-mediated indirect interactions in a marine ecosystem. *Ecology* **94**:2567–2574.

Wakeham, S. G., C. Lee, J. I. Hedges, P. J. Hernes, and M. L. Peterson. 1997. Molecular indicators of diagenetic status in marine organic matter. *Geochimica Et Cosmochimica Acta* **61**:5363–5369.

Wallis, I. R., M. J. Edwards, H. Windley, A. K. Krockenberger, A. Felton, M. Quenzer, J. U. Ganzhorn, and W. J. Foley. 2012. Food for folivores: Nutritional explanations linking diets to population density. *Oecologia* **169**:281–291.

Wang, X.-Y., Y. Miao, S. Yu, X.-Y. Chen, and B. Schmid. 2013. Genotypic diversity of an invasive plant species promotes litter decomposition and associated processes. *Oecologia* **174**:993–1005.

Ward, B. B. 2013. How nitrogen is lost. *Science* **341**:352–353.

Wardle, D. A. 2002. *Communities and Ecosystems: Linking the Aboveground and Belowground Components*. Princeton University Press, Princeton, NJ.

Wardle, D. A., K. I. Bonner, and G. M. Barker. 2002. Linkages between plant litter decomposition, litter quality, and vegetation responses to herbivores. *Functional Ecology* **16**:585–595.

Wardle, D. A., M. J. Gundale, A. Jäderlund, and M.-C. Nilsson. 2013. Decoupled longterm effects of nutrient enrichment on aboveground and belowground properties in subalpine tundra. *Ecology* **94**:904–919.

Wardle, D. A., B. J. Karl, J. R. Beggs, G. W. Yeates, W. M. Williamson, and K. I. Bonner. 2010. Determining the impact of scale insect honeydew, and invasive wasps and rodents, on the decomposer subsystem in a New Zealand beech forest. *Biological Invasions* **12**:2619–2638.

Wassen, M. J., H. O. Venterink, E. D. Lapshina, and F. Tanneberger. 2005. Endangered plants persist under phosphorus limitation. *Nature* **437**:547–550.

Weber, T. and C. Deutsch. 2012. Oceanic nitrogen reservoir regulated by plankton diversity and ocean circulation. *Nature* **489**:419–422.

Weber, T. and C. Deutsch. 2014. Local versus basin-scale limitation of marine nitrogen fixation. *Proceedings of the National Academy of Sciences* **111**:8741–8746.

Weis, J. J. and D. M. Post. 2013. Intraspecific variation in a predator drives cascading variation in primary producer community composition. *Oikos* **122**:1343–1349.

Weng, J.-K., Y. Li, H. Mo, and C. Chapple. 2012a. Assembly of an evolutionarily new pathway for α-pyrone biosynthesis in *Arabidopsis*. *Science* **337**:960–964.

Weng, J.-K., R. N. Philippe, and J. P. Noel. 2012b. The rise of chemodiversity in plants. *Science* **336**:1667–1670.

Werner, E. E. and J. F. Gilliam. 1984. The ontogenetic niche and species interactions in size structured populations. *Annual Review of Ecology and Systematics* **15**:393–425.

Werner, E. E., J. F. Gilliam, D. J. Hall, and G. G. Mittelbach. 1983. An experimental test of the effects of predation risk on habitat use in fish. *Ecology* **64**:1540–1548.

Werner, E. E. and S. D. Peacor. 2003. A review of trait-mediated indirect interactions in ecological communities. *Ecology* **84**:1083–1100.

Werren, J. H. 1980. Sex ratio adaptations to local mate competition in a parasitic wasp. *Science* **208**:1157–1159.

Wetzel, R. G. 1983. *Limnology*. Saunders College Publishing, Philadelphia, PA.

Whalen, M. A., J. E. Duffy, and J. B. Grace. 2013. Temporal shifts in top-down vs. bottom-up control of epiphytic algae in a seagrass ecosystem. *Ecology* **94**:510–520.

Whiles, M.R., M. A. Callaham, C. K. Meyer, B. L. Brock, and R. E. Charlton. 2001. Emergence of periodical cicadas (*Magicicada cassini*) from a Kansas riparian forest: Densities, biomass and nitrogen flux. *American Midland Naturalist* **145**:176–187.

White, T. C. R. 1984. The abundance of invertebrate herbivores in relation to the availability of nitrogen in stressed food plants. *Oecologia* **63**:90–105.

Whitfeld, T. J. S., V. Novotny, S. E. Miller, J. Hrcek, P. Klimes, and G. D. Weiblen. 2012. Predicting tropical insect herbivore abundance from host plant traits and phylogeny. *Ecology* **93**:S211-S222.

Whitham, T. G., W. P. Young, G. D. Martinsen, C. A. Gehring, J. A. Schweitzer, S. M. Shuster, G. M. Wimp, D. G. Fischer, J. K. Bailey, R. L. Lindroth, S. Woolbright, and C. R. Kuske. 2003. Community and ecosystem genetics: A consequence of the extended phenotype. *Ecology* **84**:559–573.

Wickings, K., A. S. Grandy, S. C. Reed, and C. C. Cleveland. 2012. The origin of litter chemical complexity during decomposition. *Ecology Letters* **15**:1180–1188.

Wilder, S. M., M. Norris, R. W. Lee, D. Raubenheimer, and S. J. Simpson. 2013. Arthropod food webs become increasingly lipid-limited at higher trophic levels. *Ecology Letters* **16**:895–902.

Williams, C. J., J. N. Boyer, and F. J. Jochem. 2009. Microbial activity and carbon, nitrogen, and phosphorus content in a subtropical seagrass estuary (Florida Bay): Evidence for limited bacterial use of seagrass production. *Marine Biology* **156**:341–353.

Williams, K. S. and C. Simon. 1995. The ecology, behavior, and evolution of periodical cicadas. *Annual Review of Entomology* **40**:269–295.

Williams, S. L., M. E. S. Bracken, and E. Jones. 2013. Additive effects of physical stress and herbivores on intertidal seaweed biodiversity. *Ecology* **94**:1089–1101.

Wilson, A. E., O. Sarnelle, and A. R. Tillmanns. 2006. Effects of cyanobacterial toxicity and morphology on the population growth of freshwater zooplankton: Meta-analyses of laboratory experiments. *Limnology and Oceanography* **51**:1915–1924.

Wilson, D. M., W. Fenical, M. E. Hay, N. Lindquist, and R. Bolser. 1999. Habenariol, a freshwater feeding deterrent from the aquatic orchid *Habenaria repens* (Orchidaceae). *Phytochemistry* **50**:1333–1336.

Wilson, J. S., M. L. Forister, L. A. Dyer, J. M. O'Connor, K. Burls, C. R. Feldman, M. A. Jaramillo, J. S. Miller, G. Rodriguez-Castaneda, E. J. Tepe, J. B. Whitfield, and B. Young. 2012. Host conservatism, host shifts and diversification across three trophic levels in two neotropical forests. *Journal of Evolutionary Biology* **25**:532–546.

Wimp, G. M., W. P. Young, S. A. Woolbright, G. D. Martinsen, P. Keim, and T. G. Whitham. 2004. Conserving plant genetic diversity for dependent animal communities. *Ecology Letters* **7**:776–780.

Winder, R. S., J. Lamarche, C. P. Constabel, and R. Hamelin. 2013. The effects of high-tannin leaf litter from transgenic poplars on microbial communities in microcosm soils. *Frontiers in Microbiology* **4** doi:10.3389/fmicb.2013.00290.

Wink, M. 1994. The role of quinolizidine alkaloids in plant-insect interactions. Pages 133–169 *in* E. A. Bernays, editor. *Insect-Plant Interactions*. CRC Press, Boca Raton, FL.

Wink, M. and L. Witte. 1991. Storage of quinolizidine alkaloids in *Macrosiphum albifrons* and *Aphis genistae* (Homoptera: Aphididae). *Entomologia Generalis* **15**:237–254.

Winnie, J. A. 2012. Predation risk, elk, and aspen: tests of a behaviorally mediated trophic cascade in the Greater Yellowstone ecosystem. *Ecology* **93**:2600–2614.

Wittstock, U. and J. Gershenzon. 2002. Constitutive plant toxins and their role in defense against herbivores and pathogens. *Current Opinion in Plant Biology* **5**:300–307.

Wondafrash, M., N. M. Van Dam, and T. O. G. Tytgat. 2013. Plant systemic induced responses mediate interactions between root parasitic nematodes and aboveground herbivorous insects. *Frontiers in Plant Science* **4** doi:10.3389/fpls.2013.00087.

Wong, C. N. A., P. Ng, and A. E. Douglas. 2011. Low-diversity bacterial community in the gut of the fruitfly *Drosophila melanogaster*. *Environmental Microbiology* **13**:1889–1900.

Woodward, G., M. O. Gessner, P. S. Giller, V. Gulis, S. Hladyz, A. Lecerf, B. Malmqvist, B. G. McKie, S. D. Tiegs, H. Cariss, M. Dobson, A. Elosegi, V. Ferreira, M. A. S. Graca, T. Fleituch, J. O. Lacoursiere, M. Nistorescu, J. Pozo, G. Risnoveanu, M. Schindler, A. Vadineanu, L. B. M. Vought, and E. Chauvet. 2012. Continental-scale effects of nutrient pollution on stream ecosystem functioning. *Science* **336**:1438–1440.

Wootton, J. T. 1995. Effects of birds on sea urchins and algae: A lower-intertidal trophic cascade. *Ecoscience* **2**:321–328.

Wotton, R. S. and B. Malmqvist. 2001. Feces in aquatic ecosystems. *Bioscience* **51**:537–544.

Wright, I. J., P. B. Reich, M. Westoby, D. D. Ackerly, Z. Baruch, F. Bongers, J. Cavender-Bares, T. Chapin, J. H. C. Cornelissen, M. Diemer, J. Flexas, E. Garnier, P. K. Groom, J. Gulias, K. Hikosaka, B. B. Lamont, T. Lee, W. Lee, C. Lusk, J. J. Midgley, M. L. Navas, U. Niinemets, J. Oleksyn, N. Osada, H. Poorter, P. Poot, L. Prior, V. I. Pyankov, C. Roumet, S. C. Thomas, M. G. Tjoelker, E. J. Veneklaas, and R. Villar. 2004. The worldwide leaf economics spectrum. *Nature* **428**:821–827.

Yang, I., E. Selander, H. Pavia, and U. John. 2011. Grazer-induced toxin formation in dino-flagellates: A transcriptomic model study. *European Journal of Phycology* **46**:66–73.

Yang, L. H. 2004. Periodical cicadas as resource pulses in North American forests. *Science* **306**:1565–1567.

Yang, L. H. 2006. Interactions between a detrital resource pulse and a detritivore community. *Oecologia* **147**:522–532.

Yang, L. H. 2008. Pulses of dead periodical cicadas increase herbivory of American bell-flowers. *Ecology* **89**:1497–1502.

Yang, Y., L. Wu, Q. Lin, M. Yuan, D. Xu, H. Yu, Y. Hu, J. Duan, X. Li, Z. He, K. Xue, J. van Nostrand, S. Wang, and J. Zhou. 2013. Responses of the functional structure of soil microbial community to livestock grazing in the Tibetan alpine grassland. *Global Change Biology* **19**:637–648.

Yates, J. L. and P. Peckol. 1993. Effects of nutrient availability and herbivory on polyphe-nolics in the seaweed *Fucus vesiculosus*. *Ecology* **74**:1757–1766.

Yelenik, S. G. and J. M. Levine. 2010. Native shrub reestablishment in exotic annual grass-lands: Do ecosystem processes recover? *Ecological Applications* **20**:716–727.

Yunev, O. A., J. Carstensen, S. Moncheva, A. Khaliulin, G. Aertebjerg, and S. Nixon. 2007. Nutrient and phytoplankton trends on the western Black Sea shelf in response to cultural eutrophication and climate changes. *Estuarine Coastal and Shelf Science* **74**:63–76.

Zagrobelny, M., S. Bak, A. V. Rasmussen, B. Jorgensen, C. M. Naumann, and B. L. Moller. 2004. Cyanogenic glucosides and plant-insect interactions. *Phytochemistry* **65**:293–306.

Zak, D. R., W. E. Holmes, A. J. Burton, K. S. Pregitzer, and A. F. Talhelm. 2008. Simulated atmospheric NO3– deposition increases soil organic matter by slowing decomposi-tion. *Ecological Applications* **18**:2016–2027.

Zalewski, M. and I. Wagner-Lotkowska, editors. 2004. *Integrated Watershed Management—Ecohydrology & Phytotechnology*. United Nations Environment Program, Osaka, Japan.

Zalucki, M. P., L. P. Brower, and A. Alonso. 2001. Detrimental effects of latex and cardiac glycosides on survival and growth of first-instar monarch butterfly larvae *Danaus plexippus* feeding on the sandhill milkweed *Asclepias humistrata*. *Ecological Entomology* **26**:212–224.

Zamzow, J. P., C. D. Amsler, J. B. McClintock, and B. J. Baker. 2010. Habitat choice and predator avoidance by Antarctic amphipods: The roles of algal chemistry and morphology. *Marine Ecology Progress Series* **400**:155–163.

Zangerl, A. R. and C. E. Rutledge. 1996. The probability of attack and patterns of constitutive and induced defense: A test of optimal defense theory. *American Naturalist* **147**:599–608.

Zastepa, A., F. R. Pick, and J. M. Blais. 2014. Fate and persistence of particulate and dissolved microcystin-LA from microcystis blooms. *Human and Ecological Risk Assessment* **20**:1670–1686.

Zehnder, C. B. and M. D. Hunter. 2007a. A comparison of maternal effects and current environment on vital rates of *Aphis nerii*, the milkweed-oleander aphid. *Ecological Entomology* **32**:172–180.

Zehnder, C. B. and M. D. Hunter. 2007b. Interspecific variation within the genus *Asclepias* in response to herbivory by a phloem-feeding insect herbivore. *Journal of Chemical Ecology* **33**:2044–2053.

Zehnder, C. B. and M. D. Hunter. 2008. Effects of nitrogen deposition on the interaction between an aphid and its host plant. *Ecological Entomology* **33**:24–30.

Zehnder, C. B. and M. D. Hunter. 2009. More is not necessarily better: The impact of limiting and excessive nutrients on herbivore population growth rates. *Ecological Entomology* **34**:535–543.

Zehnder, C. B., K. W. Stodola, R. J. Cooper, and M. D. Hunter. 2010. Spatial heterogeneity in the relative impacts of foliar quality and predation pressure on red oak, *Quercus rubra*, arthropod communities. *Oecologia* **164**:1017–1027.

Zehnder, C. B., K. W. Stodola, B. L. Joyce, D. Egetter, R. J. Cooper, and M. D. Hunter. 2009. Elevational and seasonal variation in the foliar quality and arthropod community of *Acer pensylvanicum*. *Environmental Entomology* **38**:1161–1167.

Zhen, Y., M. L. Aardema, E. M. Medina, M. Schumer, and P. Andolfatto. 2012. Parallel molecular evolution in an herbivore community. *Science* **337**:1634–1637.

Zimmer, M. and W. Topp. 2002. The role of coprophagy in nutrient release from feces of phytophagous insects. *Soil Biology & Biochemistry* **34**:1093–1099.

Zuppinger-Dingley, D., B. Schmid, J. S. Petermann, V. Yadav, G. B. De Deyn, and D. F. B. Flynn. 2014. Selection for niche differentiation in plant communities increases biodiversity effects. *Nature* **515**:108–111.

Index

Page numbers in italics refer to figures; tables are indicated with a lowercase *t*.

MONOGRAPHS IN POPULATION BIOLOGY

EDITED BY SIMON A. LEVIN AND HENRY S. HORN